Quantitative Analysis and Optimal Control of Energy Efficiency in Discrete Manufacturing System

Yan Wang · Cheng-Lin Liu · Zhi-Cheng Ji

Quantitative Analysis and Optimal Control of Energy Efficiency in Discrete Manufacturing System

Springer

Yan Wang
School of IOT Engineering
Jiangnan University
Wuxi, Jiangsu, China

Cheng-Lin Liu
School of IOT Engineering
Jiangnan University
Wuxi, Jiangsu, China

Zhi-Cheng Ji
School of IOT Engineering
Jiangnan University
Wuxi, Jiangsu, China

ISBN 978-981-15-4461-3 ISBN 978-981-15-4462-0 (eBook)
https://doi.org/10.1007/978-981-15-4462-0

This Springer imprint is published by the registered company Springer Nature Singapore Pte Ltd.
The registered company address is: 152 Beach Road, #21-01/04 Gateway East, Singapore 189721, Singapore

Preface

Recently, the problems of energy consumption and carbon emission in manufacturing have received global attention. Although the energy consumption problem in the process industry has attached much attention, it is also very prominent in the discrete manufacturing industry. To our knowledge, the effective energy consumption of the equipment is less than 30% in the discrete production line. Over 70% energy consumption costs on standby, no-load, unreasonable process parameters and routes. Hence, the research on quantitative analysis and optimal control of energy efficiency in discrete manufacturing system is a meaningful work. Most of the current relative works are focused on the analysis of the relationship between equipment energy consumption and process parameters. This is a local optimization perspective without considering the global optimization of the overall production line energy consumption.

Therefore, this book proposes some energy efficiency quantitative analysis and optimal methods for the discrete manufacturing system from the view of global optimization. To analyse and optimize the energy efficiency for discrete manufacturing systems, this book uses the real-time acquisition of energy consumption data, models the energy consumption, constructs the energy efficiency quantitative indices system and proposes a principal component quantitative analysis and a combined energy efficiency quantitative analysis based on rough set and AHP. Then, an energy quantitative analysis application system is built based on the presented analysis methods. Furthermore, this book designs several optimal control strategies including a static energy efficiency optimization based on knowledge and MOPSO, a high-dimensional dynamic energy efficiency optimization, and a process parameter optimization decision. At the last of this book, an energy efficiency optimization control software system is designed. This book is valuable for post-graduate students, teachers, engineers and individual researchers in the field of discrete manufacturing systems.

Chapter 2 focuses on the deployment method of the discrete manufacturing system's energy consumption information acquisition network. In Chap. 3, a multi-source and multi-level integrated energy consumption model and an intelligent identification method of key parameters are proposed. Chapter 4 investigates a multi-layered energy efficiency quantitative analysis index system. Chapter 5 focuses on the quantitative analysis of energy efficiency based on rough set theory and AHM. Chapter 6 is concerned with the bottleneck identification methods of the discrete manufacturing system's energy consumption. Chapter 7 proposes a discrete energy consumption analysis method based on the improved principal component analysis. In Chap. 8, a static optimization and scheduling algorithm based on knowledge and MOPSO is presented to minimize the energy consumption, machine workload and completion time. Chapter 9 studies the problem of production/energy consumption high-dimensional multi-objective dynamic optimization. A pre-response dynamic scheduling mechanism and NSGA-III-based optimization algorithm are designed to address the unexpected work events. In Chap. 10, a process parameter optimization decision model and RWA-MOPSO solution algorithm are presented to minimize the energy consumption in discrete manufacturing system. On the basis of the methods presented in the previous chapters, Chap. 11 illustrates the development steps, environments, tools and functional modules of a real energy efficiency optimization software system. Also, some industrial applications are shown to validate the effectiveness of our methods and software.

Wuxi, China
September 2019

Yan Wang
Cheng-Lin Liu
Zhi-Cheng Ji

Contents

Notations

E_c	Cutting energy consumption
E_m	Motor energy consumption
E_k	Drive link kinetic energy
E_f	Friction loss
E_a	Load loss
ΔE	Frequency conversion loss
E_e	Energy consumption of electrical control system
E_{as}	Energy consumption of each auxiliary system
$\eta(t)$	Instantaneous energy efficiency
$P_o(t)$	The effective energy change rate of the manufacturing system at a certain time t
$P_{in}(t)$	The input energy change rate
E_p	Energy efficiency of a process
E_o	Effective energy of a certain process or a certain period of time
E_{in}	Energy consumption of manufacturing system
η_{sec}	Energy efficiency under specific energy consumption
E_{ms}	Energy consumption of a manufacturing system
O_E	Effective output of a manufacturing system
E_{output}	Output energy
E_{input}	Input energy
η_{ee}	Machine energy efficiency
E_{mt}	Total energy consumption of a machine tool
$P_c(t)$	Machine tool's cutting power
$P(t)$	Total machine processing power
P_{rfo}	Fixed power attenuation during the machine tool's normal operation
$P_{sp}(t)$	Input power of the main transmission system
T_s	Start time of the manufacturing process
T_e	End time of the manufacturing process

$P_{\mathrm{u}}(t)$	Idle power of the machine tool
$P_{\mathrm{ad}}(t)$	Additional load loss of the machine tool
a_0	Additional load loss factor
a_1	Additional load loss factor
$\hat{P}_{\mathrm{c}}(t)$	Predictive cutting power of the machine tool
$\mathrm{E}\{\cdot\}$	The mean value
$\boldsymbol{\theta}$	The parameter
$\mathbf{h}(k)$	Observable data vector
$J(\boldsymbol{\theta})$	$\boldsymbol{\theta}$ criterion function
$\hat{\boldsymbol{\theta}}$	Least square estimate of parameter $\boldsymbol{\theta}$
$u(k)$	Input variables of system
$z(k)$	Output variables of system
$n(k)$	Noise variables of system
$A(z^{-1})$	Multinomials of the retardation factor z^{-1}
$B(z^{-1})$	Multinomials of the retardation factor z^{-1}
z^{-1}	Retardation factor
L	Data length
$\hat{\boldsymbol{\theta}}(k)$	Estimate value of $\boldsymbol{\theta}(k)$
$\mathbf{I}$	The unit matrix
$\mathbf{K}(k)$	The gain matrix
$\mathbf{P}(k)$	The covariance matrix
$\Gamma(k, i)$	The discount factor
$\Lambda(i)$	The weighting factor
$\mu(j)$	The forgetting factor
β	The sufficiently large positive integer
$\boldsymbol{\varepsilon}$	The sufficiently small real vector
P_{dec}	The total load irrelevant energy consumption
C_i	The power of a certain auxiliary energy consumption sub-system, which is normally a constant
g_i	A certain energy consumption sub-system is being used: 1 stands for yes, and 0 stands for no
P_{in}	Input power
P_{a}	Inverter forward loss
P_{b}	Inverter switch loss
P_{c}	Inverter recovery loss
P_{Fel}	The motor stator's iron loss
P_{f}	The friction loss
P_{w}	The wind loss
P_{cu1}	The motor stator's copper loss
P_{cu2}	The motor rotor's copper loss
P_{st}	Stray loss
P_{s}	The mechanical transmission system's energy consumption

P_{m}	The transmission system's friction loss
P_{z}	The transmission system's damping loss
P_{const}	Fixed loss
P_{var}	Variable loss
P_{imotor}	The motor's input power
P_{cu}	Copper loss
P_{Fe}	Core loss
P_{mec}	Mechanical loss
P_{ad}	Stray loss
P_{out}	Mechanical output power
ω_{m}	The angular speed
T_{en}	The electromagnetic torque
R_a	The stator resistance
B_{m}	The motor's damping coefficient
K_{T}	The motor's torque coefficient
K_{g}	The coupling's transmission ratio
B_{ls}	The ball screw's damping coefficient
μ_{v}	The viscous friction coefficient
μ_{c}	The Coulomb friction coefficient
η_{bse}	The ball screw drive efficiency
P_{pitch}	The ball screw's pitch
F_{ext}	The horizontal weight of the cutting force
M_{weight}	The machine tool's weight
i	The task number
k	The procedure number
j	The step number in the procedure
θ_l	Process parameters
O_n	The process route
M_{ik}	Processing equipment of k procedure of task i
$E_{u,M_{ik}}$	The facility's idle energy consumption
P_{idk}	The idle power of equipment no. k
z_{ik}	The total number of the facility's processing tasks
s_{ik}	The batch volume
t_{cik}	The processing time of task no. i at the machine tool no. k
t_{uik}	The idle time of the machine tool no. k
F_p,F_q	The two neighbouring structural features of the same processing part
$E_{\mathrm{non}}^{(F_p,F_q)}$	The idle energy consumption from the processing of feature p to the start of processing the next q
$E_{\mathrm{tp}}^{(F_p,F_q)}$	The idle cutting energy consumption from machining of feature F_p to machining of the next F_q
$E_{\mathrm{tc}}^{(F_p,F_q)}$	Automatic cutter changing energy consumption from machining of feature F_p to machining of the next F_q

n	Spindle speed
x, y, z	Tool coordinate
I_{spindle}	Spindle current
U_{spindle}	Spindle voltage
f	Feed rate
U_{in}	Input voltage
I_{in}	Input current
P_{stdby}	Standby power
P_{idle}	Idle power
V	The space volume of this basic hypercube
V_t	The volume of the space surrounded by this sample
n_{cmp}	The case number of the category i
$v_{d_{ij}}$	The decisional attribution i energy consumption of case no. j in the category i
$\overline{v_{d_i}}$	The average energy consumption of category i
n_{cond}	The case number of the category i
$v_{d_{ij}}$	The decisional attribution i energy consumption of case no. j in the category i
n_{num}	The number of processing parameters
Δx_j	Δx_j represents the difference after the processing parameters are normalized
α_j	The deviation after the standardization of the processing parameters
n_{obj}	The number of the selected features
β_i	The weighting factor after normalizing the attributes' importance of various features
$\lambda_{\max}$	The largest feature value
C_1	Energy consumption per ten thousand yuan production
S_1	The total amount of energy consumption
M_1	The industry added value
C_2	Energy consumption per ten thousand yuan added value
S_2	The total added value of the energy consumption
M_2	The industrial added value
C_3	Comprehensive energy consumption per unit product
N	The production output
C_4	Amount of energy saving per unit product
C_3'	The total amount of energy consumption per unit product after the energy saving optimization
C_5	The product energy utilization level
C_0	The comprehensive energy consumption per unit among the advanced-level products in China
C_6	Processing facility's energy efficiency
S_3	The total amount of energy consumption when the processing facility is cutting

S_4	The total amount of energy consumption of the processing facility throughout the entire production process
C_7	Energy transmission efficiency
S_5	The total amount after the energy has been transmitted
S_6	The total energy input amount
C_8	Energy conversion processing efficiency
S_7	The total amount of energy processing and conversion output
S_8	The total amount of energy processing and energy input
C_9	Processing technique energy efficiency
C_{10}	The comprehensive energy consumption per unit product's entire production cycle
S_9	The total amount of the energy consumption per unit product's technological process
C_{11}	Production resources scheduling energy efficiency
C_{12}	The total amount of energy consumption per unit product prior to the production resources scheduling
$C_{12}^{'}$	The total amount of energy consumption per unit product after the application of the production resources scheduling scheme
S	The decision table
D	The decision attribute set
C	The conditional attribute set
U	Object collection
V	Attribute value set
F	An information function
$\mathrm{Sig}(c)$	The weight significance of condition attribute Cin the decision table
$W_{Ai}(\mathrm{c})$	The weight of condition attributes (indexes) C
w_j	The combinational weight the item j sub-index relative to the total index
w_i	The combinational weight of the item i sub-index
w_{ij}	The weight of item j sub-index relative to i sub-index
w_{Ai}	Objective weight value
w_{Bi}	Subjective weight value
ς	The number of decision index
ξ	The number of alternative options
$r_{i,m}$	The degree of membership of u_i to c_j
c_k^i	The original value of item k sub-index in the number i option
c_k^*	The best value of the item k sub-index
W	The weight matrix
R	The quantitative analysis matrix
j_i	Quantitative analysis index for a single index c_i

TOW_{min}	Minimum difference between poles
TOW_{max}	Maximum difference between poles
L_i^k	Grey relational coefficiency
C^*	The optimal index set
A_i	The node degree
a_{ij}	Whether node i and node j are connected
B_r	The betweenness of node r
λ_{st}	The number of all the shortest routes between node s and node t
$\lambda_{st}(r)$	The number of all shortest routes between node s and node t which pass node r
e_{st}	The reciprocal of the distance between node s and node t
d_{st}	Represents the number of the edges which pass through node s and node t
W_{st}	The route energy consumption between node s and node t within a given period
N_{kind}	The kinds of components
a_{sti}	Whether there is the component mobilization of the component i between node s and node t
P_{sti}	The different transportation equipment power equipped for component mobilization
T_{sti}	The operation time of the transportation equipment for transporting i between node s and node t
C_r	The energy consumed by the equipment node r in a given period. u_i refers to the number of component i
H_i	The number of procedures of component i
β_{ijr}	The equipment coefficient
T_{ijr}	The cutting time of component i at procedure no. j
T'_{ijr}	Idle time of component i at procedure no. j
P_{ncut}	The cutting power of equipment node r
P'_{nidle}	Idle power of equipment node r
α_1	The weight functions of route energy consumption
α_2	The weight functions of node energy consumption
U_{Ci}	The substance–relation–substance set contained within node C
$U(C_i, C_j)$	The intersection of the two indexes' corresponding sets
p	The index volume of the index system
W_j	The importance weight function value endowed to each index in the index system through the expert method's subjective empowerment
X_{ij}	The energy consumption evaluation index variable value of sample no. i, section j
$k_{principal}$	The number of principal components

F	The comprehensive result of the discrete manufacturing system's energy consumption sample
m	The number of processing equipment
N	The number of workpieces to be processed
c_{ijegk}	i, e represents the workpiece number, j, g represents the procedure number
h	The device number
M	The device set
J	The workpiece set
s	The largest procedure number of all workpieces
O_{ij}	Procedure j of workpiece J_i
M_{ij}	Procedure j of workpiece J_i the available device set of procedure O_{ij} for workpiece J_i
P_{ijk}	A certain procedure of the workpiece can be processed at multiple devices
R_c	Two workpieces can be processed at one device
t_{ijk}	The processing time of procedure O_{ij} at device M_k
S_{ijk}	The start time for processing procedure O_{ij} at device M_k
E_{ijk}	The finish time for processing procedure O_{ij} at device M_k
w_k	The energy consumption of a certain procedure of a certain workpiece at device M_k
E_i	The total energy consumption of all workpieces at device M_i
$E_{\text{total-ec}}$	The total energy consumption for processing all workpieces
ω_i	Weight value of no. i component
p_{itarget}	i component's target value at the best level
$\bar{p}_i$	The historical value of no. i component's priority level
C_M	The order completion time uses the device's minimalized maximum completion time
$E_{\min-\text{total}}$	Minimalized total production energy consumption
W_T	Minimalized machine's total load
W_M	Critical machine load
G	The target component set
H	Component set of historical scheduling
$\sum t_{ij}$	The length of time to complete producing this type of component
t_{day}	The number of days prior to the completion date
t_{work}	The daily working hours
$P_i(t)$	The priority level of component no. i
$imp\%$	The MOPSO algorithm's performance upgrade rate
t_0	Initial moment
t_l	Rescheduling time point $l = 1,\ 2, \cdots$

$n(t_l)$	Rescheduling time point t_l number of workpieces with machinable operations
$J_i(t_l)$	The i of the t_l moment contains the workpieces of the machinable process
$m(t_l)$	Number of machines that can be released at t_l time
$M_k(t_l)$	k machine released at t_l time, $k = 1, 2, \cdots, m(t_l)$
$C_i(t_l)$	t_l time, the completion time of all processable parts of the workpiece $J_i(t_l)$
$e_i^l(t_l)$	Number of machinable processes included in workpiece $J_i(t_l)$ at t_l time
$DD_i(t_l)$	t_l time, the delivery date of all processable parts of $J_i(t_l)$
$S_i(t_l)$	t_l time, workpiece $J_i(t_l)$ the start time of the first process in the process
$n*(t_l)$	Rescheduling time point t_l has the number of workpieces that can be processed and processed
p^{k_r}	t_l time, the remaining processing time of the processing process on $M_k\,(t_l)$
$I_i(t_l)$	t_l time, workpiece $J_i(t_l)$ subscript of the first process in all processable processes
$c_{i(I_i(t_l)-1)}$	Process $J_i(t_l)$, the completion time of the last process in the process that was completed before the t_l time
$O_{ij}(t_l)$	Step j in the workpiece $J_i(t_l)$, $j = I_i(t_l), I_i(t_l)+1, \cdots, I_i(t_l)+e_i^l(t_l)-1, r = 1$
$M_{ij}(t_l)$	Process $O_{ij}(t_l)$ is a collection of machinable machines at t_l
$p_{ijk}(t_l)$	Process $O_{ij}(t_l)$ processing time on the machine $M_k(t_l) \in M_{ij}(t_l)$, $k = 1, 2, \cdots, \lvert M_{ij}(t_l) \rvert$, $\lvert \bullet \rvert$ indicates the size of the collection
$s_{ij}(t_l)$	Start processing time of process $O_{ij}(t_l)$
$c_{ij}(t_l)$	The completion time of the last process of the machine $M_k(t_l)$ completed before t_l
$c^{k\_last}(t_{l-1})$	The completion time of the last process of the machine $M_k(t_l)$ completed before t_l
$n_{M_k}(t_l)$	Number of operations assigned to the machine $M_k(t_l)$ at t_l, $k = 1, 2, \cdots, m(t_l)$
$O^{r,M_k(t_l)}$	The r process assigned on the machine $M_k(t_l)$, $r = 1, 2, \cdots, n_{M_k(t_l)}$
$p^{r,M_k(t_l)}$	Process $O^{r,M_k(t_l)}$ processing time
$c^{r,M_k(t_l)}$	Finishing time of process $O^{r,M_k(t_l)}$
K_i^*	The delivery period elastic coefficient of the workpiece $J_i(t_l)$, which satisfies the positive distribution of 1.5 with a variance of 0.5
E_k	Machine M_k processing energy per unit time

$A_k(t_l)$	Machine $M_k(t_l)$ in rescheduling period t_l initial release time
$X_{ijk}(t)$	Binary variable
E_{total}	Total energy consumption
E_{st}	Energy consumption when machine starts
E_{s-s}	Machine standby power consumption
E_{ie}	Energy consumption when machine is no-load
E_{ic}	Cutting energy consumption during processing
t_c	Process time
v_c	Cutting speed
f_{afeed}	The amount of feed
a_{sp}	Cutting depth
$C_{FC}, x_{FC}, y_{FC}, n_{FC}, K_{FC}$	Coefficients related to part material and tool material
t_{ot}	Auxiliary time such as clamping
L_w	The length of the machined part
T	The life of the tool used for machining
t_{ct}	The time taken by the magazine to change the tool once
Δ	The margin left in the process
d_0	The diameter of the machined part
n	The spindle speed of the machine
C_T	Constant values related to machining conditions such as parts, tools and machine tools
$x_{\text{life}}, y_{\text{life}}, z_{\text{life}}$	The life factor of the tool
$n_{\min}$	The minimum values of the spindle speed of the processing equipment
$n_{\max}$	The maximum values of the spindle speed of the processing equipment
$f_{\min}$	The minimum values of the feed allowed by the processing equipment
$f_{\max}$	The maximum values of the feed allowed by the processing equipment
$F_{\max}$	Maximum cutting force
C_F, x, y, n, K_F	The coefficient associated with the machined workpiece and the cutting conditions
r_ε	Tool radius
$R_{\max}$	The maximum value required for the surface roughness of the part
η_{total}	The total efficiency of the machine tool
$P_{\max}$	The maximum cutting power indicated on the machine nameplate
F_c	Indicates the cutting force when the machine is machined
$P_{\text{population}}$	Population of n
D	Search space dimension

V_{id}	The speed of particle i in d-dimensional space
X_{id}	The position of particle i in the d-dimensional space
P_i	Individual extremum
P_g	Group extremum
$\omega(t)$	Inertia factor
$\omega_{\max}$	The maximum values of the inertia factor
$\omega_{\min}$	The minimum values of the inertia factor
d	Particle dimensions
i_{num}	The number of particles
t	The current number of iterations
i_{num}	The maximum number of iterations
$c_1 c_2$	Acceleration factor
x_i	The approximate minimum value obtained in $i-1$ steps
u_i	A randomly generated unit vector
$P_{\text{optimization}}$	The optimal solution found by the particle swarm optimization algorithm
D_i	The crowded distance of the particles
$F(s)$	A collection of all particles with a non-dominated solution level of s
f_j^h	The hth objective function representing the jth particle, a total of p objective functions
l	The number of particles in $F(s)$set
$a_j^{\max}$	Maximum values of the corresponding j columns of the judgment matrix$A$
$a_j^{\min}$	Minimum values of the corresponding j columns of the judgment matrix$A$
p_{num}	The number of evaluation programs
q	The number of indicators

Chapter 1
Introduction

1.1 Discrete Manufacturing System

Discrete manufacturing system is an important support for the production of mechanical products and their parts, and its energy consumption is an important part of the carbon footprint of products [1]. Literatures [2, 3] show that the main energy-consuming machinery and equipments of discrete manufacturing system have large energy consumption and effective energy utilization rate of less than 30% on average. The energy-saving space is huge, and the carbon emissions caused by energy consumption are shocking. According to the literature [4], the global manufacturing industry produced nearly 50 billion tons of carbon emissions in 2001. It is expected that by 2030, global manufacturing carbon emissions will reach 100 billion tons, double the number in 2001. A set of data from MIT's research can visually illustrate the total energy consumption of a machine tool and the environmental emissions it brings [5]: a machine equipped with a corresponding auxiliary device, if the spindle power is 22 kW. The two-shift work system, in which the cutting time accounted for 57%, measured by the efficiency data of the US National Grid, the environmental emissions generated by the machine's electricity consumption for one year, and the fuel consumption is 20.7 mpg, driving 12,000 miles per year. Compared with SUVs, their CO_2 emissions are equivalent to the emissions of 61 SUVs.

Discretely manufactured products are often assembled from multiple parts through a series of discrete processes. The discrete manufacturing process is a complex process consisting of different parts processing sub-processes or parallel or series. Discrete manufacturing systems span different levels of products, workshops, tasks, manufacturing units and production equipment. Each level of energy consumption has its basic characteristics. Therefore, the discrete manufacturing system is much more complicated than the process manufacturing system. Complex characteristics such as variability in processing tasks, equipment and process diversity, complex tree characteristics of products and their manufacturing energy consumption, product energy consumption and dynamic characteristics of equipment (process) are often [6, 7]. The overall energy consumption, energy efficiency utilization and the

Y. Wang et al., *Quantitative Analysis and Optimal Control of Energy Efficiency in Discrete Manufacturing System*, https://doi.org/10.1007/978-981-15-4462-0_1

implementation of relevant at home and abroad green regulations, energy conservation and consumption reduction have become the first problem for the sustainable development of discrete manufacturing.

In recent years, China, the European Union, Japan, North America and other regions have launched plans to improve the energy efficiency of discrete manufacturing and carry out research on related mechanisms and technologies [8–10]. A large number of studies have shown that [11–16] the energy consumption of discrete manufacturing systems is closely related to equipment state and its motion properties, process parameters, process feature sequencing, product structure, production scheduling and start–stop control. In particular, for complex discrete manufacturing processes such as flexible manufacturing and mixed flow manufacturing, the energy consumption mechanism is complicated by the influence of process characteristics such as variable processing tasks, equipment state and process dynamics. The pre-unknownness of a large number of parameters in the processing parameter/thermodynamic energy consumption model, the highly nonlinear characteristics of the equipment energy consumption process and the inevitable randomness and dynamics in complex workflows pose challenges to the modelling, analysis and optimization of system energy consumption process. Therefore, the research on energy optimization of discrete manufacturing systems is of great scientific significance and engineering application value.

1.2 Energy-Related Factors and Characteristics

1.2.1 Energy-Consuming Components

The basic components of a typical discrete manufacturing system can be divided into four parts: production environment, production equipment, production objects and operators. The functional characteristics, usage status and utilization rate of these production factors will affect the energy consumption in manufacturing systems. The consumed energy in a production process can be divided into direct energy and indirect energy [17], where direct energy is the energy consumed by various processes (such as casting, processing, painting and inspection) for manufacturing products, and the indirect energy is the energy required to maintain the production environment in the workshop (such as lighting, heating and ventilation). Taking a typical machining manufacturing system of complex discrete manufacturing as an example, the mechanical manufacturing system is composed of nine types of processes such as cutting, pressure processing, casting, welding, special processing, heat treatment, covering, assembling and packaging. Energy-consuming components can be further divided into direct and indirect energy-consuming components according to their manifestations. Direct energy consumption from production machines depends on the operating state of the equipments. The process characteristics of the parts, the scheduling of the production system and the start–stop control strategy of the equipment constitute the indirect energy-consuming components.

1. **Direct energy consumption of machines**. The production equipment of discrete manufacturing systems is mostly CNC machine tools with machine–electric–liquid multi-source energy fusion characteristics. There are many energy-consuming components in a machine tool, including auxiliary system, main transmission system, feed system and other energy-consuming subsystems. It is a multi-component and multi-source energy-consuming system. The basic energy consumption characteristics of the motion unit are determined by its own structure and physical and motion properties [18], such as the structural design and motion control mode of each moving component. The energy consumption of each main component is shown in Table 1.1. During the operation of the production equipment, the energy loss caused by the interaction and mutual influence of the various moving parts makes the energy loss law quite complicated, and the energy consumption exhibits instantaneous dynamic changes in different operation stages such as starting, no-load and processing. The CNC machine tools mainly consume electric energy. The change of the whole machine power during the operation process depends not only on the performance of its internal component system, the movement composition of the components and the operating parameters, but also the processing object and the processing conditions.
2. **Production object process characteristics indirect energy consumption**. From the perspective of production objects, the same production equipment, different part materials, tool types and machining parameters will form different machining forces/torques, affecting the load of the moving unit of the processing equipment [19], thus consuming different energy.
3. **Auxiliary production equipment indirect energy consumption**. In addition to processing equipment, the energy consumption generated by other auxiliary processing equipment, such as lighting equipment, workshop handling equipment, workshop cooling and heating equipment.
4. **Production management indirect energy consumption**. In the process of processing, affected by the production planning scheduling and production resource scheduling scheme, the energy consumption difference of different batches of products during processing.

1.2.2 Energy Consumption Hierarchy and Characteristics

According to the different energy consumption levels of discrete manufacturing systems, the energy consumption of discrete manufacturing processes is analysed from three aspects: equipment layer, product layer and process layer:

Table 1.1 Function and energy consumption description of energy-consuming components in a CNC machine tool [18]

Name		Component composition	Functional description	Energy consumption statement
Processing power system	Spindle drive	Spindle motor and mechanical transmission components	Drive the tool or workpiece for high-speed rotation and provide cutting power	Cutting energy consumption $E_\mathbf{c}$ Motor energy consumption $E_\mathbf{m}$ Drive link kinetic energy $E_\mathbf{k}$ Friction loss $E_\mathbf{f}$ Load loss $E_\mathbf{a}$
	Feed drive	Feed shaft motor and mechanical transmission components	Electric tool or workpiece linear motion along the feed axis	
Electrical control system	Numerical control device	Computer machine display	CNC program processing and display	Frequency conversion loss ΔE Electrical control system energy consumption $E_\mathbf{e}$
	Spindle control	Spindle drive/inverter	Converting NC instruction of spindle into electrical signal	
	Feed axis control	Feed axis drive/inverter	Convert CNC commands of the feed axis to electrical signals	
Auxiliary system	Heat exhaling system	Fan	Electric control cabinet cooling	Energy consumption of each auxiliary system $E_\mathbf{as}$
	Cooling system	Cooling pump	Provide coolant	
	Chip removal system	Chip removal motor	Chip removal	
	Tool change system	Storage motor	Tool change	

1. **Equipment layer energy consumption**

Discrete workshop equipment layer has many types of energy consumption and large quantity. Workshop energy-consuming equipment contains trampolines, grinding machines, etc. Production and transportation energy-consuming equipment includes hanging towers, forklifts, etc., workshop production environment equipment such as refrigeration and air conditioning, heating and lighting. The energy consumption of each equipment in the discrete manufacturing workshop is mainly electric energy. As the main driving force of workshop processing equipment and auxiliary processing equipment, electric energy is an indispensable energy medium in the whole processing.

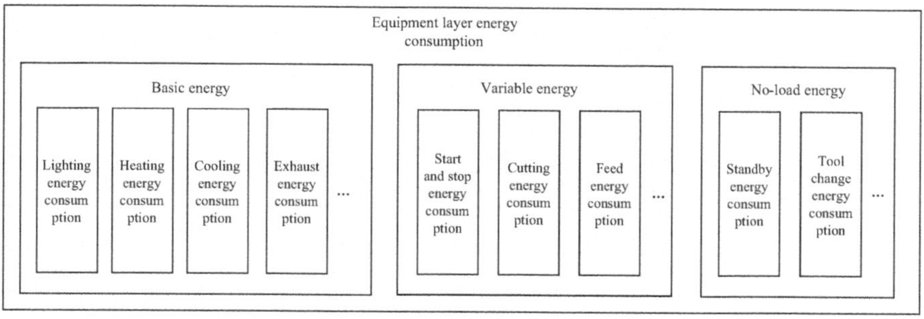

Fig. 1.1 Discrete manufacturing system equipment layer energy consumption

The energy consumption of the equipment layer is divided into three parts: basic energy, variable energy and no-load energy, as shown in Fig. 1.1. Basic energy includes energy consumption in system equipment such as lighting, hydraulics and cooling. Variable energy includes machine tool equipment start-up, cutting, tool change, feed and other processing energy consumption; during the machining process, the no-load energy generated by the machine tool standby due to the machining gap is no-load energy consumption, such as waiting for no-load, tool-less no-load energy consumption and no-load energy consumption during processing. Variable energy is the main body of the energy of the device layer. The greater the proportion of variable energy, the higher the energy efficiency of the device layer.

The operating state of machining equipment is divided into four categories: start-up, no-load between work steps, cutting and stop. When the equipment is switched between different states, the energy consumption also changes, as shown in Fig. 1.2. Among them, cutting energy consumption is the effective energy consumption part of the total energy consumption of machining equipment.

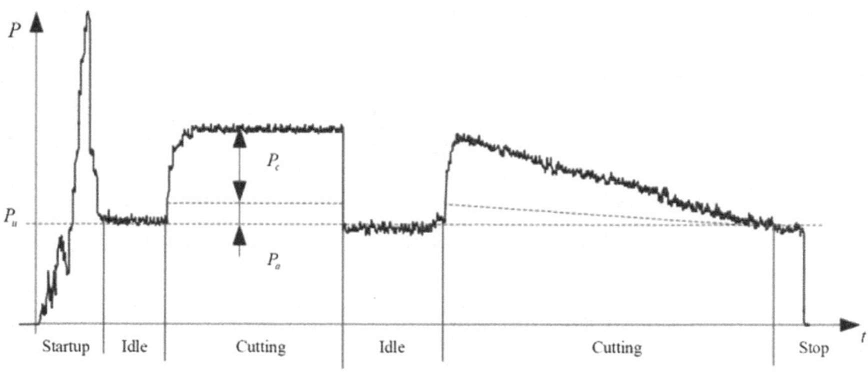

Fig. 1.2 Input power profile

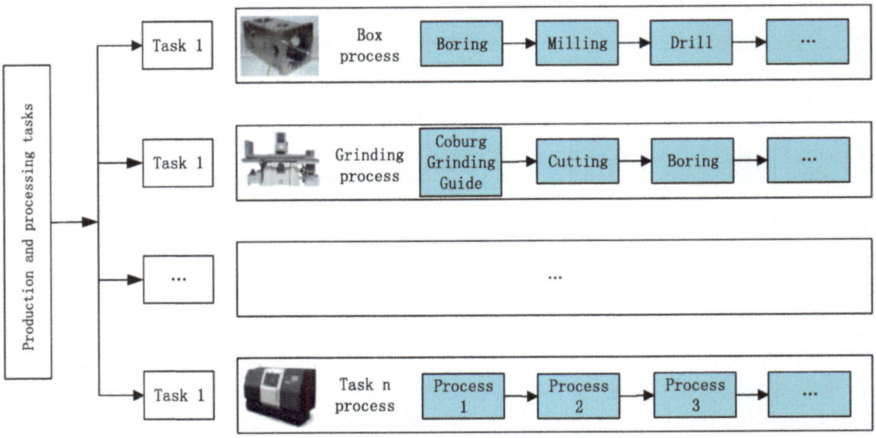

Fig. 1.3 Discrete manufacturing system task layer

2. **Process layer energy consumption**

Discrete manufacturing system processing technology has a variety of characteristics, the same product has a number of different processing routes, each process consists of several processing steps, the same process is processed on different equipment, and its energy consumption is generally different [20]. As shown in Fig. 1.3, the process layer contains all the processing routes of the product production process, which indirectly reflects the energy efficiency level of the equipment corresponding to all processed parts in the product processing process.

Considering the machining process of the workpiece, each machining task unit can be regarded as a machining process of a workpiece, and multiple machining operations are required to complete each machining task. Manufacturing system task layer energy consumption is determined by various factors such as process parameters, task plan and scheduling plan during production and processing. However, there are mainly two kinds of energy in the task layer of manufacturing system: First, the process energy consumption characteristics determined by factors such as processing parameters and process schedules, different process parameters and different process routes will bring about differences in energy consumption of processing equipment; second, the production planning and processing plan determine the energy consumption characteristics of the scheduling. Task-level energy consumption is defined as the energy consumption of the equipment required to complete a specific processing task for a discrete manufacturing system.

3. **Product layer energy consumption**

The discrete manufacturing system production process is the entire processing flow from raw material to product, including the processing of raw materials, transportation, product processing, assembly and rework [21], as shown in Fig. 1.4.

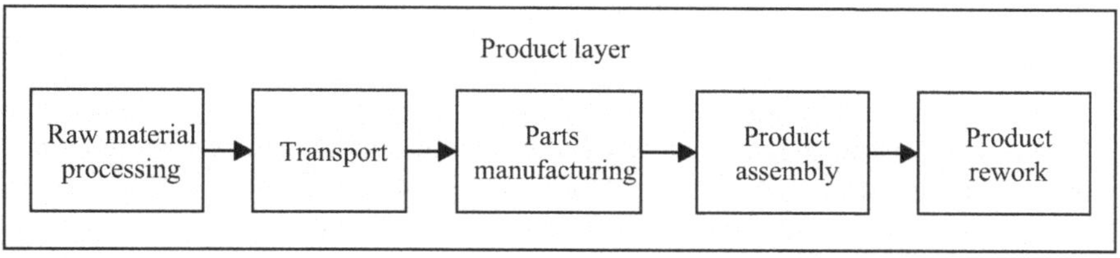

Fig. 1.4 Discrete manufacturing system product layer

Discrete workshop product layer energy consumption difference. The manufacturing environment of discrete manufacturing systems is complex and varied, including the changing production environment, numerous production equipments and different energy consumption characteristics. Therefore, the same type of energy-consuming equipment in different processing environments and different machines needs different energy for processing different products. When producing the same product, different task sequences or fine-tuning processing parameters can also lead to differences in energy consumption of the equipment. Therefore, the discrete manufacturing system is a changing, fluctuating process, and there is room for optimization. The product layer is the top layer of the manufacturing system, as shown in Fig. 1.4. The processing of the product includes raw material preparation, parts manufacturing, workpiece assembly and product remanufacturing. The product layer contains all the processes of discrete system production and processing, and the product layer energy consumption includes energy consumption of all types of equipment in the system.

1.3 Definition of Energy Efficiency in Discrete Manufacturing Systems

For discrete manufacturing systems dominated by machining, electrical energy is the main energy consumer. This book defines its energy efficiency from the perspective of thermodynamics, which is divided into two types: instantaneous energy efficiency and process energy efficiency.

Definition 1.1 Instantaneous energy efficiency $\eta(t)$ refers to the ratio of the effective energy change rate $P_o(t)$ of the manufacturing system at a certain time t to the input energy change rate $P_{in}(t)$.

$$\eta(t) = \frac{P_o(t)}{P_{in}(t)} \tag{1.1}$$

Definition 1.2 Process energy efficiency E_p refers to the ratio of the effective energy E_o of a certain process or a certain period of time to the energy consumption of the system E_{in}.

$$E_p = \frac{E_o}{E_{in}} = \frac{\int P_o(t)\mathrm{d}t}{\int P_{in}(t)\mathrm{d}t} \tag{1.2}$$

In the subsequent chapters of this book, the instantaneous energy efficiency defined by Formula (1.1) for the real-time energy efficiency of the device layer, the energy efficiency of the device layer step energy or the statistical energy efficiency within a certain time period using the process energy efficiency defined by Formula (1.2), both the task layer and the product layer are defined by process energy efficiency, and the time interval is the task or product processing start time to the task or product processing end time.

1.4 Research Situation at Home and Abroad

With the increasing shortage of energy shortages and environmental degradation problems, how to use energy more efficiently from the input and output of manufacturing systems to improve their utilization efficiency, that is, the problem of efficient energy production, has received widespread attention worldwide. The so-called energy-efficient manufacturing is the whole process of manufacturing resources from the input, conversion to output of the system, more efficient use of resources, reducing energy consumption and environmental impact and improving system energy efficiency [22]. In order to be able to achieve a high point in the research of energy-efficient manufacturing, the EU, Japan and other regions have invested in research aimed at reducing energy consumption and improving energy efficiency in manufacturing industries [23]. According to the study of National Institute of Standards and Technology (NIST) in 2009 [24], in order to achieve energy-efficient manufacturing of manufacturing systems, it will be necessary to conduct scientific and in-depth research in specific directions such as data collection, modelling theory and methods, interoperability of related terms/standards and support tools. In 2009, Intelligent Manufacturing System (IMS), in the blueprint for the future intelligent manufacturing system 2020, "energy efficient manufacturing" is listed as one of the five research areas of sustainable manufacturing, products and services, standardization, robots and key technologies of smart materials [25]. In recent years, most of the research on energy efficient manufacturing has focused on the field of process (process) industrial systems and has achieved certain research results, while the research on energy efficient manufacturing for discrete manufacturing systems is rare. With the deeper understanding of the scientific research and application significance of energy-efficient manufacturing in complex discrete manufacturing processes in industry and academia, the research on energy-efficient manufacturing of complex discrete manufacturing systems is emerging at home and abroad [22, 26].

At present, research on energy-efficient manufacturing focuses on three aspects of energy system modelling, energy efficiency analysis and evaluation, and energy system optimization scheduling, both for process industry and discrete manufacturing, and has achieved certain research results. From the research method, the previous research is based on the mechanism model, lacking the global and system modelling and analysis of the manufacturing life process, and the research results are also constrained by the limitations of the model. With the increase in the complexity of the discrete manufacturing process, it is difficult to establish a mechanism model, which has been widely recognized by researchers. However, it is worth noting that the development of industrial informationization, data acquisition and storage technology has accumulated huge amounts of data related to manufacturing processes, energy systems, equipment status, product information, decision management, etc. [27, 28]. Hence, it is convenient for research on energy consumption and energy efficiency in manufacturing systems.

1.4.1 Modelling of Energy Process in Discrete Manufacturing Systems

Discrete manufacturing systems are much more complex than process manufacturing systems. There are often a series of complex characteristics such as variability in processing tasks, equipment and process diversity, complex tree characteristics of products and their manufacturing energy consumption, product energy consumption and equipment (process) dynamic and variable correlation characteristics [7, 18]. Therefore, its energy efficiency modelling problem is much more complicated.

Manufacturing system energy consumption modelling and analysis are mainly three categories: manufacturing equipment energy modelling and analysis, process energy modelling and analysis, manufacturing system energy modelling and analysis.

1. **Manufacturing equipment energy consumption model**

The energy consumption of mechanical equipment is a multi-disciplinary problem. There are machining fields, such as cutting parameter selection, cutting tool selection, machining materials, which involve servo drive technology and motor control technology, and also cover sorting and automatic processing during machine tool processing. Automatic tool changing system (ATCS), hydraulic control technology, etc., are typical combination of machine–electric–liquid integration [29]. The manufacturing equipment has many energy sources and complicated energy flow, and the research on energy consumption modelling and analysis has received the most attention. Considering the multi-source energy consumption characteristics and motion properties of manufacturing equipment, the energy consumption model is also diversified, such as specific energy model [30], associated energy consumption model of machine tools and their energy consumption components [31], dynamic energy consumption model of machine tools and their main drive systems [32].

Literature [33] established the machine energy consumption model by analysing the energy consumption characteristics of each device in the generation process and predicted the cutting energy efficiency through the energy consumption model; literature [34] decomposes the energy consumption of the product processing process, establishes the energy consumption model based on the spindle drive system and combines the model to explore the energy consumption under different processing schemes. Literature [35], the intelligent sensor network is used to monitor and collect the energy consumption of the machine tool in the workshop generation process, and establish the mathematical model of the energy consumption of the machine tool. On this basis, the energy efficiency optimization of each device is carried out. Literature [36], the multi-source energy consumption characteristics of CNC machine tools are studied. The main drive system and feed system energy consumption characteristics related to machine tool load are studied. The energy consumption integration model of main drive system and feed system is given. Combined with this model, an online monitoring method for machine tool energy consumption is proposed, which realizes real-time perception of equipment energy consumption and effectively reduces the "information island" of equipment. Literature [37, 38] establishes the energy flow model of the main drive system of processing equipment based on the integration of machine tool motor and mechanical transmission system, based on the energy flow process of electromechanical system and the energy flow of each transmission component. Literature [39], Professor Kordonowy of MIT University of the USA proposed a calculation model based on the rated power of equipment to predict its energy efficiency, using statistical methods to divide the energy consumption of machine tools into variable (processing) energy consumption and constant (fixed) energy consumption. Literature [40] proposed an energy consumption estimation model based on processing code to accumulate the energy consumption of the whole process to obtain the total energy consumption.

2. **Process energy modelling**

The research on process energy consumption modelling mainly investigates the relationship between tools, processing materials, processing parameters, product structure characteristics and energy consumption, and establishes a series of process energy consumption mechanism models and empirical models [41, 42].

Literature [43], based on the energy consumption modelling of a single mechanical device, the difference in energy consumption between different processing equipments for processing different raw materials was studied. Literature [19], the polynomial fitting method is used to derive the functional relationship between the specific cutting energy (SCE) and the combined processing parameters such as spindle speed, feed rate and depth of cut. By analysing the function, the optimal cutting parameters can be found. Literature [44], Yan et al. optimized the machining parameters of the milling process and determined the optimal set of cutting parameters by the weight grey correlation analysis method and the response surface method to minimize the processing consumption.

3. **System energy modelling**

Modelling system energy consumption, mainly from the perspective of product manufacturing life cycle process modelling and analysis. Mose and Weinert [45] analysed the effects of factors such as production location, transport weight, transportation distance and transportation mode on specific energy consumption of products; Hu et al. [46] proposed an energy-related association modelling method for mechanical product manufacturing process based on design features; Wang et al. [47] analysed the multi-element features of system energy consumption and proposed a multi-energy characterization model for object-oriented Petri nets; Li et al. proposed a multi-element influence on the modelling of energy consumption characteristics in the dynamic environment of machine shop and the dynamic change of production environment. The proposed time-based object-oriented Petri net (CTOPN) was proposed. In terms of energy consumption feature modelling method [48], Dietmair and Verl [30] used statistical discrete event modelling to model the energy consumption behaviour of machine shops. Our research [49] established an ontology-based energy consumption knowledge model from the perspective of multiple sources of energy consumption.

1.4.2 Energy Efficiency Quantitative Analysis and Prediction

In the light of establishing system energy efficiency analysis and evaluation index system, the manufacturing system energy efficiency will be predicted well. On the basis of completeness, conciseness, importance, hierarchy and comparability, establishing the energy efficiency evaluation index of manufacturing system is the basis of energy-saving optimization of discrete manufacturing system, which has aroused the research boom at home and abroad. The U.S. Department of Energy (DOE) has established the "Industrial Evaluation Centre" [50]. The evaluation centre analyses and evaluates the energy consumption and energy efficiency of the manufacturing system production site with the joint efforts of 29 well-known universities in the USA, with a view to reducing the energy consumption of manufacturing systems and processing and improving energy efficiency.

The complex characteristics of energy consumption in discrete manufacturing systems make it difficult to accurately evaluate the energy efficiency of discrete manufacturing systems. It includes the following: (1) The complexity of the products processed by discrete manufacturing systems makes it difficult to express the effective output of discrete manufacturing systems; at the same time, the relationship between effective output and the energy consumption process of the manufacturing system is difficult to describe, increasing the difficulty of modelling energy efficiency evaluation. (2) Discrete manufacturing systems involve a wide variety of energy-consuming equipment such as manufacturing equipment, auxiliary equipment, transfer equipment, assembly equipment and energy consumption characteristics. The processing technology and processing parameters are ever-changing, which makes it possible to establish a complete energy consumption basic database and process energy consumption. The database of energy efficiency evaluation such as basic database and basic energy consumption database is a long way to go [7]. (3) The detailed evaluation of energy efficiency of discrete manufacturing systems also needs to solve the boundary problem of energy efficiency evaluation and the problem of energy efficiency evaluation index and evaluation parameter systemization. In summary, the problem of fine evaluation of energy efficiency in discrete manufacturing systems is complex and is a technical problem to be solved.

1. **Energy efficiency simulation and prediction**

Literature [51], for the increase of energy costs, new environmental legislation and concerns about energy security, a set of plant-level energy monitoring and control system for manufacturing execution system (MES), PC data analysis and decision support is proposed. Manage the framework, and analyse the energy consumption and energy efficiency of the manufacturing equipment, and display it on the MES system in a visual way. Literature [52] proposed a fuzzy modelling method for enterprise energy consumption process based on fuzzy extended Petri nets, which can quantitatively and qualitatively analyse the energy consumption of manufacturing process in manufacturing enterprises. It is an effective method for enterprises to carry out energy efficiency analysis and evaluation of manufacturing systems. Literature [53] combines Discrete Event Simulation (DES) and Life Cycle Assessment (LCA) to analyse the utilization rate of manufacturing resources and product life processes in the context of sustainable manufacturing. The system module AUTOMOD has established a comprehensive evaluation simulation model, and the simulation results can output the most productive and energy-saving production methods. Literature [54] propose a multi-granularity state diagram model for discrete manufacturing systems, simulate and control energy consumption in manufacturing processes and manage and evaluate energy consumption in discrete manufacturing systems.

2. **Energy efficiency quantitative analysis**

The process of evaluation and analysis of discrete manufacturing systems is a process in which all indicators obtain an analysis result through data processing and give a system comprehensive index value. The indicator data processing method is the most important part of the entire evaluation analysis process. The existing research methods mainly include the following:

(a) **Qualitative analysis**: There are mainly Delphi method, expert analysis method [55], etc. This method reflects the knowledge that exists in the human brain and can take advantage of the rich experience of industry experts, but it is also subject to this and is easily influenced by subjective ideas. Chu et al. [54] proposed a manufacturing resource allocation model based on the analysis of aircraft structural parts manufacturing process, where the expert experiences and historical data are used to evaluate the ability of personnel, the complexity of parts, the reliability of equipment, the reliability of machining tools, and the correlation between resources and parts. The expert analysis method has the advantages of simple process and evaluates the comprehensive results through the evaluation of industry experts. It is easy to realize in the actual production process. However, due to the influence of human experience level and knowledge reserve, there may be cases where expert decision-making may not be agreed.

(b) **Operational research method**: Use statistical, mathematical models and algorithms to mine the hidden information in the data to solve some complex problems. Common methods include grey correlation analysis method, principal component analysis method and multi-objective decision method. Feng et al. [56] applied an improved analytic hierarchy process in the analysis process. According to the evaluation matrix, the relative weights of the energy-influencing factors are obtained, and the grey correlation degree is used as the measurement index to determine the influencing factors or optimal solutions. This method avoids the disadvantages of ambiguity in the quantitative analysis process. Tang et al. [57] fusion super data envelopment analysis (DEA) model and grey fuzzy theory. A new super-DEA model for measuring weights is proposed, and a grey relational projection model for device program ordering is established, solved the grey attribute of the device program and its difficulty in quantitative evaluation of device program information. Finally, the device scheme is evaluated by this method, and the effectiveness of the proposed method is verified.

(c) **Machine learning method**: Simulate or implement human learning behaviours, acquire new knowledge or skills, and give evaluation analysis results. There are mainly particle swarm optimization algorithms, neural network algorithms, leapfrog algorithms, etc. Lu and Mu [58] applied the genetic algorithm to BP neural network training to evaluate the aging performance of stealth coating and established the basic model of genetic BP neural network, and in order to predict the aging performance of coated samples in high-altitude environment, the proposed method is effective and accurate by comparing experimental data with model output data.

(d) **Compound analysis method**: The composite superposition of multiple analysis methods can not only avoid the deficiencies of the single method, but also take advantage of the respective methods to achieve more accurate analysis. For example, Aqueveque et al. [59] combines the characteristics of entropy and expert analysis with the characteristics of entropy and expert analysis, combines the weights of the two, improved fuzzy evaluation methods, and establishes the relationship between energy quality and energy efficiency. Lee et al. [60] used fuzzy analytic hierarchy method to replace the fuzzy number in the two stages to reflect the ambiguity of human thinking and used the data envelopment analysis method to measure the relative efficiency of energy technology to high oil price with economic point of view. Decision-makers allocate decision data for limited resources.

3. **Discrete manufacturing system bottleneck identification**

Energy efficiency optimization of manufacturing systems is a key way to reduce energy costs, increase profitability, and enhance competitive advantage. However, the energy consumption process of discrete manufacturing systems shows complexity, diversity, dynamics, etc. [61]. The factors that lead to the level of energy efficiency are difficult to determine. Thus, the improvement of resource adjustment and control of manufacturing systems is impended, and the realization of energy efficiency optimization of enterprises is restricted. Hence, how to develop an effective optimization plan is an indispensable part of energy efficiency research in manufacturing systems.

The bottleneck of energy efficiency of the system restricts the energy efficiency of the production process and determines the overall energy efficiency level of the manufacturing system. Therefore, the identification and analysis of bottlenecks is an important way to improve the energy efficiency of the system and optimize the energy efficiency of the system. The existing researches have different definitions of bottleneck points, and the bottlenecks are mainly defined by the following aspects.

(a) **Equipment bottleneck**

The manufacturing equipment is regarded as the bottleneck point of the system [62], and the equipment with the worst working ability of the equipment is generally defined as the bottleneck point of the production system. Cao et al. [63] proposed a method of growth pruning neural network for the dynamic bottleneck analysis of manufacturing process. The method integrates multiple indicators, such as production load, utilization rate and buffer queue length, define the bottleneck of the device through a comprehensive bottleneck index and establishes a bottleneck determination mechanism to discover the bottleneck device; secondly, predict the possible bottlenecks through the model obtained by training, Finally, the feasibility and effectiveness of the method are verified by simulation. Zhai [64] for the production process with the goal of "minimizing the delay time," the workpieces whose completion time of the last process is greater than the corresponding delivery date are classified into the candidate bottleneck set, and these workpieces affect the level of the processing target. Among them, the most serious elements in the collection have the greatest impact on the job target, called the bottleneck artefact. Brundage et al. [65] introduced two bottlenecks: shutdown energy bottlenecks and power rating bottlenecks. The machine with the largest energy consumption per unit of energy consumption on the continuous production line is the bottleneck of the shutdown energy. The equipment with the largest power consumption per unit of energy consumption on the continuous production line is the rated power bottleneck. Rated power bottlenecks provide plant managers with the information they need to replace individual machines on the production line. Improve downtime energy bottlenecks and power rating bottlenecks to minimize energy waste and minimize energy waste compared to other machines on the line.

(b) **Process bottleneck**

The process factor of the manufacturing process is an angle to study the bottleneck of the job shop. Zhang and Chiong [66] proposed a multi-objective genetic algorithm based on the inefficiency of production system scheduling tasks, combining the local improvement strategies of two targets. In each iteration, select a bottleneck machine, improve the total weighted delay time by adjusting the processing sequence of related operations on the machine, which has certain reference value for future research on energy-efficient production scheduling. Zou et al. [67] used processing load, production capacity and yield rate as bottleneck indicators, and established a decision tree model for identifying bottlenecks, using decision tree algorithm to identify bottleneck processes in the system. Then, the decision tree model is established through the production process data to obtain the optimal solution of the process parameter combination. The optimized process parameters greatly reduce the rework rate of the production process, improve the yield rate of the product and solve the bottleneck problem. Wang et al. [68] in the Extended Flexible Job Shop Scheduling Problem (EFJSP) analysed time constraints between tasks, including serial, parallel and flexible relationships and then introduced a way to identify the bottleneck task that most significantly affects the completion time. First, the EFJSP solution (corresponding to the scheduling plan) is transferred to the Activity On Edge (AOE) network (edge active network), and then the bottleneck identification index is proposed according to the critical path of the AOE network, and the task with the largest bottleneck index value is used as the bottleneck.

(c) **Indicator bottleneck**

Han [69] proposed the decision-making unit with input and output data restrictions based on data envelopment analysis (DEA), and used one-year data on energy efficiency for monthly processing. Weighted fusion of DEA data for energy efficiency was realized with the help of analytic hierarchy process. Evaluate the relative effectiveness of energy efficiency data of different technologies to find out the direction of energy saving and consumption reduction. Beginning with the network characteristics of manufacturing system, Li et al. [70] established a dynamic identification method of multiple bottlenecks in production workshop based on Coupled Map Lattices (CML). First, based on the production and processing data, establish a networked model of manufacturing plant manufacturing nodes. Then, according to the characteristics of the node, the relationship between the nodes and the propagation path of the disturbance in the network structure, the definition of the bottleneck is expanded to complete the quantitative description of the bottleneck of the production workshop and predict the possible bottleneck. Zhang and Wu [71] designed a statistical method to evaluate the bottleneck characteristic value of each machine (machine bottleneck [MBN]). It reflects the impact of the processing tasks of each machine on the overall mission program performance. Based on the MBN value, the bottleneck machine is identified, and the proposed hybrid coding genetic algorithm reduces the potential search space by performing more searches on these bottleneck machines.

1.4.3 Manufacturing System Energy Efficiency Optimization Control

Energy Efficiency-Oriented Multi-Link Integrated Multi-objective Optimization of Discrete Manufacturing System is a typical NP-hard problem. The connotation of multi-objective optimization is to consider the economic indicators such as time, cost and quality while pursuing the highest energy efficiency, often with multivariate, multi-dimensional, multi-constraint, nonlinear and dynamic characteristics. It is a complex decision process involving many parameters such as process parameter optimization, machine tool optimization and production scheduling optimization. Since the energy efficiency optimization model of discrete manufacturing systems involves high-dimensional and multi-variable decision-making, the workload of optimization calculation increases exponentially with the increase of the number of dimensions and variables. How to improve the efficiency of optimization method is also a hot topic.

Energy efficiency optimization control of discrete manufacturing systems can be realized in two ways: First, technical parameters are energy saving, and overall control of system energy consumption is achieved by optimizing equipment operating conditions and processing parameters; The second is to manage energy conservation, optimize the production system production planning and generate energy-saving production plans from the process layer and the system layer.

1. **Technical parameters energy saving**

Energy saving of technical parameters mainly includes upgrading of production equipment, optimization of processing technology (standard use of tools, rational arrangement of equipment processing order), etc. [72]. Among them, the optimization and upgrading of production equipment are quickly effective and good effective, but in the actual production environment, during the life cycle of the equipment, the enterprise cannot update the equipment for cost reasons. In actual production, enterprises often reduce energy consumption by optimizing equipment processing parameters, and energy consumption optimization through this method makes the process simple and convenient and practical. For example, the literature [73] compares the effects of process parameters and material removal rates on energy consumption in production processes through numerical control during NC machining, Optimization of cutting parameters and energy efficiency using multi-objective genetic algorithm. Literature [74], the carbon emission model of the cutting process is constructed for the carbon emission of the process, and the purpose of energy saving, environmental protection and green manufacturing is achieved by rationally optimizing the cutting parameters of the mechanical equipment. Literature [75], in the cutting process of raw materials, the Taguchi method was used to optimize the turning parameters, and the combination of cutting parameters with the best energy consumption was obtained. The experimental results verified that the machining energy consumption increased with the increase of cutting speed. Literature [76], the particle swarm algorithm is used to optimize the machining process parameters of the machine tool. The experiment proves that the energy consumption of the equipment is reduced by about 64%. Ultimately achieve the goal of energy saving and emission reduction. Literature [77], the machine tool processing energy consumption and roughness are optimized. The turning regression equation is established by grey correlation method, and the influence of feed rate on energy consumption and roughness during machining is analysed. Literature [78], the response surface method is used to establish the parameter model of milling parameters and energy consumption, analyse historical energy consumption data, determine the main parameters affecting the energy consumption of the workshop and make real-time predictions of system energy consumption according to the workshop resources and mission plan.

2. **Production management energy saving**

Energy conservation in production management mainly optimizes the layout of the workshop, optimizes the scheduling and then achieves energy conservation through various organizational measures, such as shortening the no-load time of equipment through reasonable scheduling methods so as to reduce the no-load energy consumption, rationally setting the workshop layout to reduce transportation energy consumption, etc. On the production management energy saving of discrete manufacturing process, the literature [79] analyses and calculates the energy consumption utilization rate of discrete manufacturing system based on the historical data of workshop production scheduling and establishes the workshop energy consumption evaluation system; statistical methods are used to evaluate and analyse the energy consumption of the workshop and predict the energy consumption value of the workshop in the future; literature [80] acquisition of basic production data of discrete manufacturing workshop was completed through workshop informatization construction. On this basis, a simulation model of energy consumption-oriented scheduling was established. The validity of the model was verified by aluminium die casting workshop, and the energy consumption value of workshop was calculated. Literature [81], a flexible workshop energy-saving scheduling model with minimum energy consumption as the scheduling target is established under the premise of meeting the delivery requirements, and the sliding window mechanism is used to dynamically optimize the shop scheduling. Literature [82], the method based on process and equipment coding separation is used to solve the multi-objective optimization problem of flexible job shop and improve the utilization rate of workshop equipment. Literature [83] et al., based on the energy consumption of the production workshop, optimize the energy consumption and completion time objectives of the production auxiliary process, establish a multi-objective planning model for the workshop and study the problem of shop scheduling.

In summary, the current research on energy efficiency optimization of discrete manufacturing systems has yielded fruitful results, but it also has the following shortcomings: (1) The optimization of energy consumption for processing equipment parameters belongs to local optimization, and the optimization result is greatly affected by external influence factors. (2) At present, the research on production scheduling of discrete manufacturing systems is mainly based on the optimization of processing energy consumption and processing time. It is necessary to further consider the quality of products, customer satisfaction and environmental impact. (3) The traditional optimization method lacks the reuse of the production history scheduling data of the workshop, and it is easy to have multiple calculations for the same problem, increasing the waste of manpower and material resources.

1.5 Content Structure of This Article

In this book, the complex discrete manufacturing system is taken as the research object, considering the multi-source, dynamic uncertainty of energy consumption in the production process, and the complex correlation between process parameters and equipment state, research on key scientific issues involved in system energy integration and knowledge modelling, energy consumption prediction and analysis and energy intelligent optimization decision. The theory of energy efficiency optimization control based on discrete manufacturing system is proposed, and the corresponding software platform is developed to solve the research difficulties of real-time energy consumption monitoring, energy efficiency quantitative evaluation and bottleneck identification, process parameters and production scheduling optimization decision, and to explore the maximum energy saving space of the system. The research content of this book is shown in Fig. 1.5.

In this chapter: It mainly expounds the research background and research significance of this book, describes the energy efficiency optimization problem of discrete manufacturing system and introduces the research status of the subject at home and abroad.

Chapter 2: Based on the energy consumption of discrete manufacturing systems, a deployment method for energy consumption information collection network of discrete manufacturing system is proposed, and aiming at the problem that the effective processing energy consumption of the equipment layer is difficult to obtain directly, combined with the recursive identification algorithm, an indirect acquisition method for effective processing energy consumption is presented.

Chapter 3: Aiming at the randomness and dynamics of discrete manufacturing systems in the production process, a modelling method for manufacturing system energy consumption is proposed. Based on the mechanism to establish a multi-source

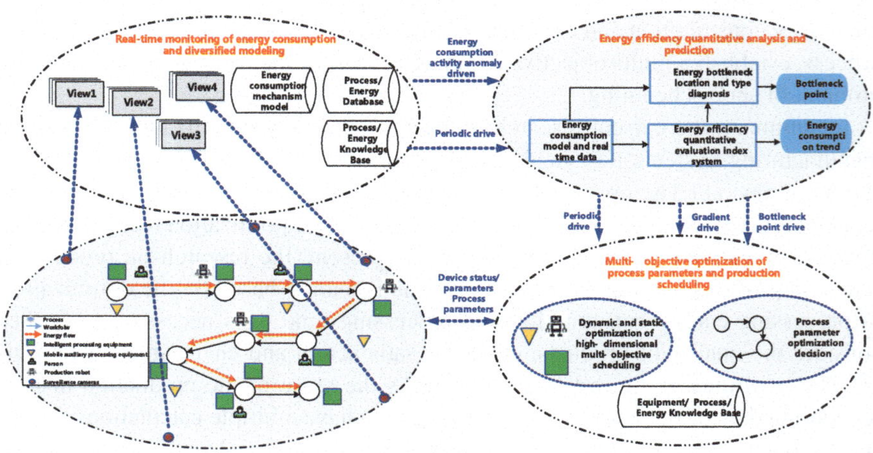

Fig. 1.5 Research content and research ideas

and multi-level energy integration mathematical model, and further consider the dynamic evolution and uncertainty of the manufacturing process, the ontology-based energy consumption knowledge model construction method is proposed.

Chapter 4: A number of indicators are summarized from the product layer, device layer, system layer, and process layer, establish a three-tiered indicator system for energy efficiency quantitative analysis of discrete manufacturing systems including economic energy efficiency, product energy efficiency, equipment energy efficiency, task process energy efficiency and four secondary indicators.

Chapter 5: A combined energy efficiency quantitative analysis method, in the process of weight determination, the method of rough set-attribute level model combination is used to fully integrate the index weights of objective and subjective aspects to overcome the greyness and uncertainty in the process of energy efficiency quantitative analysis.

Chapter 6: Establishing a discrete manufacturing process energy network model from the perspective of complex networks, a method for identifying energy consumption bottlenecks in discrete manufacturing processes based on complex networks is proposed to realize energy consumption bottleneck identification in manufacturing processes in complex environments.

Chapter 7: A discrete energy analysis method based on improved principal component analysis is proposed. Ensure that information loss is minimal, through the dimension reduction process and improving the weight determination and non-dimensionalization method of traditional principal analysis, the system energy efficiency evaluation of each stage is realized.

Chapter 8: Aiming at the static multi-objective optimization scheduling problem for energy efficiency, a flexible job shop optimization scheduling model with minimum machine completion time, task layer product energy consumption and total machine load as optimization targets is established. A variable particle swarm optimization algorithm based on knowledge fusion and public key block for variable neighbourhood search is proposed.

Chapter 9: Considering three dynamic events of emergency insertion, machine failure and machine repair, a high-dimensional multi-objective flexible job shop scheduling model with performance indicators including scheduling efficiency, machine load, rescheduling stability and energy consumption is constructed. Research on dynamic pre-reaction scheduling method is based on NSGA-III for high-dimensional multi-objective discrete manufacturing system.

Chapter 10: Under the constraints of the actual machining environment, a multi-objective optimization model for processing parameters of discrete manufacturing systems with cutting speed, feed rate and depth of cut as optimization variables and energy efficiency as the optimization goal is established. A random walk multi-objective particle swarm optimization algorithm is proposed to solve the process parameter optimization problem.

Chapter 11: Based on the ISOK industrial operation platform, integrated energy efficiency real-time monitoring, quantitative analysis, process parameter optimization decision, static and dynamic optimization scheduling algorithm. Introduce the development of energy efficiency optimization control software system, and verify the application on multi-variety flexible bearing processing production line.

References

1. Cai L, Wang Y, Ji Z (2016) Prediction of energy efficiency of NC machine tools based on recursive method with discounted measurements. J Syst Simul 28(8):1725–1731
2. Rohdin P, Thollander P (2006) Barriers to and driving forces for energy efficiency in the non-energy intensive manufacturing industry in Sweden. Energy 31(12):1836–1844
3. Huang W (2013) Research on energy consumption monitoring system of main transmission system of CNC machine tools. Chongqing University, Chongqing
4. Quadrelli R, Peterson S (2007) The energy–climate challenge: recent trends in CO_2 emissions from fuel combustion. Energy Policy 35(11):5938–5952
5. MIT (2009) Environmentally benign manufacturing [EB/OL], 2009-10-08
6. Liu F, Xu Z, Dan B et al (1995) Energy characteristics of machining systems and its application. China Machine Press, Beijing
7. Liu F, Zhou X, Cai W (2015) The complexity and strategy for establishing product energy consumption allowance in discrete manufacturing industry. J Mech Eng 51(19):138–145
8. European Commission (2016) 2030 climate & energy framework. https://ec.europa.eu/clima/policies/strategies/2030/index_en.htm
9. Innovation T (2016) Horizon 2020. https://ec.europa.eu/programmes/horizon2020/
10. USA Clean Energy Manufacturing Innovation Institute (2016). http://www.energy.govenergy-department-announces-70-million-innovation-institute-smart-manufacturing
11. Verl A, Westkamper E, Abele E (2011) Architecture for multilevel monitoring and control of energy consumption. In: The 18th CIRP international conference on life cycle engineering, pp 347–352
12. Zhou LR, Li JF, Li FY et al (2016) Energy consumption model and energy efficiency of machine tools: a comprehensive literature review. J Clean Prod 112(5):3721–3734
13. Li C-B, Li L-L, Tang Y et al (2019) A comprehensive approach to parameters optimization of energy-aware CNC milling. J Intell Manuf 30(1):123–138
14. Cai W, Liu F, Zhou XL (2016) Fine energy consumption allowance of workpieces in the mechanical manufacturing industry. Energy 114:623–633
15. Velchev S, Kolev I, Ivanov K (2014) Empirical models for specific energy consumption and optimization of cutting parameters for minimizing energy consumption during turning. J Clean Prod 80:139–149
16. Eisele C, Schrems S, Abele E (2011) Energy-efficient machine tools through simulation in the design process. Springer
17. Rahimifard S, Seow Y, Childs T (2010) Minimising embodied product energy to support energy efficient manufacturing. CIRP Ann Manuf Technol 59(1):25–28
18. Liu F, Xu ZJ, Dan B (1995) Mechanical characteristics of mechanical processing systems and their applications. Machinery Industry Press, Beijing
19. Draganescu F, Gheorghe M, Doicin CV (2003) Models of machine tool efficiency and specific consumed energy. J Mater Process Technol 141(1):9–15
20. Lim J, Chae MJ, Yang Y et al (2016) Fast scheduling of semiconductor manufacturing facilities using case-based reasoning. IEEE Trans Semicond Manuf 29(1):22–32
21. Wang RD, Sun XS, Yang X et al (2016) Cloud computing and extreme learning machine for a distributed energy consumption forecasting in equipment-manufacturing enterprises. Cybern Inf Technol 16(6):83–97
22. Tridech S, Cheng K (2011) Low carbon manufacturing: characterization, theoretical models and implementation. Int J Manuf Res 6(2):110–121

23. Park CW, Kwon KS, Kim WB et al (2009) Energy consumption reduction technology in manufacturing: a selective review of policies, standards, and research. Int J Precis Eng Manuf 10(5):151–173
24. Rachuri S, Ram D, Narayanan A et al (2009) NIST workshop on sustainable manufacturing: metrics, standards, and infrastructure. In: NIST workshop sustainable manufacturing, Gaithersburg, 13–15 Oct 2009
25. Intelligent Manufacturing System (2010) Supporting global research for 2020 manufacturing vision. http://www.ims2020.net
26. Wang JF, Li SQ, Liu JH (2013) A survey on energy efficient discrete manufacturing system. J Mech Eng 49(11):89–97
27. Zheng L, Jang PY, Qiao LH et al (2010) Challenges and frontiers of manufacturing systems. J Mech Eng 46(21):124–136
28. Bunse K, Vodicka M (2010) Managing energy efficiency in manufacturing processes-implementing energy performance in production information technology systems. IFIP Adv Inf Commun Technol 328:260–268
29. He C (2013) Study on energy-saving of running main driving system of CNC machine tool and its application. Chongqing University, Chongqing
30. Dietmair A, Verl A (2009) A generic energy consumption model for decision making and energy efficiency optimization in manufacturing. Int J Sustain Eng 2(2):123–133
31. Cai W, Liu F, Dinolov O et al (2018) Energy benchmarking rules in machining systems. Energy 142:258–263
32. Pellegrinelli S, Valente A, Tosatti LM (2012) An integrated setup planning and pallet configuration approach for highly automated production systems with energy modelling of manufacturing operations. Procedia CIRP 3:49–54
33. Chen Y, Liu B, Shen Y et al (2016) The energy efficiency of China's regional construction industry based on the three-stage DEA model and the DEA-DA model. KSCE J Civ Eng 20(1):34–47
34. Li XL, Xing KY, Zhou MC et al (2018) Modified dynamic programming algorithm for optimization of total energy consumption in flexible manufacturing systems. IEEE Trans Autom Sci Eng 99:1–15
35. Gahm C, Denz F, Dirr M et al (2016) Energy-efficient scheduling in manufacturing companies: a review and research framework. Eur J Oper Res 248(3):744–757
36. Hu SH (2012) Energy consumption characteristics of multiple-component of modern CNC machine tools. Chongqing University, Chongqing
37. Liu F, Li CB, Cao HJ et al (2009) Green manufacturing technology connotation and system framework based on product life cycle. J Mech Eng 5(12):115–120
38. Liu F, Xu ZJ (1990) Energy transmission mathematical models of main driving system of machine tools. J Chongqing Univ (Nat Sci Ed) 13(2):8–14
39. Kordonowy D (2003) A power assessment of machining tools. Massachusetts Institute of Technology, Cambridge
40. Avram O, Xirouchakis P (2011) Evaluating the use phase energy requirements of a machine tool system. J Clean Prod 19(6–7):699–711
41. Gutowski T, Branham M, Dahumus J (2009) Thermodynamic analysis of resources used in manufacturing processes. Environ Sci Technol 43(5):1584–1590
42. Matibenga PT, Rajemi MF (2011) Calculation of optimum cutting parameters based on minimum energy footprint. CIRP Ann Manuf Technol 60(1):149–152
43. Li Y, Yan H, Yan W et al (2014) A framework for characterising energy consumption of machining manufacturing systems. Int J Prod Res 52(2):314–325
44. Yan JH, Li L (2013) Multi-objective optimization of milling parameters the trade-offs between energy, production rate and cutting quality. J Clean Prod 52(1):462–471
45. Mose C, Weinert N (2015) Process chain evaluation for an overall optimization of energy efficiency in manufacturing—the welding case. Robot Comput Integr Manuf 24:44–51
46. Hu KL, Peng C, Evans S et al (2017) Minimising the machining energy consumption of a machine tool by sequencing the features. Energy 121:292–305

47. Wang Q, Wang X, Yang S (2014) Energy modeling and simulation of flexible manufacturing systems based on colored timed petri nets. J Ind Ecol 18(4):558–566
48. Li Y, Hi Y, Wang Y et al (2015) A modeling method for hybrid energy behaviors in flexible machining systems. Energy 86:164–174
49. Xu B, Wang Y, Ji ZC (2017) Knowledge network model of the energy consumption in discrete manufacturing system. Mod Phys Lett B 1740100
50. IAC. Industrial assessment centers. http://eere.energy.gov/industry/bestpractices/iacs.html
51. Vikhorev K, Greenough R, Brown N (2013) An advanced energy management framework to promote energy awareness. J Clean Prod 43:103–112
52. Ma FM, Wang J (2008) Enterprise energy consumption process simulation method for energy efficiency evaluation. Comput Integr Manuf Syst 14(12):2361–2368
53. Muroyama A, Mani M, Lyons K et al (2011) Simulation and analysis for sustainability in manufacturing processes. In: Proceedings of the ASME. 2011 international design engineering technical conferences & computers and information in engineering conference IDETC/CIE 2011, Washington, 28–31 Aug 2011
54. Chu W, Li Y, Liu C et al (2014) A manufacturing resource allocation method with knowledge-based fuzzy comprehensive evaluation for aircraft structural parts. Int J Prod Res 52(11):3239–3258
55. Zhu HW, Chen ZH, Zeng M (2011) The study on synthetical evaluation system of e-government system performance. In: International conference on information management, vol 3, no 20, pp 541–544
56. Feng Q, Wu B, Jiang W et al (2012) The evaluation of design for energy efficiency of buildings based on analysis hierarchical process and grey theory. Appl Mech Mater 6(3):2–7
57. Tang LB, Guo D, Wu J et al (2014) Program evaluation and its application to equipment based on super-efficiency DEA and gray relation projection method. J Syst Eng Electron 25(6):1037–1042
58. Lu YL, Mu JY (2015) Evaluation model of aging properties of stealth coatings based on genetic BP neural network. Acta Armamenta 36(8):1580–1586
59. Aqueveque P, Wiechmann EP, Henriquez JA et al (2014) Energy quality and efficiency of an open pit mine distribution system: an evaluation. In: Industry applications society meeting. IEEE, pp 1–7
60. Lee SK, Mogi G, Hui KS (2013) A fuzzy analytic hierarchy process (AHP)/data envelopment analysis (DEA) hybrid model for efficiently allocating energy R&D resources: in the case of energy technologies against high oil prices. Renew Sustain Energy Rev 21:347–355
61. Zhang YX (2011) Research on performance optimization of parallel discrete event simulation for complex system application. National University of Defense Technology, Hunan
62. He WJ, Lu JS, Li XL (2013) Research on production scheduling based on the production of logistics bottleneck. Light Ind Mach 31(1):101–110
63. Cao ZC, Qiu MH, Liu M (2016) Dynamic bottleneck analysis for semiconductor wafer fabrication system based on growing and pruning neural networks. Acta Electron Sin 44(7):1636–1642
64. Zhai YN, Wang JQ, Chu W et al (2015) A study of large-scale job shop scheduling problem based on TOC. Mech Sci Technol Aerosp Eng 34(8):1222–1228
65. Brundage MP, Chang Q, Li Y et al (2014) Energy efficiency management of an integrated serial production line and HVAC system. IEEE Trans Autom Sci Eng 11(3):789–797
66. Zhang R, Chiong R (2016) Solving the energy-efficient job shop scheduling problem: a multi-objective genetic algorithm with enhanced local search for minimizing the total weighted tardiness and total energy consumption. J Clean Prod 112:3361–3375
67. Zou FL, Leng S, Lian PF et al (2016) An improved method on bottleneck based on decision tree. Mod Manuf Eng 6:121–128
68. Wang Y, Tang L, Zhou Y et al (2019) Bottleneck identification of extended flexible job shop scheduling problem. In: 2018 6th international symposium on computational and business intelligence (ISCBI), Basel. IEEE, pp 23–27

69. Han YM, Geng ZQ, Liu YQ (2014) Energy efficiency evaluation based on data envelopment analysis integrated analytic hierarchy processing ethylene production. Chin J Chem Eng 37(12):1279–1284
70. Li XJ, Sun WL, Yuan YP et al (2016) Multi-bottleneck identification for job-shop network in disturbance environment. J Xi'an Jiaotong Univ 50(12):64–72
71. Zhang R, Wu C (2009) Bottleneck identification procedures for the job shop scheduling problem with applications to genetic algorithms. Int J Adv Manuf Technol 42(11–12):1153–1164
72. Mashaei M, Lennartson B (2013) Energy reduction in a pallet-constrained flow shop through on–off control of idle machines. IEEE Trans Autom Sci Eng 10(1):45–56
73. Duflou JR, Sutherland JW, Dornfeld D et al (2012) Towards energy and resource efficient manufacturing: a processes and systems approach. CIRP Ann Manuf Technol 61(2):587–609
74. Zhou Z, Yao B, Xu W et al (2017) Condition monitoring towards energy-efficient manufacturing: a review. Int J Adv Manuf Technol 91(9):3395–3415
75. Yang J, Dong R, You WL (2018) Analysis of multi objective optimization model for machining process route for high efficiency and low carbon. Mod Manuf Technol Equip 257(4):150–154
76. Lin WW (2016) Research and application on the key technologies of low carbon operation optimization in a mechanical machining system. Huazhong University of Science and Technology, Wuhan
77. Meeran S, Morshed MS (2012) A hybrid genetic tabu search algorithm for solving job shop scheduling problems: a case study. J Intell Manuf 23(4):1063–1078
78. Yoon HS, Lee JY, Kim HS et al (2015) A comparison of energy consumption in bulk forming, subtractive, and additive processes: review and case study. Int J Precis Eng Manuf-Green Technol 1(3):261–279
79. Claudia E, Marija J (2015) Energy efficiency, robustness, and makespan optimality in job-shop scheduling problems. Ai Edam Artif Intell Eng Des Anal Manuf 30(3):300–312
80. Xiong J, Xing L, Chen Y (2013) Robust scheduling for multi-objective flexible job-shop problems with random machine breakdowns. Int J Prod Econ 141(1):112–126
81. Bo H, Jiang R, Zhang G (2014) Search strategy for scheduling flexible manufacturing systems simultaneously using admissible heuristic functions and nonadmissible heuristic functions. Comput Ind Eng 71(1):21–26
82. Jia ZH, Leung YT (2015) A meta-heuristic to minimize makespan for parallel batch machines with arbitrary job sizes. Eur J Oper Res 240(3):649–665
83. Wang L, Deng J, Wang SY (2016) Survey on optimization algorithms for distributed shop scheduling. Control Decis 31(1):1–11

Chapter 2
Perceiving the Energy Consumption Information of Discrete Manufacturing Systems

2.1 Introduction

The discrete manufacturing system is a complex giant. For instance, the machine tool manufacturing system used in the equipment manufacturing industry includes not only such processes as the transportation, processing, inspection and assembling of the components, but also those of energy consumption, waste emission and information transmission (of processing, scheduling and quality control). The energy consumption problems involved are the overdue number of energy-consuming equipment, the excessive variety of the energy consumed and the huge difference in energy usage. Such equipments include numerical control machine tools, common machine tools, bridge crane and platform lorry. The energy consumed involves primary energy sources (e.g. coal and natural gas) and secondary energy sources (e.g. electricity, steam and coal gas). Additionally, machine tool could be viewed as a small system, comprising main transmission system, feeding transmission system, hydraulic pressure system, cooling and lubrication system, as well as peripheral equipment like fans, lighting and human–machine interface. Therefore, the discrete manufacturing system is an organic complex strongly influenced by the interconnection of energy, material and information flows [1]. Movements of these relevant "triple flows" are shown in Fig. 2.1.

The perceived energy consumption information of the discrete manufacturing system includes overall power consumption of equipments directly related with energy consumption, effective manufacturing energy consumption and the real-time operation state of equipments; and information indirectly related to energy consumption, such as processing data, data of product or component structural features, status of processing status, manufacturing scheduling or scheduling data. The collection and transmission of various information can be achieved via implementing wireless sensors, RFID and smart data collection terminals in the manufacturing workshops or production lines, as well as integrating multiple dispersed information sub-system. Of the above, the most difficult is the perception of effective processing energy consumption. Because its direct assessment requires installation of cutting force sensors

Y. Wang et al., *Quantitative Analysis and Optimal Control of Energy Efficiency in Discrete Manufacturing System*, https://doi.org/10.1007/978-981-15-4462-0_2

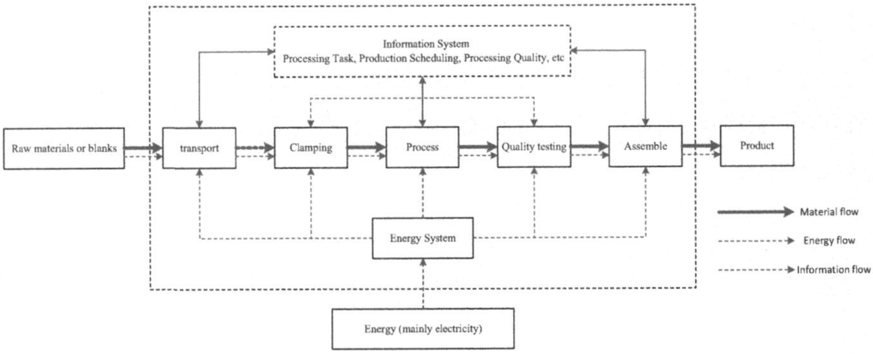

Fig. 2.1 Movements of the "triple flows" in the discrete manufacturing system

on the manufacturing facility, which in turn affects the normal operation and manufacturing precision of the equipment. Therefore, normally we use soft measuring methods to obtain effective processing energy consumption.

2.2 Web Implementation of Energy Consumption Information Collection in Discrete Manufacturing System

The first chapter has described and analysed the equipment level, task level and auxiliary production level of the discrete manufacturing system. Here, we will ascertain the information collection nodes at each level, concerning the total energy consumption of the equipment, processing tasks and auxiliary production, to get a transparent and detailed overview of the entire workshop's energy consumption. The data collection process is mainly to conduct statistical analysis of the required data collected from bottom facilities. It enforces real-time monitoring of the target system. And by transmitting the facility's real-time operation status to the workshop's scheduling centre, it helps the dispatcher to provide more reasonable processing task allocation. In this section, hardware network deployment can be divided into the collection, the communication and the data processing layers. These three layers refer respectively to: data collection at the facility and hardware layer, workshop data communication and network transmission, and data processing at the back office database. It adopts the mode of decentralized management and centralized supervision, as shown in Fig. 2.2.

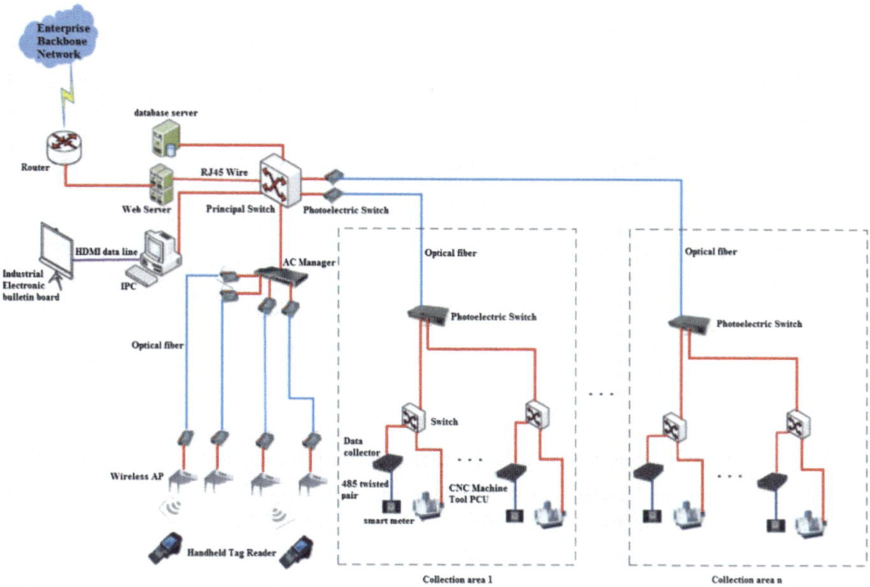

Fig. 2.2 Topological graph of the discrete manufacturing system's acquisition network

2.2.1 *The Discrete Manufacturing System's Energy Consumption Acquisition Network*

1. **The total energy consumption acquisition of the discrete manufacturing system's manufacturing facility**

A smart meter is used to collect the total energy consumption of the facilities. Its built-in RS485 interface outputs acquisition signals, first transmitted to the data acquisitor via RS485 and the USB module, then to the photoelectric switches in the form of TCP/IP protocols. The optical signals thus generated are transmitted to the planning and scheduling office, wherein they are transformed into network signals via photoelectric transceivers, and finally transmitted to the data acquisitor. The relevant configuration software installed on upper-computer processes and categorizes the collected parameter variables and then saved in the local SQL database installed in the data acquisitor. The Salien acquisition software on the data acquisitor then synchronize these SQL data to the Oracle database on the servers. The actual acquisition process is shown in Fig. 2.3.

2. **Energy consumption acquisition of the discrete manufacturing system's main spindle and feed shaft**

Facilities of the discrete manufacturing system are connected via a wired network. Their energy consumption information is collected by the Original Equipment Manufacturer (OEM) of Siemens. Once the machine tool activates, the OPC client program

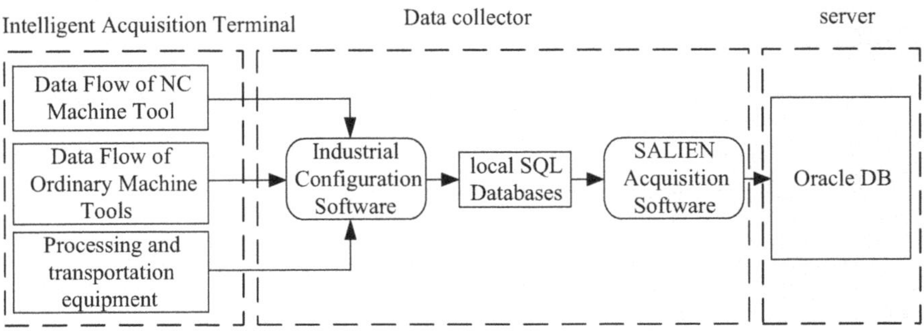

Fig. 2.3 Data acquisition flow chart of the total energy consumption of discrete manufacturing system's production facilities

written on Siemens's OEM software development platform is embedded into the PCU of numerical control machine tools and operates as the HMI software activates. In accordance with the address of a variable, the OPC client makes a fetch request to the OPC server. The OPC server answers the request and transports the data to the photoelectronic switches via PCU's in-built Ethernet interface under TCP/IP protocols. The optical signals thus generated are passed to the planning and scheduling office, then transformed into network signals via photoelectric switches and transported to the upper computer, wherein the configuration software processes, classifies and saves these data into the local SQL database, before synchronizing them to the Oracle database on the server by the Salien acquisition software. The actual acquisition and transmission process is shown in Fig. 2.4.

3. **The processing information acquisition network of the discrete manufacturing system**

Once semi-finished products or components arrive at the manufacturing workshop, the workshop administrator will warehouse this batch by writing the processing information and parameters of this batch on tags and then pasting them on corresponding

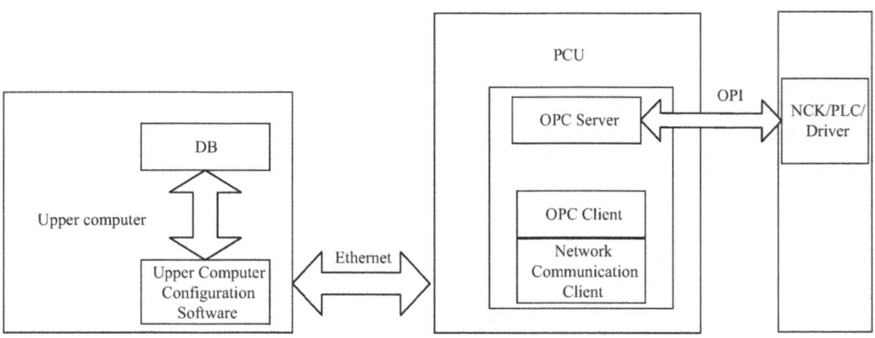

Fig. 2.4 Energy consumption data acquisition of machinery production equipment's main spindle and feed shaft

products or components. Workers are operating each processing machine equipped with a hand-held tag reader. Before the machine undertakes a certain procedure for a certain component, the operator shall first read the information tag on the component with this hand-held tag reader, and the component's processing parameters will be displayed by the information extraction software on the hand-held machine. Once the processing is completed, the worker will edit the processing completion information and write additional notes onto the tag with the reader's software. The hand-held reader's reading distance is set within 20 cm. Once the hand-held machine reads the tag information, the relevant installed software will display the tag information, while transmitting it to the upper computer via a wireless network and saving it in local SQL database, and then synchronizing it into the server's Oracle database by the Salien acquisition software. Employed on-site is the high-frequency anti-metal tag, which is magnetically mounted. As for those circular shaft parts, the tag is attached to one side of the component with a rubber band for strengthened anchorage. High-frequency hand-held machine is used for responding to the tag. The tag information acquisition process is shown in Fig. 2.5.

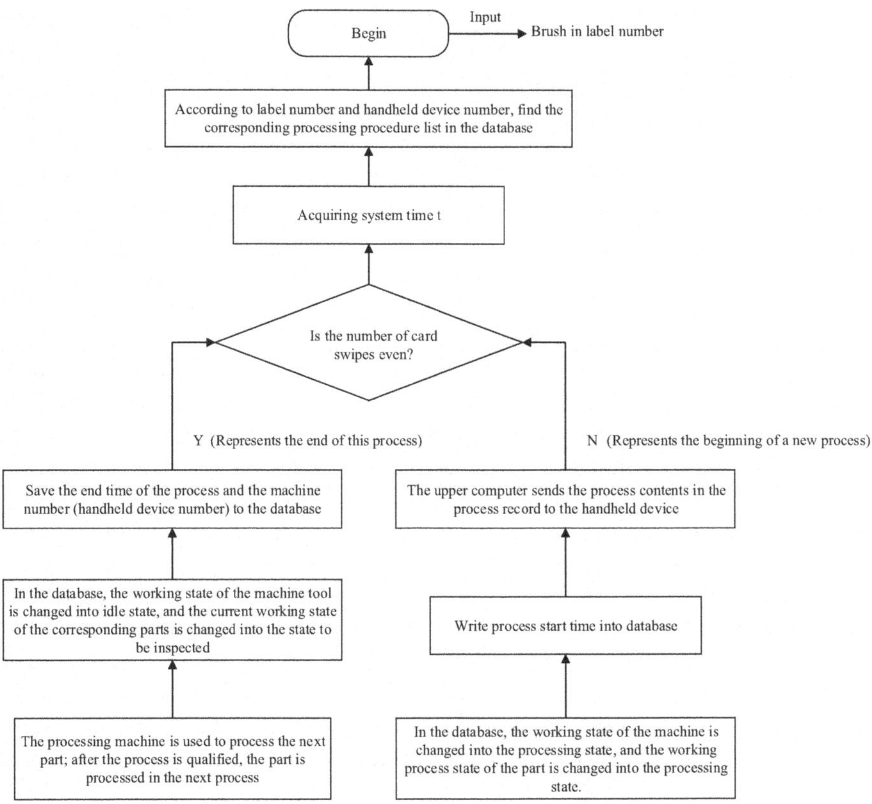

Fig. 2.5 Tag processing information extraction flow chart

2.2.2 Sensors for the Discrete Manufacturing System's Energy Consumption Acquisition

1. Smart meter: The smart meter is chosen in accordance with the harsh environment of the workshop, where the working temperature ranges from −10 to 55 °C. It can run assessment of three-phase parameters (including three-phase currents, three-phase voltage, three-phase active power, in-phase active power, three-phase reactive power, in-phase reactive power, three-phase apparent power, in-phase apparent power, three-phase power factors, in-phase power factors, frequency), active energy and reactive energy, while capable of collecting and transmitting data as well. It is equipped with an RS-485 interface which adopts the international standard of MODBUS-RTU protocol, thus capable of transmitting the acquired data first to the SQL Server database at the local data acquisitor, then to the back office database server via the workshop's intranet.
2. Hand-held machine and e-tag: Various lathes are distributed in the processing workshop, such as millers, lathe tools, drillers, grinders, and MC processing centre. Since it is a metallic environment, anti-metallic e-tag and reader are necessary. Adopted on-site is ADS-883 anti-metallic tag of 900 MHz. There are two types of tags: component tags and machine tags. The former is bound with components; the latter, with equipment. Magnetic anti-metallic tags can be stuck to components during transportation and removed during processing. As for round shaft components, one side of the tag can be attached the components with a rubber band for reinforcement. The operators approach the e-tag with hand-held machines, and by clicking the "read" button on the screen, the reading operation beings. The extracted tag data is connected to the middleware program in the terminal PC via WIFI network. Successful network connection ensures the data be to processed and analysed by the middleware program, before finally being saved to the Oracle database at the server.
3. Data communication and network transmission: The workshop is more than 10,000 m^2 large. All the processing and transportation facilities are metallic products. Given the huge dimension and multiplicity of large-scale metallic equipment, it is important to avoid data transmission problems caused by lengthy wires and strong currents' interference. Therefore, the optical cable is chosen as the main communication material, while data transmission of the hand-held tag reader uses wireless industrial AP (with in-built directional antenna) as transmission hotspot. The work field uses three first-level wireless AP and AC administration servers to form wireless Intranet. The wireless APs are connected respectively to the AC administration server, which runs collective management of these wireless APs and charges them via POE, to avoid unnecessary power cord arrangement. The seamless relay roaming function thus achieved allows an automatic switch from one AP of weak signals to a wireless AP of stronger signals.

Additionally, data acquisitors and switches loaded with local database and configuration software are also required. Communication between the data acquisitor and the numerical control machine tools saves the data collected by the machine tool in the local SQL Server's database, before transmitting them to the Oracle database via the workshop's Intranet.

2.3 Indirect Acquisition of the Effective Processing Energy Consumption, Based on Recursive Method with Discounted Measurements

There are two methods to define the energy efficiency of the discrete manufacturing system: effective energy definition and effective output definition. The effective energy approach uses the effective output-input ratio of the manufacturing system as the definition of energy efficiency. The effective output is defined as the ratio between the energy consumed by the manufacturing system and the system's effective output, and is in the form of specific energy consumption (SEC) shown in Formula (2.1).

$$\eta_{\mathrm{SEC}} = \frac{E_{\mathrm{ms}}}{O_{\mathrm{E}}} = \frac{(E_{\mathrm{input}} - E_{\mathrm{output}})}{O_{\mathrm{E}}} \tag{2.1}$$

where O_{E} refers to effective output, E_{o} refers to the output energy consumption and E_{i} refers to the input energy consumption.

Here, we use the processing equipment machine tool of the discrete manufacturing system in the equipment manufacturing industry as an example to study the energy efficiency forecast. The machine tool's energy efficiency is defined by its effective energy, namely the ration between its effective energy and its total energy during a certain period or when running a certain processing procedure [2], shown in Formula (2.2).

$$\eta_{\mathrm{ee}} = \frac{E_{\mathrm{c}}}{E_{\mathrm{mt}}} = \frac{\int_{T_s}^{T_e} P_{\mathrm{c}}(t)\mathrm{d}t}{\int_{T_s}^{T_e} P(t)\mathrm{d}t} = \frac{\int_{T_s}^{T_e} P_{\mathrm{c}}(t)\mathrm{d}t}{\int_{T_s}^{T_e} P_{\mathrm{rfo}}\mathrm{d}t + \int_{T_s}^{T_e} P_{\mathrm{sp}}(t)\mathrm{d}t} \tag{2.2}$$

where E_{c} means the cutting energy consumption, E means the total energy consumption of the machine tool, $P_{\mathrm{c}}(t)$ means the cutting power, P_{rfo} means fixed power attenuation during the machine tool's normal operation, $P_{\mathrm{sp}}(t)$ means the input power of the main transmission system, T_{s} means the start time of processing, T_{e} means the end time of processing.

Formula (2.1) shows that, to solve the energy consumption, we need to know such information as the cutting power, the fixed power attenuation, the input power of the main transmission system, the processing start time and the processing end time. Of the above parameters: (1) the fixed power attenuation and the main transmission system's input power is acquired from the power sensors and the smart meters installed; (2) the start time and end time of the processing is obtained by using the hand-held machine to scan the e-tag. Acquisition of the cutting power in Formula (2.2) could use either the direct or the indirect approach. The direct approach is to install a torque transducer or force transducer on one end of the main spindle, and calculate the cutting power by measuring the torque or cutting force. Installing these transducers affect the precision and solidness of the machine tool; moreover, torque and force transducers are very expensive. Using the indirect method, which estimates the cutting power through the combination of the main transmission system's power balance equation and load loss function, can avoid these disadvantages of the direct method.

Hu et al. [3] proposed an on-line monitoring method to calculate the energy efficiency of the equipment (machine tool). Based on the machine tool's energy consumption model, an energy efficiency model is built for its assessment. But this method, which uses the least square identification algorithm to solve the additional load loss factor, does not take into account the fact that excessive data volume would cause data saturation for the identification algorithm and results in unsatisfactory identification. Professor Kordonowy from MIT proposes a calculation model which predicts an equipment's power efficiency based upon its rated power [4]. Using statistical methods, this model divides the machine tool's energy consumption into variable (processing) and invariable (fixed) ones. This method, though handy, does not take into account the additional load loss during the equipment's operation, so the acquired data deviates a lot from reality. Documentary [5] research indicates that the additional load loss during the equipment's operation takes up 15–20% of the cutting power, the energy consumption of which cannot be omitted. Therefore, this chapter proposes a method to predict the equipment's energy efficiency, which adopts the recursive method with discounted measurements. Based on the main equipment transmission system's power balance equation and the additional load loss function analysed in the last chapter, here we deduce an estimation model of the cutting power. Furthermore, considering that the additional load loss factor in the model cannot be measured directly, we adopt the recursive method with discounted measurements to identify the additional load loss factor, thence estimating the cutting power and indirectly obtain the effective processing energy consumption of the manufacturing facility.

The energy flow of the main transmission system comprises three parts, namely the machine tool's idle power, the machine tool's cutting power and the additional load loss caused by chips [2]. The main transmission system's power balance equation is as follows:

$$P_{\mathrm{sp}}(t) = P_{\mathrm{u}}(t) + P_{\mathrm{c}}(t) + P_{\mathrm{ad}}(t) \tag{2.3}$$

where $P_{\mathrm{u}}(t)$ represents the machine tool's idle power and $P_{\mathrm{ad}}(t)$ represents the machine tool's additional load loss. Its additional load loss is composed of the electric loss of electric motor and the mechanical transmission loss caused by cutting work. Literature [6] research reveals that the additional load loss is the quadratic function of the machine tool's cutting power, and there is a formula:

$$P_{\mathrm{ad}}(t) = a_0 P_{\mathrm{c}}(t) + a_1 P_{\mathrm{c}}^2(t) \tag{2.4}$$

where a_0 and a_1 represent the additional load loss factors.

From Formulas (2.3) and (2.4), we get:

$$P_{sp}(t) = P_u(t) + (1 + a_0)P_c(t) + a_1 P_c^2(t) \tag{2.5}$$

Formula (2.5) could be rewritten as:

$$\hat{P}_c(t) = \frac{-(1 + a_0) + \sqrt{(1 + a_0)^2 + 4a_1(P_{sp}(t) - P_u(t))}}{2a_1} \tag{2.6}$$

In Formula (2.6), the main transmission system's input power $P_{sp}(t)$ and the machine tool's idle power $P_u(t)$ can be measured by the aforementioned method; the additional load loss factor a_0 and a_1, which cannot be measured directly, can be estimated by using the parameter identification method. The traditional least square identification algorithm would gradually become incapable of being amended as the identification data grows to the extent of data saturation, which affects the algorithm's amenability and tractability, and leads to an imprecise estimation of parameters. Therefore, this chapter introduces the discount factor and proposes to calculate the additional load loss factor by using the recursive method with discounted measurements (RDM).

2.3.1 The Basic Ordinary Least Squares Techniques

Definition 2.1 Suppose the mean value of a random sequence is the linear function of the parameter $\boldsymbol{\theta}$

$$E\{z(k)\} = \mathbf{h}^{\mathrm{T}}(k)\boldsymbol{\theta} \tag{2.7}$$

where $\mathbf{h}(k)$ is **the observable data vector**, a realization which makes criterion function using the random sequence:

$$J(\boldsymbol{\theta}) = \sum_{k=1}^{L} [z(k) - \mathbf{h}^{\mathrm{T}}(k)\boldsymbol{\theta}]^2 \tag{2.8}$$

The minimum parameter estimate $\overset{\wedge}{\boldsymbol{\theta}}$ thus achieved is called the least squares estimate of the parameter $\boldsymbol{\theta}$.

Suppose the mathematical model of a single-input single-output system is:

$$A(z^{-1})z(k) = B(z^{-1})u(k) + n(k) \tag{2.9}$$

where $u(k)$ and $z(k)$ are respectively the input and the output variables of the model, $n(k)$ is the model noise, $A(z^{-1})$ and $B(z^{-1})$ are the multinomials of the retardation factor z^{-1}.

$$\begin{cases} A(z^{-1}) = 1 + a_1 z^{-1} + a_2 z^{-2} + \cdots + a_{na} z^{-na} \\ B(z^{-1}) = b_1 z^{-1} + b_2 z^{-2} + \cdots + b_{nb} z^{-nb} \end{cases} \tag{2.10}$$

Rewriting the model Formula (2.9) into the least square format:

$$z(k) = \mathbf{h}^{\mathrm{T}}(k)\boldsymbol{\theta} + n(k) \tag{2.11}$$

where

$$\begin{cases} \mathbf{h}(k) = [-z(k-1), \ldots, -z(k-n_a), u(k-1), \ldots, u(k-n_b)]^{\mathrm{T}} \\ \boldsymbol{\theta} = [a_1, a_2, \ldots, a_{na}, b_1, b_2, \ldots, b_{nb}]^{\mathrm{T}} \end{cases} \tag{2.12}$$

Assume that $k = 1, 2, \ldots, L$ (L is the data length), from Formula (2.11) we can have the following linear equation group:

$$z_L = \mathbf{H}_L \boldsymbol{\theta} + \mathbf{n}_L \tag{2.13}$$

$$\begin{cases} z_L = [z(1), z(2), \ldots, z(L)]^{\mathrm{T}} \\ \mathbf{n}_L = [n(1), n(2), \ldots, n(L)]^{\mathrm{T}} \\ \mathbf{H}_L = \begin{bmatrix} \mathbf{h}^{\mathrm{T}}(1) \\ \mathbf{h}^{\mathrm{T}}(2) \\ \vdots \\ \mathbf{h}^{\mathrm{T}}(L) \end{bmatrix} = \begin{bmatrix} -z(0) & \ldots & -z(1-n_a) & \ldots & u(1-n_b) \\ -z(1) & \ldots & -z(2-n_a) & \ldots & u(2-n_b) \\ \vdots & \vdots & \vdots & \vdots & \vdots \\ -z(L-1) & \ldots & -z(L-n_a) & \ldots & u(L-n_b) \end{bmatrix} \end{cases} \tag{2.14}$$

Based on Formula (2.14), the criterion function formula can be written into the quadratic form:

$$J(\boldsymbol{\theta}) = (z_L - \mathbf{H}_L\boldsymbol{\theta})^{\mathrm{T}}(z_L - \mathbf{H}_L\boldsymbol{\theta}) \tag{2.15}$$

where $\mathbf{H}_L\boldsymbol{\theta}$ stands for the model's output and $J(\boldsymbol{\theta})$ is used to measure the approximate condition of the model's and the system's outputs. By minimalizing $J(\boldsymbol{\theta})$, the model parameter estimate $\hat{\boldsymbol{\theta}}$ thus derived makes the model's output a better approximate to the its real figure.

Suppose $\hat{\boldsymbol{\theta}}$ make $J(\boldsymbol{\theta})|_{\hat{\boldsymbol{\theta}}} = \min$, then we have

$$\left.\frac{\partial J(\boldsymbol{\theta})}{\partial \boldsymbol{\theta}}\right|_{\hat{\boldsymbol{\theta}}} = \left.\frac{\partial}{\partial \boldsymbol{\theta}}(z_L - \mathbf{H}_L\boldsymbol{\theta})^{\mathrm{T}}(z_L - \mathbf{H}_L\boldsymbol{\theta})\right|_{\hat{\boldsymbol{\theta}}} = 0 \tag{2.16}$$

Deriving the parameter estimate at the moment k:

$$\hat{\boldsymbol{\theta}}(k) = \left(\sum_{i=1}^{k} \mathbf{h}(i)\mathbf{h}^{\mathrm{T}}(i)\right)^{-1} \left(\sum_{i=1}^{k} \mathbf{h}(i)z(i)\right) \tag{2.17}$$

Let $\mathbf{R}(k) = \sum_{i=1}^{k} \mathbf{h}(i)\mathbf{h}^{\mathrm{T}}(i)$, from Formula (2.17) we have

$$\mathbf{R}(k)\,\hat{\boldsymbol{\theta}}(k) = \mathbf{R}(k-1)\,\hat{\boldsymbol{\theta}}(k-1) + \mathbf{h}(k)z(k) \tag{2.18}$$

Bringing $\mathbf{R}(k) = \mathbf{R}(k-1) + \mathbf{h}(k)\mathbf{h}^{\mathrm{T}}(k)$ into Formula (2.18), we reorganize it into:

$$\hat{\boldsymbol{\theta}}(k) = \hat{\boldsymbol{\theta}}(k-1) + \mathbf{R}^{-1}(k)\mathbf{h}(k)\left[z(k) - \mathbf{h}^{\mathrm{T}}(k)\,\hat{\boldsymbol{\theta}}(k-1)\right] \tag{2.19}$$

Let $\mathbf{P}(k) = \mathbf{R}^{-1}(k) = \left[\sum_{i=1}^{k} \mathbf{h}(i)\mathbf{h}^{\mathrm{T}}(i)\right]^{-1}$, then we have

$$\mathbf{P}(k) = \left[\mathbf{P}^{-1}(k-1) + \mathbf{h}(k)\mathbf{h}^{\mathrm{T}}(k)\right]^{-1} \tag{2.20}$$

Using the matrix inversion formula, and let $\mathbf{K}(k) = \mathbf{P}(k)\mathbf{h}(k)$, thus derive the recursive identification algorithm of the least square:

$$\begin{cases} \hat{\boldsymbol{\theta}}(k) = \hat{\boldsymbol{\theta}}(k-1) + \mathbf{K}(k)\left[z(k) - \mathbf{h}^{\mathrm{T}}(k)\,\hat{\boldsymbol{\theta}}(k-1)\right] \\ \mathbf{K}(k) = \mathbf{P}(k-1)\mathbf{h}(k)\left[\mathbf{h}^{\mathrm{T}}(k)\mathbf{P}(k-1)\mathbf{h}(k) + 1\right]^{-1} \\ \mathbf{P}(k) = \left[\mathbf{I} - \mathbf{K}(k)\mathbf{h}^{\mathrm{T}}(k)\right]\mathbf{P}(k-1) \end{cases} \tag{2.21}$$

where $\hat{\boldsymbol{\theta}}(k)$ is the estimate value of k moment's parameter, $\mathbf{I}$ is the unit matrix, $\mathbf{K}(k)$ represents the gain matrix and $\mathbf{P}(k)$ is the covariance matrix.

The covariance matrix in Formula (2.21) is used to sum up the data information volume. As time passes, old data keep accumulating in $\mathbf{P}(k)$, until to a certain extent when new data cannot be inserted. Thus, the least square identification algorithm gradually loses the power to amend itself, and the derived identification parameter fails to be precise. To solve this problem, this chapter introduces the concept of the discount factor.

2.3.2 RDM Algorithm

Based on the least square algorithm, the discount method [7] introduces the forgetting factor and the weighting factor. Introduction of the forgetting factor reduces the occupancy of the old data in the algorithm and increases the volume of the new data information, thereby amending the identification algorithm. Introduction of the weighting factor weighs data of different levels of credibility. The discount method takes into account the functions of both the weighting factor and the forgetting factor. And its combination of these two factors brings together their respective advantages.

Relationship between the discount factor $\Gamma(k, i)$ and the weighting factor plus the forgetting factor is as follows:

$$\Gamma(k, i) = \Lambda(i) \prod_{j=i+1}^{k} \mu(j) \tag{2.22}$$

where $\Lambda(i)$ is the weighting factor and $\mu(j)$ is the forgetting factor.

Given the introduction of the discount factor, discussions in Sect. 2.3.1 reveal that the criterion function of the identification model takes

$$J(\boldsymbol{\theta}) = \sum_{i=1}^{k} \Gamma(k, i)\left[z(i) - \mathbf{h}^{\mathrm{T}}(i)\boldsymbol{\theta}\right]^2 = (z_L - \mathbf{H}_L\boldsymbol{\theta})^{\mathrm{T}}\boldsymbol{\Gamma}_L(z_L - \mathbf{H}_L\boldsymbol{\theta}) \tag{2.23}$$

where $\boldsymbol{\Gamma}_L = \begin{bmatrix} \Gamma(L,1) & & \mathbf{0} \\ & \Gamma(L,2) & \\ & & \ddots & \\ \mathbf{0} & & & \Gamma(L,L) \end{bmatrix}$, $\mathbf{H}_L\boldsymbol{\theta}$ represents the model's output and $J(\boldsymbol{\theta})$ is used to measure how much does the model's output approximate the system's actual output.

Through the criterion function $J(\boldsymbol{\theta})$ of the minimalized formula, the derived model parameter estimate $\hat{\boldsymbol{\theta}}$ makes the model's output approximates its real figure better. Then, the parameter estimate at the moment k:

$$\hat{\boldsymbol{\theta}}(k) = \left(\sum_{i=1}^{k} \mathbf{h}(i)\mathbf{h}^{\mathrm{T}}(i)\right)^{-1} \left(\sum_{i=1}^{k} \mathbf{h}(i)z(i)\right) \tag{2.24}$$

Let $\mathbf{P}(k) = \left[\sum_{i=1}^{k} \mathbf{h}(i)\mathbf{h}^{\mathrm{T}}(i)\right]^{-1}$, then we have

$$\mathbf{P}(k) = \left[\mathbf{P}^{-1}(k-1) + \mathbf{h}(k)\mathbf{h}^{\mathrm{T}}(k)\right]^{-1} \tag{2.25}$$

Let $\mathbf{K}(k) = \Lambda(k)\mathbf{P}(k)\mathbf{h}(k)$, deducing the following recursive method with discounted measurements:

$$\begin{cases} \hat{\boldsymbol{\theta}}(k) = \hat{\boldsymbol{\theta}}(k-1) + \mathbf{K}(k)\left[z(k) - \mathbf{h}^{\mathrm{T}}(k)\hat{\boldsymbol{\theta}}(k-1)\right] \\ \mathbf{K}(k) = \mathbf{P}(k-1)\mathbf{h}(k)\left[\mathbf{h}^{\mathrm{T}}(k)\mathbf{P}(k-1)\mathbf{h}(k) + \frac{\mu(k)}{\Lambda(k)}\right]^{-1} \\ \mathbf{P}(k) = \frac{1}{\mu(k)}\left[\mathbf{I} - \mathbf{K}(k)\mathbf{h}^{\mathrm{T}}(k)\right]\mathbf{P}(k-1) \end{cases} \tag{2.26}$$

where $\hat{\boldsymbol{\theta}}(k)$ is the estimate of k moment's parameter, $\mathbf{I}$ is the unit matrix, $\mathbf{K}(k)$ represents the gain matrix and $\mathbf{P}(k)$ is the covariance matrix. Introduction of the discount factor allows the RDM identification algorithm to keep reducing the old data, while increasing the occupancy of new data and the number of adjustable parameters. In a real-life application, with regard to the specific object, a well-chosen discount factor will effectively enhance the amendment and identification precision of the algorithm.

2.3.3 The Discount Recursive Load Loss Factor Identification

Using Formula (2.5) to derive the additional load loss factor of the identification model, Formula (2.4) can be transposed into the following form:

$$P_{\mathrm{sp}}(k) - P_{\mathrm{u}}(k) = \left[P_{\mathrm{c}}(k)P_{\mathrm{c}}^{2}(k)\right]\begin{bmatrix} a_0 + 1 \\ a_1 \end{bmatrix} \tag{2.27}$$

Formula (2.27) can be arranged in the form of Formula (2.11):

$$z(k) = \mathbf{h}^{\mathrm{T}}(k)\boldsymbol{\theta} \tag{2.28}$$

$$\begin{cases} z(k) = \left[P_{\mathrm{sp}} - P_{\mathrm{u}}\right] \\ \mathbf{h}(k) = \left[\mathbf{P}_{\mathrm{c}}, \mathbf{P}_{\mathrm{c}}^{2}\right]^{\mathrm{T}} \\ \boldsymbol{\theta} = [u_0, u_1]^{\mathrm{T}} \end{cases} \tag{2.29}$$

where $\mathbf{P}_{\mathrm{c}} = \begin{bmatrix} P_{\mathrm{c}1} \\ \vdots \\ P_{\mathrm{c}l} \end{bmatrix}$, $\mathbf{P}_{\mathrm{c}}^{2} = \begin{bmatrix} P_{\mathrm{c}1}^{2} \\ \vdots \\ P_{\mathrm{c}l}^{2} \end{bmatrix}$, $\begin{cases} u_0 = a_0 + 1 \\ u_1 = a_1 \end{cases}$.

Using the recursive method with discounted measurement Formula (2.26) to estimate the load loss function, setting the initial value of the covariance matrix $\mathbf{P}(k)$ as $\mathbf{P}(0) = 10^{\beta}\mathbf{I}$, with β as the sufficiently large positive integer, the estimated initial value as $\hat{\boldsymbol{\theta}}(0) = \boldsymbol{\varepsilon}$ and $\boldsymbol{\varepsilon}$ as the sufficiently small real vector. The values of the weighting factor and the forgetting factor are set between (0, 1], namely $0 < \Lambda(k), \mu(k) \leq 1$. After experiments and analyses, here we choose $\mu(k) = 0.95$, $\Lambda(k) = 0.6$.

The acquisition procedure of the effective processing energy, based on RDM, or the recursive method with discounted measurements, is shown in Fig. 2.6.

From the energy efficiency estimation procedure in Fig. 2.7, we know that indirect acquisition of the manufacturing facility's effective processing energy consumption requires information of relevant parameters, such as the load-irrelevant energy consumption P_{rfo}, the input power of the main spindle P_{sp} and the idle power P_u. Table 2.1 lists the details.

2.3.4 *Acquisition of Relevant Parameters*

As Table 2.1 illustrates, the required data comprises three parts, which are:

1. **Basic data**: using the smart meter to acquire the equipment's idle power P_u and fixed energy consumption P_{rfo}; using the power sensor to measure the main transmission system's input power P_{sp}.
2. **Measuring data**: the processing start time T_s and the processing end time T_e are obtained by the hand-held machine's reading the e-tag on the manufactured parts.
3. **Identification estimate data**: using the recursive method with discounted measurements to solve the additional load loss factors a_0 and a_1, thereby solving the machine tool's cutting power.

Analyses in this chapter reveal that, with regard to the facility's energy consumption information and processing energy consumption information in the manufacturing workshop, the acquisition network deployment designed for collecting the energy consumption information in a discrete manufacturing workshop can obtain the basic data and measuring data listed in Table 2.1. And the recursive method with discounted measurements proposed above can help identify the additional load loss factors a_0 and a_1. Once the main transmission system's input power P_{sp}, idle power P_u and the additional load loss factors a_0 and a_1 have been obtained, we can use Formula (2.6) to estimate the cutting power $P_c(t)$ at the moment t, then solving the numerator in the energy efficiency Formula (2.2). The fixed energy consumption and the main transmission system's input power in the denominator of the energy efficiency Formula (2.2) can be obtained by the acquisition terminals of such smart hardware as the smart meter. Then, we can calculate the energy efficiency of the machine tool.

2.4 Software Design of the Discrete Manufacturing System's Energy Consumption Information Acquisition

The software part comprises four sections: firstly, the local database installed on the data acquisitor, which uses SQL database in case of data loss accidents caused

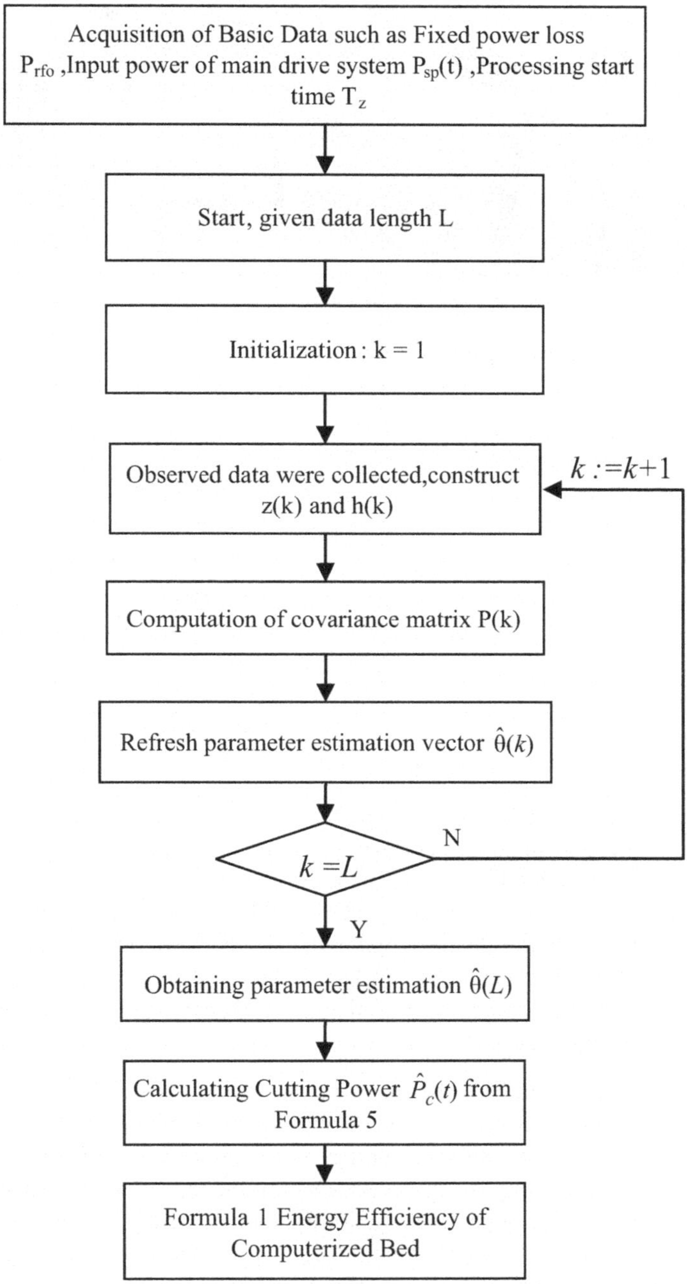

Fig. 2.6 Indirect acquisition procedure of the effective processing energy consumption, based on RDM parameter estimate

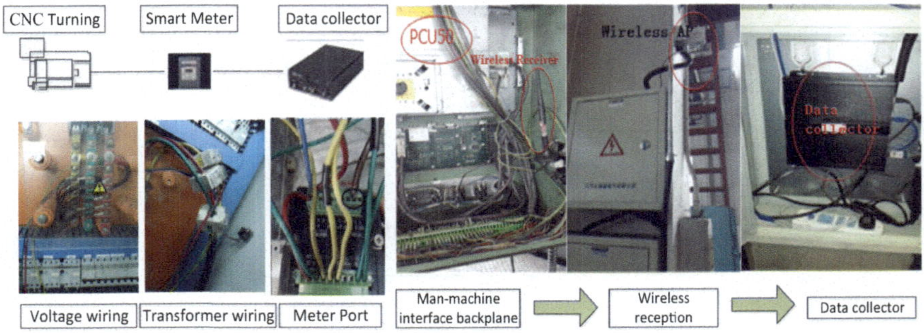

Fig. 2.7 Deployment of data acquisition facility

Table 2.1 Variables relevant to the manufacturing facility's energy consumption

Basic data/parameter	Meaning
T_s	Processing start time
T_e	Processing end time
a_0, a_1	Additional load loss factor
P_u	Idle power
P_{sp}	Main transmission system's input power
P_{rfo}	Fixed energy consumption
P_c	Cutting power

by power cut or network problems; secondly, the configuration design, which uses KingView software as the upper computer to configure parameters with the lower computer or smart meter for data acquisition at certain frequencies, and mutual data storage with SQL Server; thirdly, the Original Equipment Manufacturer (OEM) software of Siemens, under the development environment of which we can use VB or C++ designed application interface to access the necessary parameters of machine tools, to display data and graphics, as well as to export data; and fourthly, the Oracle database at the back office.

1. **SQL Server database SQL Server**. As a product of Microsoft, it is a relational database management system. The database is an important tool for data storage, statistical analysis and data loading, widely applied in data-analysis-related industries. The ever-improving development of the SQL Server database series, together with its wide-openness, flexibility, high security and easy operability, makes SQL Server a primary database choice for more and more enterprises, making it their memory database storing raw data.
2. **KingView monitoring software**. In the configuration software, we can create a new smart meter data acquisition project, configure the parameters of meter and configuration software, add variables that need to be fetched and complete

data storage via mutual interaction with SQL Server database. The workshop's Intranet then transmits the data to the back office Oracle database.

3. **Siemens's OEM software** [8, 9]. The OEM software development kit provided by Siemens offers secondary development for machine tools of specific processing application. After installing OEM, VB 6.0 and VC++6.0 software on a personal computer, we can use the NCDDE server interface provided by the OEM software to gain access to variables in NCU (e.g. axis coordinates), data block variables in PLC, MMC103 variables as well as files. A system structure based on PC brings convenience to the development of data acquisition software. Users can code their own programs with the OEM development kit provided by Siemens, and the .exe file thus generated is precisely the application file of OEM. With some relevant configuration files added, it can be copied to the hard disk of the human–machine interface and be called by specific soft keys of MMC103. We can call pictures to check whether necessary data are collected and whether the acquired data are correct.
4. **Oracle database as the server**. This design scheme uses Oracle10g as the back office database, given the fact that Oracle database can be run across platforms, is more secure than the SQL Server database and can support all industrial standards with a completely open strategy.

2.5 Experiments and Application Analysis

2.5.1 Realization and Display of the Discrete Manufacturing System's Energy Consumption Acquisition

Implementation of the data acquisition is the precondition for data display. After data acquisition has been implemented, we can adopt the B/S structure and use C# language in the Microsoft VS2010 environment to start relevant development operation in Fig. 2.7, to process and display the acquired data into forms like graphs, tables or curves. After deploying the workshop's hardware scheme and designing the software, we can acquire the data required in real time.

After the wiring terminals on the smart meter have been connected, we can start relevant configurations using the KingView software. Besides, the RS-485 interface on the meter's terminal can transmit data to the local SQL Server database via RS-485 USB cable. Then, the data would be transmitted to the back office Oracle database via the workshop's Intranet. The actual acquisition and transmission process is shown in Fig. 2.8.

After installing the OEM data on the PC, we can use VB and C++ to write relevant programs to acquire the facility's processing information, as shown in Fig. 2.9.

The hand-held machines installed on the workshop's facilities can scan the e-tags on the parts being processed, and thus acquire the manufacturing information of these parts. These include details like the components' name, fabrication information,

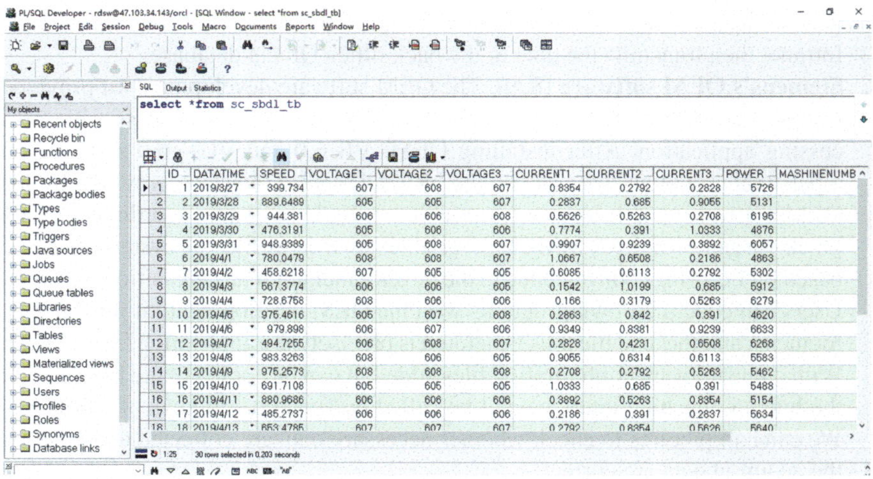

Fig. 2.8 Acquiring the data of the equipment's electric quantity

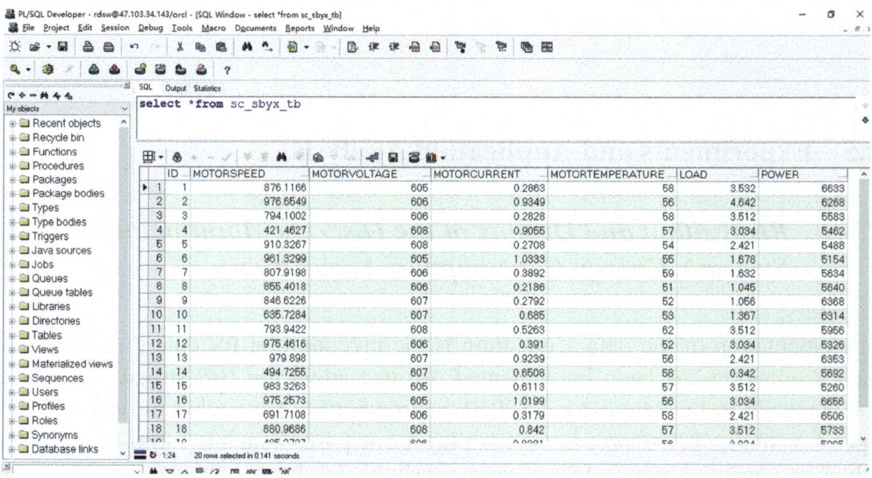

Fig. 2.9 Acquiring the facility's operation data

processing start time or end time, and they can be accessed directly in the system, as shown in Fig. 2.10.

The above data can be acquired and displayed directly. Some other data need to be processed before being displayed, such as the daily power consumption and hourly power consumption. The data, once being processed by the system's back office, would show clearly and directly power consumption fluctuation of each equipment, as shown in Fig. 2.11.

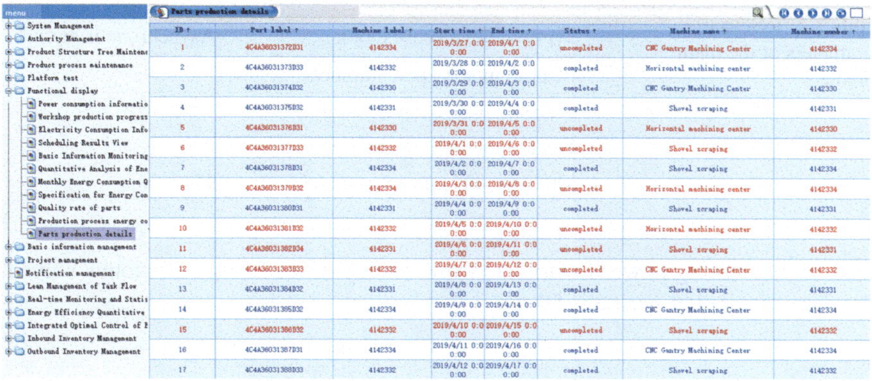

Fig. 2.10 Acquiring the components' processing data

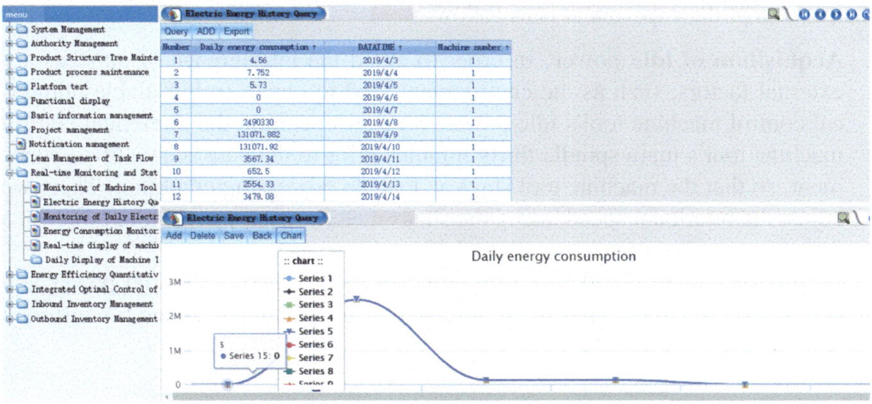

Fig. 2.11 Daily power consumption of the manufacturing system's facilities

2.5.2 *Validating the Indirect Method of Effective Processing Power Consumption Acquisition*

1. **Summary of the experiment**

After acquiring relevant data of energy consumption, we can undertake cutting experiment on machine tools to validate the effectiveness of the method proposed. To compare algorithms, this cutting experiment adopts the same manufacturing experiment environment as that of the document [10]. The relevant technical parameters of the main spindle are listed in Table 2.2.

By installing a smart meter at the numerical control machine tools' incoming line terminal and installing power sensor at the front of the main spindle motor inverter, we achieve real-time monitoring and statistics of the machine tool's energy consumption. Meanwhile, by installing a torque transducer at the main cutting spindle, we can

Table 2.2 Relevant parameters of the main spindle

Types of the electric motor	Asynchronous motor
Electric motor's pole numbers	2
Electric motor's rated power (kW)	5.5
Electric motor's rated slip ratio	5%
Main spindle's speed range (r/min)	45–2100

measure the cutting power to validate the accuracy of the energy efficiency solved. The data acquisition period is 50 ms.

2.5.3 *Process of the Experiment*

The process of the experiment is as follows:

1. **Acquisition of idle power**. In order to avoid the interference of unnecessary external factors, such as the environment, and to obtain truly reliable numerical control machine tool's idle power, we start operating the numerical control machine tool's main spindle thirty minutes prior to the data acquisition experiment, so that the machine tool stays at a stable processing condition. Then, we adjust the rotational speed, using the smart meter to measure the idle power at a different speed, shown here in Fig. 2.12.
2. **Acquisition of the load loss function**. We record, firstly, the idle power at a certain rational speed and then the different processing schemes for different

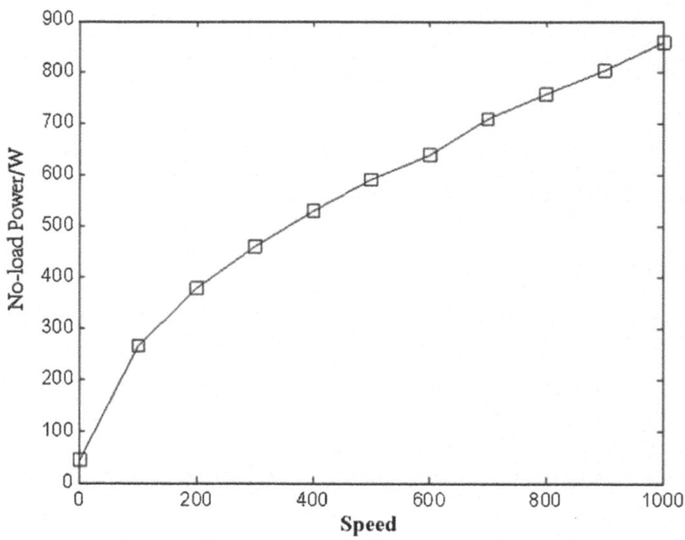

Fig. 2.12 Relationship between the idle power and the rotational speed

Table 2.3 Experiment parameters

Main spindle's rotational speed (r/min)	Feed volume (mm/r)	Cutting depth (mm)
100	0.2	0.8
200	0.4	0.8
400	0.4	0.1
400	0.4	0.4
600	0.4	0.4
600	0.4	0.8
800	0.4	0.4
800	0.4	0.8
1000	0.4	0.4
1000	0.4	0.8

cutting parameters at the same main spindle rotational speed. The specific experiment parameters are listed in Table 2.3, with the data length L set at the value of 5000. Based on the recursive parameter method with discounted measurements proposed in Sect. 2.3, we solve the additional load loss factor. The result of the load loss function identification is shown in Table 2.3.

From Fig. 2.13, we can see that the additional load loss factor solved by the recursive method with discounted measurements, compared with that solved by the traditional least square identification, shrinks faster and fluctuates less. In Fig. 2.13a, the recursive discounted method shrinks after 300th iteration, while the least square algorithm shrinks by the 2500th iteration. In Fig. 2.13b, the least square algorithm still fluctuates at the 1000th iteration, when the recursive discounted method already stays stable. Based on the traditional least square algorithm, the recursive method with discounted measurements introduces the weighting factor and the forgetting factor. The former gives different credibility to data at different moments. The latter gradually shrink the old data, increases the new data volume and strengthens the algorithm's ability to amend itself. Introduction of these two factors helps to improve the overall performance of the algorithm. The Identification result leads to $u_0 = 1.1603$, $u_1 = 5.2 \times 10^{-5}$. Thence, we solve the additional load loss factor as:

$$\begin{cases} a_0 = u_0 - 1 = 0.1603 \\ a_1 = u_1 = 5.2 \times 10^{-5} \end{cases} \tag{2.30}$$

From Formula (2.4), we solve the additional load loss function as:

$$P_{\mathrm{ad}} = 0.1603 P_{\mathrm{c}} + 5.2 \times 10^{-5} P_{\mathrm{c}}^2 \tag{2.31}$$

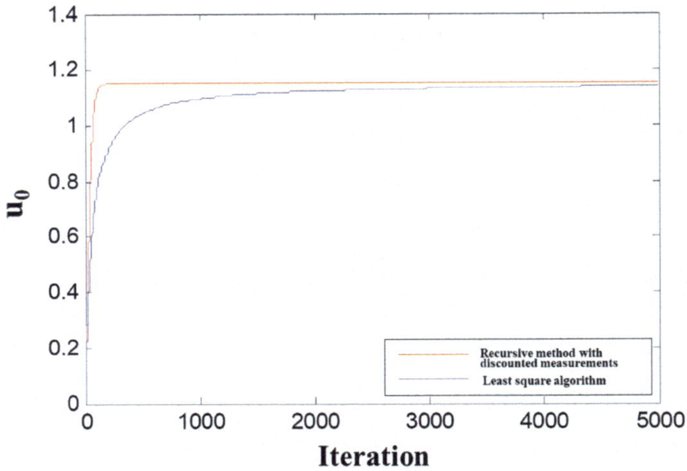

(a) Identification effect of the load loss function u_0

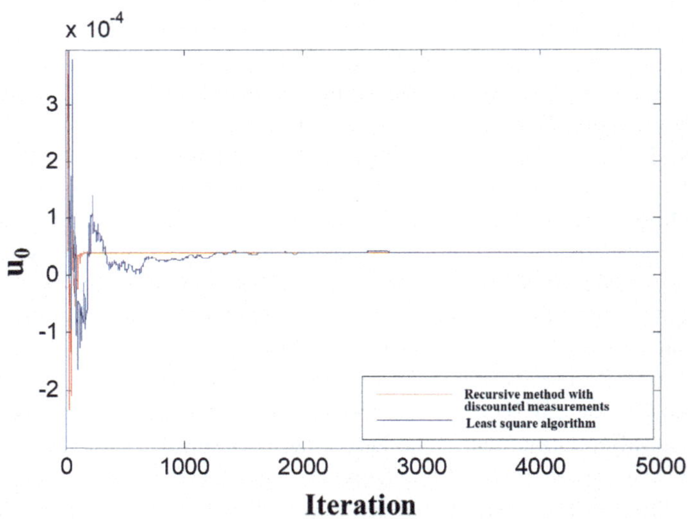

(b) Identification effect of the load loss function u_1

Fig. 2.13 Load loss coefficient identification result. **a** Identification effect of the load loss function u_0. **b** Identification effect of the load loss function u_1

3. **Estimation of the cutting power and calculation of power efficiency**. We choose to conduct the cutting experiment on a numerically controlled lathe in a machine tool processing workshop. We choose the processing parameters of Table 2.4 to undertake cylindrical turning operation of a 45# steel bar of 80 mm long and 60 mm in diameter. We choose six groups of cutting experiment results,

Table 2.4 Manufacturing facility's energy efficiency of effective processing energy consumption cutting power estimate total value

Cutting parameters (S, f, a_{sp})	Idle power P_u (W)	Load-irrelevant loss P_{rfo} (W)	Additional load loss P_{ad} (W)	Cutting power estimate value P_c (W)	Cutting power measured value P_{cm} (W)	Energy efficiency estimate value η (%)	Real energy efficiency value (%)	Estimate error E_{err} (%)	Least square identification error (%)
(400, 0.153, 2)	420	630	72	395	400	26.04	26.37	−1.25	−2.50
(400, 0.198, 2)	420	630	116	590	577	33.41	32.67	2.25	3.11
(400, 0.243, 2)	420	630	198	947	955	43.14	43.51	−0.84	−1.57
(800, 0.153, 2)	500	630	161	794	780	37.90	37.32	1.79	1.92
(800, 0.198, 2)	500	630	261	1179	1205	45.88	46.89	−2.16	−2.90
(800, 0.243, 2)	500	630	466	1890	1920	55.10	55.98	−1.56	−2.08

listed here in Table 2.4. From Table 2.4, we can see that, when the cutting parameters remain the same, the ratio proposed in this chapter, namely that between the effective processing energy consumption cutting power estimate and the machine tool's energy efficiency value, better approximates the real value than the estimated value solved by previous reference [11]. Additionally, the ratio's error remains within ±5%.

2.6 Conclusion

Based on the energy consumption composition of the discrete manufacturing system, this chapter proposes the method to implement the discrete manufacturing system's energy consumption information acquisition network. It ascertains the energy consumption information that needs to be acquired from nodes at the facility level, task level and secondary processing equipment. It selects relevant hardware network and network transmission protocols to collect the energy consumption information, which would be finally transmitted to the back office database server. Furthermore, concerning the difficulty to directly obtain the effective processing energy consumption at the equipment level, this chapter proposes an indirect acquisition method to obtain the effective processing energy consumption, using the recursive method with discounted measurements. Based on the facility's main transmission system power balance equation and the additional load loss function, this chapter constructs an estimation model of effective processing energy consumption cutting power. It uses the recursive method with discounted measurements to identify the additional load loss factor, which would be applied to estimate the effective processing energy consumption and equipment's energy efficiency. Experiments and simulation results demonstrate that, compared with the least square estimation method, the method proposed in this chapter can achieve higher identification precision.

References

1. Yin J (2012) Carbon-flow dynamic modeling and its application based on first-order hybrid Petri nets. Chongqing University, Chongqing
2. Liu F, Wang QL, Liu GJ (2013) Content architecture and future trends of energy efficiency research on machining systems. J Mech Eng 49(19):88–92
3. Hu SH, Liu F, He Y et al (2012) An on-line approach for energy efficiency monitoring of machine tools. J Clean Prod 27:133–140
4. Kordonowy D (2003) A power assessment of machining tools. Massachusetts Institute of Technology, Cambridge
5. Hu SH, Liu F, He Y et al (2010) Characteristics of additional load losses of spindle system of machine tools. J Adv Mech Des Syst Manuf 4(7):1221–1231
6. Ma J, Ge X, Lei S et al (2014) Assessment of cutting energy consumption and energy efficiency in machining of 4140 steel. Int J Adv Manuf Technol 74(9–12):1701–1708

7. Xiao DY (2014) System identification theory and application. Tsinghua University Press, Beijing
8. Du YB, Li CB (2014) Implementing energy-saving and environmental-benign paradigm: machine tool remanufacturing by OEMs in China. J Clean Prod 66:272–279
9. Newman S, Nassehi A, Imani-Asrai R et al (2012) Energy efficient process planning for CNC machining. CIRP J Manuf Sci Technol 5(2):127–136
10. Apostolos F, Alexios P, Georgios P et al (2013) Energy efficiency of manufacturing processes: a critical review. Procedia CIRP 7:628–633
11. IAC (2010) Industrial assessment centers. http://eere.energy.gov/industry/bestpractices/iacs.html

Chapter 3
Energy Consumption Model of the Discrete Manufacturing System

3.1 Introduction

For the discrete manufacturing system, its energy consumption and its facility condition have closely connections with its other factors, such as motion properties, technical parameters, a ranking of technical features, product structure, production scheduling and commitment control. Especially, for the complex discrete manufacturing processes like flexible manufacturing and flow manufacturing, their energy consumption mechanism is so complicated due to the influence of technological progress like the variable processing tasks, equipment condition and ever-changing techniques. What challenges the modelling, analysis and improvement of the system energy consumption progress include: the unpredictability of voluminous parameters in the processing parameter/thermodynamic energy consumption model, the highly non-linear feature of the facility energy consumption progress and the inevitable randomness and dynamism of complex workflow.

Modelling the manufacturing system's energy consumption involves three layers—the facility level, the task level and the system level, which led to three types of modelling: manufacturing facility's energy consumption modelling, technical energy consumption modelling and the manufacturing system's energy consumption modelling. The issue of modelling the manufacturing facility's energy consumption has aroused the widest research interest. Given the fact manufacturing facility has multiple energy sources and keeps moving, the energy consumption models are quite diverse as well. For instance, there are specific energy model [1], correlated energy consumption model of the equipment and its energy-consuming components [2], and dynamic energy consumption model of the equipment and its main transmission system [3]. The technical energy consumption modelling mainly studies the correlation between cutters, processing materials, processing parameters, products' structural features and energy consumption. A series of mechanism model and experience model on technical energy consumption have been established [4, 5]. The system energy consumption modelling mainly assumes the perspective of the entire life cycle of product manufacturing. Mose et al. [6] analysed the impact of factors

Y. Wang et al., *Quantitative Analysis and Optimal Control of Energy Efficiency in Discrete Manufacturing System*, https://doi.org/10.1007/978-981-15-4462-0_3

like place of origin, shipping weight, transport distance and transportation methods on the product's specific energy consumption. The existing single mechanism model, which emphasizes describing the facility's energy consumption, does not give a comprehensive overview of the correlation between the layers and the constituents of the energy consumption network, nor does it construct a multi-dimensionally integrated monitoring model which systematically combines material flow, energy flow and information flow. Besides, given the random and dynamic features of the discrete manufacturing system's processing progress, a purely mathematical model can hardly construct an accurate model of the manufacturing system and its facility's energy-consuming process. Therefore, this chapter first constructs a mechanism-based, multi-source and multi-level mathematical model of energy consumption integration. Then, it takes into consideration the dynamic evolution and uncertainty of the manufacturing process and proposes an ontological method for constructing the energy consumption knowledge model.

3.2 The Dynamic Correlation Between the System's Energy Consumption and the Manufacturing Facility

Since the products of the discrete manufacturing system are assembled from multiple parts, and since the energy consumption of producing all these parts comes from the facility, the energy consumption of product manufacturing is closely related with the manufacturing facility. The manufacturing facility for producing multiple parts may involve different energy-consuming system and energy-consuming components. In other words, various energy consumption sources are involved, which led to different kinds of energy consumption. Figure 3.1 uses the numerical control machine tool—one typical mechanical processing facility—as an example and classifies the energy sources according to the correlation between each energy source and its processing status. In reality, the numerical control machine tool may contain one part or more of these energy consumption sources.

Based on the energy consumption types of the manufacturing facility, we can classify the constituents of the facility's energy consumption into load-irrelevant energy consumption, facility's main transmission system energy consumption and the feed system energy consumption. The latter two jointly constitute the load relevant effective processing energy consumption.

3.2.1 Load-Irrelevant Energy Consumption

Load-irrelevant energy consumption mainly includes the energy consumed by subsystems of hydraulic pressure, cooling and chip removal. This part of energy consumption is related only with the corresponding motor's rated power and working

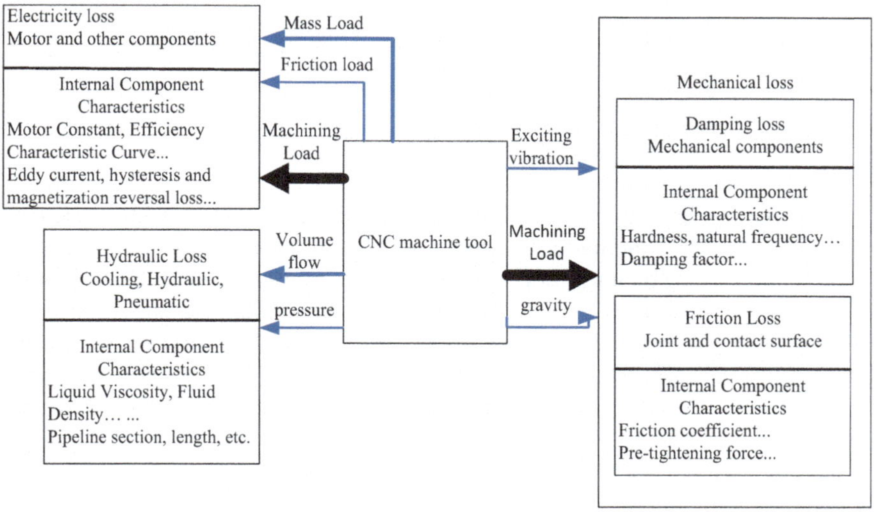

Fig. 3.1 Energy consumption forms of the numerical control manufacturing facility

hours; therefore, its energy consumption model is:

$$P_{\text{dec}}(t) = \sum_{i=1}^{n} g_i(t) \cdot C_i \tag{3.1}$$

where P_{dec} is the total load-irrelevant energy consumption, C_i is the power of a certain auxiliary energy consumption sub-system, which is normally a constant. g_i represents whether a certain energy consumption sub-system is being used: 1 stands for yes, and 0 stands for no. Here is the representation:

$$g_i(t) = \begin{cases} 0, & \text{use} \\ 1, & \text{no use} \end{cases} \tag{3.2}$$

3.2.2 *The Facility's Main Transmission System Energy Consumption*

The facility's main spindle system energy is transmitted through the inverter to the motor and finally to the cutting region for processing parts. During this process, the energy consumption is mainly that caused by the inverter and the main spindle motor themselves, as well as that caused by friction during the mechanical transmission progress. The energy flow is shown in Fig. 3.2.

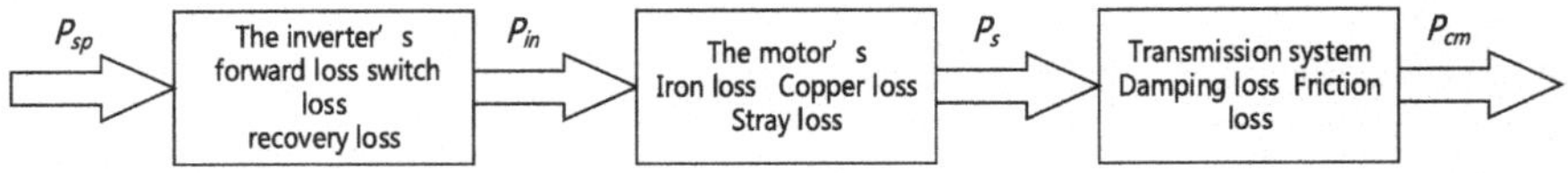

Fig. 3.2 Main and auxiliary transmission system's energy flow graph

The input power of the main transmission system $P_{\mathrm{sp}}(t)$, after being consumed by the inverter itself, the motor and the transmission system, finally transmits the energy to the cutting area for processing parts. The inverter's loss includes forward loss, switch loss and recovery loss. The motor's loss includes iron loss, copper loss and stray loss. The mechanical transmission loss includes the friction loss and the damping loss generated during the transmission process. We construct its energy consumption model into three parts: the inverter's energy consumption model [7, 8]:

$$P_{\mathrm{sp}}(t) = P_{\mathrm{in}} + P_a + P_b + P_c \tag{3.3}$$

The motor's energy consumption model:

$$P_{\mathrm{in}} = P_{\mathrm{Fel}} + P_{\mathrm{f}} + P_{\mathrm{w}} + P_{\mathrm{cu1}} + P_{\mathrm{cu2}} + P_{\mathrm{st}} + P_{\mathrm{s}} \tag{3.4}$$

The mechanical transmission system's energy consumption model:

$$P_{\mathrm{s}} = P_{\mathrm{m}} + P_z + P_{\mathrm{cm}} \tag{3.5}$$

We study the main transmission system's energy flow from the perspective of input and out and arrive at this model based on Formulas (3.3), (3.4) and (3.5):

$$P_{\mathrm{sp}}(t) = P_{\mathrm{Fel}} + P_{\mathrm{f}} + P_{\mathrm{w}} + P_{\mathrm{cu1}} + P_{\mathrm{cu2}} + P_{\mathrm{st}} + P_{\mathrm{m}} + P_z + P_{\mathrm{c}}(t) + P_a + P_b + P_c \tag{3.6}$$

where the inverter' loss P_a represents the forward loss, P_b the switch loss, P_{c} the recovery loss, P_{Fel} the motor stator's iron loss, P_{f} the friction loss, P_{w} the wind loss, P_{cu1} the motor stator's copper loss, P_{cu2} the motor rotor's copper loss, P_{st} the motor's stray loss, P_{m} the transmission system's friction loss, P_z the transmission system's damping loss and $P_{\mathrm{c}}(t)$ the machine tool's cutting power. The motor stator's iron loss P_{Fel}, the wind loss P_{w} and the friction loss P_{f} does not change along with the load, hence are called fixed loss P_{const}. The motor stator's copper loss P_{cu1}, rotor's copper loss P_{cu2}, stray loss P_{st}, the mechanical transmission system's friction loss P_{m} and the damping loss P_z changes along with the load, hence are called variable loss P_{var}. Therefore, the main transmission system's energy flow model Formula (3.6) could also be represented as:

$$P_{\mathrm{sp}}(t) = P_{\mathrm{const}} + P_{\mathrm{var}} + P_{\mathrm{c}}(t) + \Delta P \tag{3.7}$$

$$\Delta P = P_a + P_b + P_c \tag{3.8}$$

$$P_{\text{const}} = P_{\text{Fel}} + P_{\text{f}} + P_{\text{w}} \tag{3.9}$$

$$P_{\text{var}} = P_{\text{cu1}} + P_{\text{cu2}} + P_{\text{st}} + P_{\text{m}} + P_z \tag{3.10}$$

Besides, previous research [9] demonstrates that the inverter's loss normally takes up 8% of the motor's total power that the inverter's energy consumption is in proportion to the motor's energy consumption. Therefore, we can convert the inverter's loss into the idle power, the cutting power and the additional load loss. Thus, Formula (3.7) can be simplified into:

$$P_{\text{sp}}(t) = P_{\text{const}} + P_{\text{var}} + P_{\text{c}}(t) \tag{3.11}$$

The processing facility normally has five operation conditions: launching, standby, idle, cutting, and shut-down. The machine tool is launched at the moment when the facility's general power is switched on. As the control system and the peripheral equipment are turned on, the machine tool enters the standby status, at which moment the main spindle, the feed shaft and the transmission system are launched but not operating. According to Formula (3.11), when $P_{\text{c}}(t) = 0$, the main spindle and the feed system operates, and the current condition of the machine tool is called "idle". According to Fig. 3.2, the input power of an idle machine tool is the idle power of the entire system P_{u}. When $P_{\text{c}}(t) \neq 0$, the machine tool's current condition is called cutting condition, the main spindle system. Meanwhile, the feed system and the transmission system start operating, and apart from the losses caused by the idle motor and the transmission system, there is also the variable energy consumption caused by the components' cutting processing, which is called the machine tool's additional load loss P_{ad}. Therefore, the variable loss P_{var} is the sum of the idle loss and the additional load loss during the cutting operation P_{ad}. From Formula (3.11), we get the facility's main transmission system power balance equation:

3.2.3 Energy Consumption Model of the Feed System

The feeding transmission system drives the cutters or the workbench to coordinate with the main transmission system's parts processing. Each actuating motor controls one axial movement. The number of actuating motors depends on the complexity of the equipment to process the parts. The feed system also has energy consumption and energy flow during its operation, which involves energy input, storage, consumption and output as well.

Similar to the main transmission system, the feed transmissive energy flow can be shown in the following block graph:

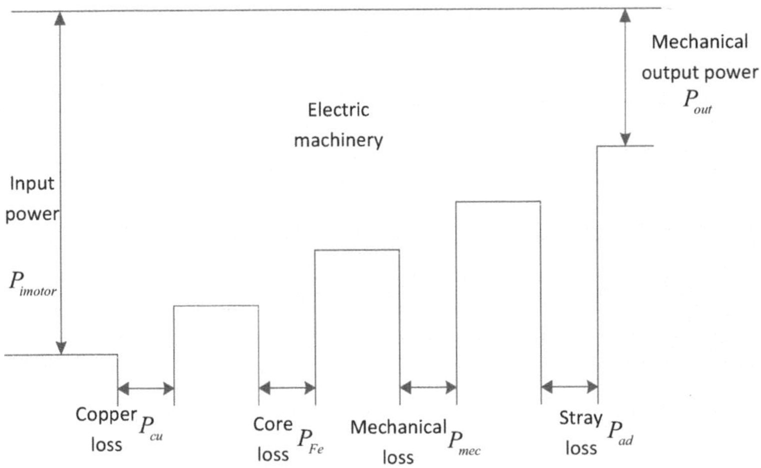

Fig. 3.3 Feed system's energy flow graph

The energy flow of the feed system is similar to that of the main transmission system, with losses like the motor's copper loss and iron loss. From Fig. 3.3, we know that

$$P_{\mathrm{imotor}} = P_{\mathrm{cu}} + P_{\mathrm{Fe}} + P_{\mathrm{mec}} + P_{\mathrm{ad}} + P_{\mathrm{out}} \tag{3.13}$$

On the other hand, the motor's input power can be represented as the electromagnetic torque T_{en} multiplied by the angular speed ω_{m}. The electromagnetic torque T_{en} involves the motor's mechanical loss, stray loss, mechanical output torque and iron loss torque. Therefore, Formula (3.13) could also be written as the sum of the motor's electromagnetic power and copper loss:

$$P_{\mathrm{imotor}} = P_{\mathrm{cu}} + \omega_{\mathrm{m}} T_{\mathrm{en}} \tag{3.14}$$

Document [10] has provided us with a dynamical model analysis of the motor's feed system and the motor's vector control, from which we derive the following feed system's power model:

$$P_{i\mathrm{motor}} = 3R_a(B'_{\mathrm{m}}\omega_{\mathrm{m}} + K'_{\mathrm{eq}}M_{\mathrm{weight}} + T_{\mathrm{c}} + T'_0)^2 + K_{\mathrm{T}}\omega_{\mathrm{m}}(B'_{\mathrm{m}}\omega_{\mathrm{m}} + K'_{\mathrm{eq}}M + T_{\mathrm{c}} + T'_0) \tag{3.15}$$

where $B'_{\mathrm{m}} = \frac{1}{K_{\mathrm{T}}}\left(B_{\mathrm{m}} + \frac{B_{\mathrm{ls}}}{K_{\mathrm{g}}^2} + \frac{P^2\mu_{\mathrm{v}}}{4\pi^2 K_{\mathrm{g}}^2 \eta_{\mathrm{bse}}}\right)$, $K'_{\mathrm{eq}} = \frac{P\mu_{\mathrm{c}}}{2\pi K_{\mathrm{g}} K_{\mathrm{T}} \eta_{\mathrm{bse}}}$, $T'_0 = \frac{T_0}{K_{\mathrm{T}}}$, $T_{\mathrm{c}} = \frac{P F_{\mathrm{ext}}}{2\pi K_{\mathrm{g}} K_{\mathrm{T}} \eta_{\mathrm{bse}}}$, R_a represent the stator resistance, B_{m} the motor's damping coefficient, K_{T} the motor's torque coefficient, K_{g} the coupling's transmission ratio, B_{ls} the ball screw's damping coefficient, μ_{v} the viscous friction coefficient, μ_{c} the Coulomb friction coefficient, η_{bse} the ball screw drive efficiency, P_{pitch} the ball screw's pitch, F_{ext} the horizontal weight of the cutting force, and M_{weight} the machine tool's weight.

3.3 The Dynamic Correlation Between Processing Energy Consumption and the Manufacturing Techniques

The manufacturing techniques include cutting processing, press working, foundry, welding, special processing, heat treatment, superstratum assembling and packaging. Each type of techniques could be further divided. For instance, the cutting processing can be classified into the common techniques of turning, milling, planning, drilling and grinding. A complex product manufacturing progress which comprises multiple parts and components necessarily involves multiple crafts of manufacturing. Different types of craft have different energy consumption which follows different laws. This leads to the issue of the impact of the manufacturing system's energy consumption. The discrete manufacturing system's energy consumption has a complex dynamic correlation with technological routes and parameters, which include cutting depth, feed speed, main spindle rotation speed, feed volume, depth of cut, parts' feature arrangement.

(1) **The functional relationship between the crafting elements and the cutting power consumption**

Setting a certain procedure i of task P_{ik} as the node, we establish a node craft parameter and cutting power consumption model. We set this processing procedure's facility as M_{ik}, the process parameters as $\{\theta_1, \theta_2, \ldots, \theta_l\}$, and the process route as $\{\mathrm{O}_1, \mathrm{O}_2, \ldots, \mathrm{O}_n\}$. Then, we have the energy consumption model:

$$E_{c,M_{ik}} = \sum_{j=1}^{S} \int_{t_{cj}} f_1(\theta_{ik,j}, O_{ik,j})\mathrm{d}t \tag{3.16}$$

where i represents the task number, k the procedure number, j the step number in the procedure.

(2) **The functional relationship between the production plan, scheduling and idle energy consumption.**

When we examine the processing facility M_{ik} during the task flow i, we observe that different production plan and scheduling scheme would result in the different idle time of the machine tool during the processing period, hence the different idle energy consumption. The facility's idle energy consumption could be represented as

$$E_{u,M_{ik}} = f_2(P_{idk}, z_{ik}, s_{ik}, t_{cik}, t_{uik}) \tag{3.17}$$

where P_{idk} represents the idle power of equipment no. k, z_{ik} the total number of the facility's processing tasks, s_{ik} the batch volume, t_{cik} the processing time of task no. i at the machine tool no. k, t_{uik} the idle time of the machine tool no. k.

(3) **The functional relationship between the manufacturing energy consumption and the parts' feature arrangement**

The manufacturing facility's dry running progress includes idle cutting, automatic cutter shift, main spindle speed change, feed shaft speed change, coolant fluid's switch on/off, chip removal's switch on/off. According to statistics, the energy consumption of the machine tool's idle cutting, automatic cutter shift and main spindle speed change take up more than 90% of the total idle cutting power [Jia Shun, 2014]. When the parts' feature processing arrangement changes, the machine tools' cutting route, cutter's changing frequency, as well as the main spindle's speed change frequency and range will all change as well, which in turn affect the idle cutting time and energy consumption.

$$E_{\text{non}}^{(F_p,F_q)} = E_{\text{tp}}^{(F_p,F_q)} + E_{\text{tc}}^{(F_p,F_q)} + E_{\text{src}}^{(F_p,F_q)}$$

where F_p and F_q are the two neighbouring structural features of the same processing part, $E_{\text{non}}^{(F_p,F_q)}(0 \leq p \leq n, 1 \leq q \leq n+1, p \neq q)$ represents the idle energy consumption from the moment when the feature F_p has been finished to the moment when the next feature F_q starts to be processed. $E_{\text{tp}}^{(F_p,F_q)}$ and $E_{\text{tc}}^{(F_p,F_q)}$ are, respectively, the idle cutting energy consumption and automatic cutter changing energy consumption from the moment when feature F_p finishes processing to the moment when the next feature F_q is processed.

3.4 The Layer-Structured System's Energy Consumption Integration Model

3.4.1 The Discrete Manufacturing System's Energy Consumption Hierarchical Structure

The manufacturing progress can be classified according to the technological processes: the working step level, the procedure level, the parts level and the product level, as shown in Fig. 3.4.

(1) **The working step level** is the most basic unit during the manufacturing process. It means that part of the procedure continuously undertaken when the processing surface and the cutting tool remain the same, and when the feed volume used for the cutting parameters and the cutting speed remains relatively the same. The working step level's energy consumption depends on the manufacturing facility's direct energy consumption. From Sects. 3.2 and 3.3, we know that the

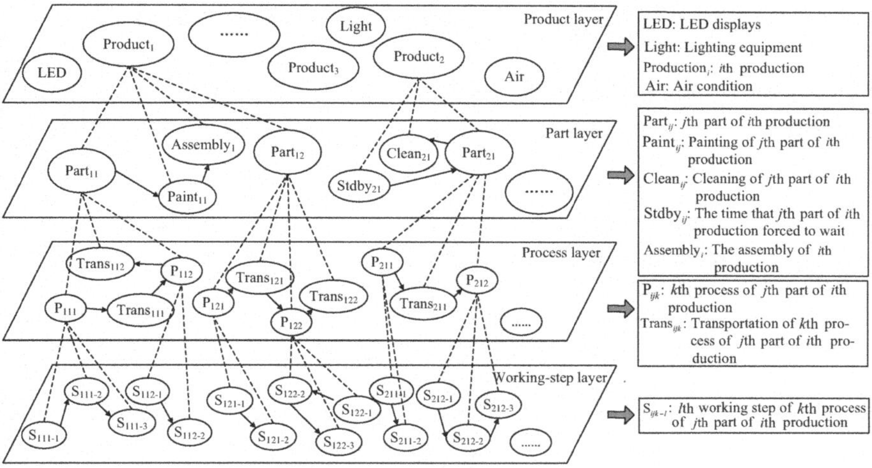

Fig. 3.4 Complex network structure of energy consumption in the workshop

facility's direct energy consumption depends on the facility's operation state and technological parameters.

(2) **The procedure level** means that part of the technological process continuously undertaken when one or one group of workers work on one or simultaneously on some parts at one working place. It also includes quality control and parts' transportation. The manufacturing energy consumption at the procedure level mainly depends on the manufacturing facility's energy consumption, idle power between working steps and auxiliary processing energy consumption of multiple working steps.

(3) **The parts level** refers to the fact that a finished part needs not only to go through multiple procedures, but also to be transported between different procedures and processing facilities. The energy consumption at the parts level mainly comprises the energy consumption of multiple procedure facilities, the parts' processing features arrangement, the production plan scheduling as well as transportation energy consumption.

(4) **The product level** refers to the fact that multiple parts are assembled into one final product. The product's energy consumption includes its components' energy consumption, assembling energy consumption and the possible auxiliary energy consumption caused by transportation. We also need to take into account the shared architectural technology service energy consumption, such as lighting and air conditioner. The parts' energy consumption comprises the energy consumption of those multiple parts which jointly make up the product. The assembling energy consumption includes the equipment and facility's energy consumption, such as that of the logistic line and the robots. The transportation energy consumption includes that caused by short-distance transfer equipment such as forklift truck and bridge crane.

3.4.2 Energy Consumption Model at the Working Step Level

Energy consumption of a working step is the sum of energy that consumed in the four periods:

$$E_{ijk-l} = E_1 = E_{\mathrm{st}} + E_{\mathrm{s\text{-}s}} + E_{\mathrm{ie}} + E_{\mathrm{c}} \tag{3.18}$$

According to the novel definition, energy efficiency should be

$$\eta_1 = \frac{E_{\mathrm{c}}}{E_1} = \frac{\int_{t_{\mathrm{c}}} P_{\mathrm{in}}(t)\,\mathrm{d}t}{\int_{t_{\mathrm{T}}} P_{\mathrm{in}}(t)\,\mathrm{d}t} \tag{3.19}$$

where

$$t_{\mathrm{T}} = t_{\mathrm{st}} + t_{\mathrm{sy}} + t_{\mathrm{ie}} + t_{\mathrm{c}} \tag{3.20}$$

3.4.3 Energy Consumption Model at the Procedure Level

Energy consumption of this layer includes every working step, which belongs to the same process, and transportation equipment. Energy consumption of a process could be described as:

$$E_{\mathrm{P}_{ijk}} = \sum_{l=1}^{N_{ijk}} E_{\mathrm{S}_{ijk-l}} + E_{\mathrm{Trans}_{ijk}} \tag{3.21}$$

Transportation is included because transportation is considered as the final step of each process in this book. It transports semi-processed workpieces to intermediate buffer prepared for the next process, especially large parts. Hence, the total energy consumption of this layer is obtained:

$$E_2 = \sum_{i=1}^{N_1} E_{1_i} + E_{\mathrm{ts}} \tag{3.22}$$

It is easy to find that the main part of E_2 is the sum of energy consumed in each working step. And the formula that calculates energy consumption for transport one part is:

$$E_{\mathrm{ts}} = (P_{\mathrm{ts}} \times t_{\mathrm{ts}})/Q \tag{3.33}$$

where Q is the number of workpieces at one transportation.

Therefore, energy efficiency in the process layer is:

$$\eta_2 = \frac{\sum_{i=1}^{N_1} E_{c_i}}{E_2} \tag{3.34}$$

3.4.4 Energy Consumption Model at the Parts Level

In this layer, the actual production situation is taken into consideration. Firstly, machine tools are forced to wait when something wrong with the supply of components. Then, the energy consumption of painting and cleaning is also considered.

$$E_{\text{Part}_{ij}} = \sum_{k=1}^{N_{ij}} \left(E_{\text{P}_{ijk}} + E_{\text{Trans}_{ijk}}\right) + E_{\text{Paint}_{ij}} + E_{\text{Clean}_{ij}} + E_{\text{Stdby}_{ij}} \tag{3.35}$$

In this part, the total energy consumption of a part is divided into four components: process energy, painting energy, cleaning energy and standby energy, which described as follows:

$$E_3 = \sum_{i=1}^{N_2} E_{2_i} + E_{\text{pt}} + E_{\text{cn}} + E_{\text{s-p}} \tag{3.36}$$

In this study, the energy consumption of painting and cleaning is fixed. Notice that standby energy here is different from the energy in the step layer.

Usually, a part needs to experience many processes before it is finished, just like Fig. 3.5 shows.

In Fig. 3.5, M1 and M2 represent machine tools and B1 is the intermediate buffer. Raw parts are loaded on the M1 from B0. When they complete service at M1, these parts are moved to buffer B1. Then, they are transported to M2, from which they are sent to buffer B2 as finished parts. When the former buffer of a machine is empty, that machine cannot get a part or latter buffer is full that machine cannot release its part, machine tools are forced to wait. This part of energy consumption is the standby energy mentioned above. It is easy to calculate because standby power and corresponding time are known.

Hence, energy efficiency in this layer could be obtained from Eq. (3.16).

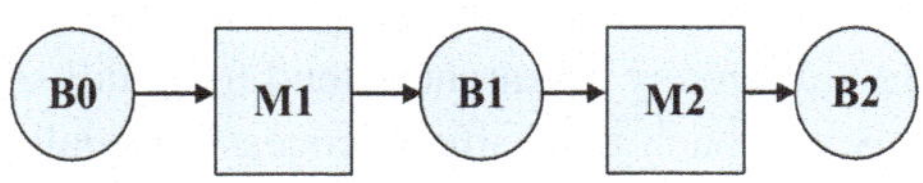

Fig. 3.5 Workflow of workpieces

$$\eta_3 = \frac{\sum_{i=1}^{N_2} \sum_{j=1}^{N_{1_i}} E_{c_{ij}}}{E_3} \tag{3.37}$$

3.4.5 Energy Consumption Model at the Product Level

Analysis of the product layer focuses on the energy consumed while finishing a product. This energy consists of two parts: producing all the parts and assembling. Besides, the related supporting facility is also taken into account. The energy consumption of a product is:

$$E_{\text{Product}_i} = E_{\text{Assembly}_i} + \frac{E_{\text{LED}} + E_{\text{Light}} + E_{\text{Air}}}{KM} + \sum_{j=1}^{N_i} E_{\text{Part}_{ij}} \tag{3.38}$$

Equation (3.15) given below calculates the energy used to finish a product.

$$E_4 = \sum_{i=1}^{N_3} E_{3_i} + E_{\text{ae}} + \frac{E_{\text{fy}}}{KM} \tag{3.39}$$

where K and M are the number of kinds of finished products and the number of a given kind of finished products, respectively, in a specific period.

Therefore, the energy efficiency of a finished product is:

$$\eta_4 = \frac{\sum_{k=1}^{N_3} \sum_{i=1}^{N_{2k}} \sum_{j=1}^{N_{1_i}} E_{c_{ijk}}}{E_4} \tag{3.40}$$

3.4.6 Experiment Analysis

(1) Experiment Environment and Parameters

This integrated energy consumption model is verified in a machine tool plant. Take bearing grinder as an example, the hierarchy of this production is shown in Fig. 3.6.

(2) Energy Consumption Data Acquisition and Processing

In this study, data acquisition is the foundation. To obtain machining data during production, many researchers tend to install sensors on the spindle [11, 12]. However, this approach may affect processing quality. So, it is not suitable in the workshop or production line. Today, open CNC system with open interfaces makes it possible to obtain machining process data collected from internal sensors of machine

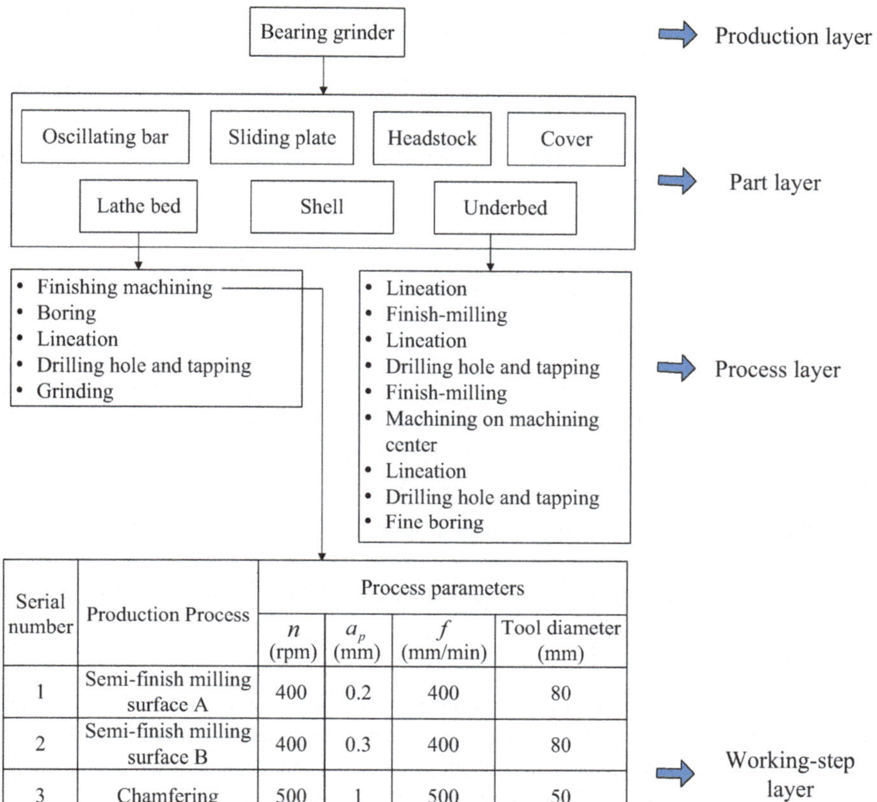

Serial number	Production Process	Process parameters			
		n (rpm)	a_p (mm)	f (mm/min)	Tool diameter (mm)
1	Semi-finish milling surface A	400	0.2	400	80
2	Semi-finish milling surface B	400	0.3	400	80
3	Chamfering	500	1	500	50
4	Semi-finish milling side surface	200	0.5	200	20
5	Rough drill $\Phi 8 \times 36$	120	25	120	6.8
6	Semi-finish boring	40	1	40	6.8

Fig. 3.6 Hierarchy of bearing grinder

tools. Therefore, almost all of the status data shown on the display screen could be obtained without extra hardware devices. In this book, SINUMERIK 840D system is taken as a case. This CNC system is based on PC, and its Man Machine Communication (MMC) system is running on Windows NT. On the one hand, application programme written in VB and VC is embedded in the 840D system to read data with the help of Human Machine Interface (HMI) under the environment of Original Equipment Manufacturer (OEM). On the other hand, the input voltage and current of the whole machine are read by smart metres in this novel data acquisition method. The parameters collected through this method are shown in Table 3.1.

Table 3.1 Parameters collected from CNC system

Parameter	Meaning	Data source
n	Spindle speed	Internal sensors of CNC machine tools
x, y, z	Tool coordinate	
$I_{spindle}$	Spindle current	
$U_{spindle}$	Spindle voltage	
f	Feed rate	
U_{in}	Input voltage	Smart metres
I_{in}	Input current	
P_{in}	Input power	

Obviously, the signal collected in the actual production usually has lots of interference (Fig. 3.7(a)). Hence, signal needs to be filtered to make it more smooth and accurate. The filtering method is shown below:

① Delete the first data of queue. Length of the queue is N, which is an odd.
② Read a new data and put it into the end of the queue.
③ Sort the data in the queue in ascending order and then average the three data in the middle of the queue.
④ Output the result (Fig. 3.7(b)).

NC gantry machining centre, whose numerical control system is 840D, is taken as an example. This process is implemented to produce a machine bed. The material of this part is HT200 grey cast iron. Production data of this process are recorded with a sampling period of 5 s, and the filtering result is shown in Fig. 3.8.

(3) **Identification of the Manufacturing Equipment's Operation Status**

As discussed above, a machining process includes four periods, and each period has a different energy-cost characteristic. So, it is particularly important to identify operating state of machine tools with production data that are acquired. Hu [11]

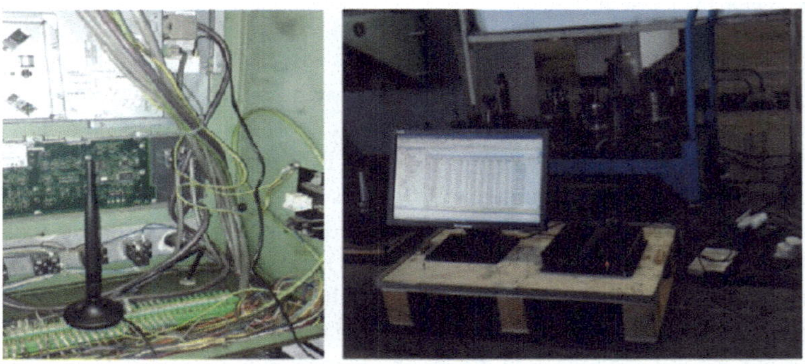

(a) Installation of wireless transceiver (b) Installation of IPC

Fig. 3.7 Energy consumption at the working step level (文中没有找到出处)

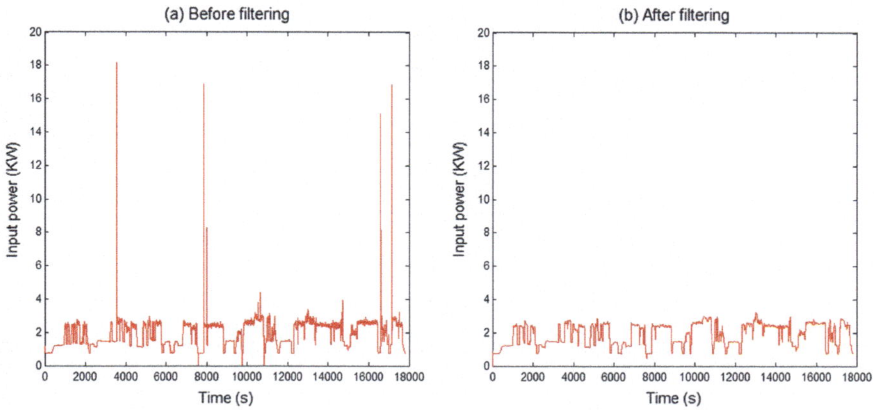

Fig. 3.8 Data filtering treatment effect

proposed a method to identify three operating states: startup phase, idle phase, cutting phase. He did not consider the standby phase. In this chapter, a feasible method is presented to identify four operating states of machine tools:

- **The startup state.**

Input power is calculated with filtered input current I_{in} and input voltage U_{in}. Then, this real-time variable is used to compare with a constant z, the zero-drift value of machine tool. If several sequential values are all greater than the constant, the machine is considered as startup, and the corresponding flag is set as STATE = 100 while 000 represents downtime.

- **Standby state.**

Standby state needs to satisfy three conditions: (a) spindle speed n is zero; (b) input power P_{in} is larger than a set value s_1; (c) recent data changes little. That is, the standard deviation of these data is smaller than a set value *sd*. In this case, set the flag as STATE = 101 and current input power as standby power P_{stdby}.

- **Idle state.**

Identification of idle state is easier because spindle is rotating and the input power of spindle is almost invariable in this state. Flag of this state is set as STATE = 110, and current input power is also recorded as idle power of spindle, namely P_{idle}.

- **Cutting state.**

Because power during the cutting period P_{cut} is greater than idle power P_{idle}, cutting state could be identified if the difference between them exceeds the setting value s_2. Of course, the condition of that spindle rotation is also needed. Set the flag as STATE = 111.

The corresponding algorithm flow is shown in Fig. 3.9.

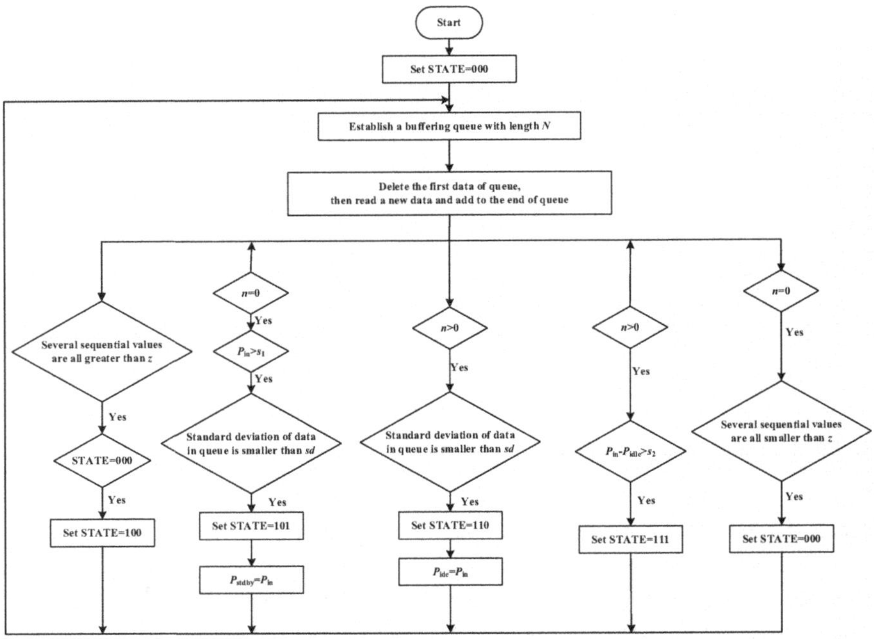

Fig. 3.9 Manufacturing equipment status recognition process

(4) **Analysis of the System's Multi-layer Energy Consumption**

The standby power and idle power of this machine tool are 750 W and 1300 W, respectively. Besides, startup energy could be obtained from production data, which is 13944 J. Table 3.2 shows part of these data during the cutting period and energy efficiency of each working step. The low energy efficiency is mainly because the first working step includes startup period and long standby period to adjust the cutting tool.

Table 3.2 Part of input power during machining

Serial number	Production process	η (%)	Input power (W)		
			Group 1	Group 2	Group 3
1	Semi-finish milling surface A	9.53	1800	1865	1796
2	Semi-finish milling surface B	46.29	1832	1920	1841
3	Chamfering	31.98	1930	1998	1952
4	Semi-finish milling side surface	29.33	1993	2021	2001
5	Rough drill $\Phi 8 \times 36f$	43.12	2476	2523	2411
6	Semi-finish boring	52.04	1906	1857	1968

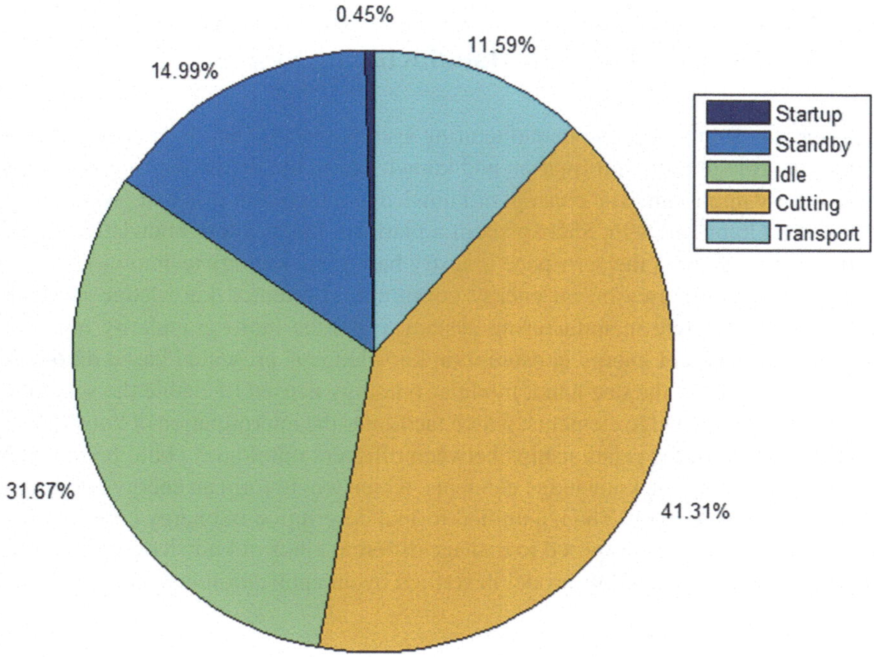

Fig. 3.10 Proportion of four periods

Working steps performed on the same machine tool compose a process, as shown in Fig. 3.10. Because of its big size, chosen parts are transported by crane whose rated power is 10 KW and one part at a time. According to production data collected during machining, the energy efficiency of this process is $\eta_{\text{process}} = 45.48\%$.

To finish a part, workpieces need to be processed on several different machine tools. In the workshop, we studied, painting and cleaning are performed by workers. Production data on processing a part are recorded, and the proportion of each period is shown in Fig. 3.10. Hence, the energy efficiency of this part is shown below: $\eta_{\text{part}} = 41.31\%$.

A finished production is composed of some different parts. Besides, the energy consumption of support facilities is also considered, like lights and central air-conditioning. In a machine tool plant, energy to assemble is mainly used on cranes for its big size. The method used here is similar to other layers and would not be explored here.

3.5 Ontologically-Based Energy Consumption Knowledge Model of the Discrete Manufacturing System

With the development of the manufacturing system towards intelligence and digitization, the role of data, information and knowledge in the manufacturing system is increasingly important. The concept of knowledge integration was first proposed by Robert M. Grant[2] in 1996. Shehab[3] built a platform-independent knowledge-based engineering system in the aerospace industry based on ontology technology.

This chapter focuses on an energy consumption-oriented knowledge network model of the discrete manufacturing system. First, the multi-granularity modular hierarchical model of energy consumption knowledge is presented based on ontology technology. On the one hand, modular ontology is used to enable the semantic annotation of knowledge elements, which facilitates the interpretation of knowledge; on the other hand, the relationships between different ontologies could be mapped to energy consumption knowledge elements, which would form an energy consumption knowledge network. Then, a unified formal description of energy consumption knowledge elements is proposed to manage different kinds of knowledge uniformly. Finally, the feasibility of this model is verified by an application instance.

3.5.1 Ontology-Based Modular Multi-granularity Hierarchical Model

The discrete manufacturing system is large and complex for its multiple energy consumption sources. In this case, managing the large volumes of information is a big challenge for enterprises, let alone a mass of knowledge generated from this information. Therefore, ontology technology is introduced to knowledge management. The ontology technology defined a standard representation of terms and concepts to interpret knowledge for various users and fill in the semantic gap between different people. However, since the discrete manufacturing system contains various ontologies and terminologies, it is hard to construct and manage ontologies effectively. To address this problem, the ontological modularity strategy is proposed to build domain ontologies. The modular ontology technique is a top-down construction method, which starts from the most fundamental and general concepts.

Combined with the energy consumption characteristics of discrete manufacturing and modular ontology technology, this book built an ontology-based modular multi-granularity hierarchical model of energy consumption knowledge, as shown in Fig. 3.11. This model describes relationships between ontology and ontology, knowledge and knowledge, ontology and knowledge with three layers from different granularity.

At the top of this model is the root ontology layer. Root ontology is also called modular reference ontology. Here, energy-related knowledge in discrete manufacturing can be divided into five modules: task knowledge module, resources knowledge module,

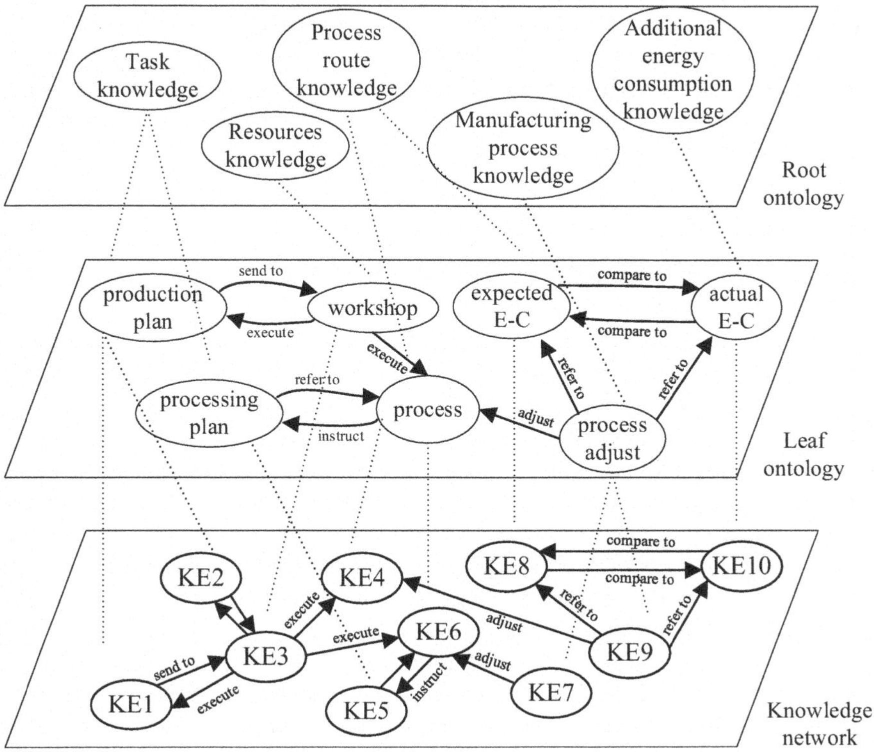

Fig. 3.11 Modular multi-granularity hierarchical model

process route knowledge module, manufacturing process knowledge module and additional energy consumption knowledge module. Modular ontology is main to describe the most fundamental and general concepts and terms in each module. The advantage of modular ontology is that it fundamentally clears the relation of different ontologies and facilitates the management, reuse and maintenance of ontologies and knowledge.

At the middle is the leaf ontology layer. Leaf ontologies are the domain-specific ontologies which not only inherit the semantics and properties of root ontologies, but also have their own properties. They describe specific concepts, terminology and the logical relationship between them in each module. Leaf ontology layer has more than one level. It is generated from the corresponding root ontology, grows down with the continuous specification of ontologies and forms ontology hierarchical tree eventually. Leaf ontology layer includes not only the concepts and terminology, but also the relationship between them. In the ontology theory, the relationship between ontologies is defined as "object property" of ontologies. Typically, a leaf ontology inherits the properties from its upper ontology. On the one hand, leaf ontology layer has multiple levels; on the other hand, properties connect different ontologies in the leaf ontology layer. Hence, an ontology network is built.

The knowledge network layer is the bottom of the model. Ontologies represent concepts, terminology and properties, while knowledge is the instance of these ontologies. Typically, the knowledge element inherits the property of the ontology. Then, a bigger knowledge network will be formed on the basis of the ontology network in leaf ontology layer.

3.5.2 Construction of Knowledge Ontologies

Studer[5] proposed that an ontology is a formal, explicit specification of a shared conceptualization. This definition is widely recognized by other researchers. According to the modular multi-granularity hierarchical model built above, knowledge ontology in discrete manufacturing is divided into five modules: resource-related knowledge ontology, task-related knowledge ontology, process-route-related knowledge ontology, additional energy-related consumption knowledge ontology and manufacturing-process-related knowledge ontology. As an example, this section constructs the manufacturing-process-related knowledge ontology with Protégé. Untagged object properties in Fig. 3.12 are property "has subclass", which indicates that ranges of this property are part of the domain.

As shown in Fig. 3.13, manufacturing-process-related knowledge ontology focuses on the description of the production activities and the monitoring and record of energy consumption during the manufacturing, such as production schedule, actual energy consumption and change of plan. Actual energy consumption knowledge is formed from the combination of energy consumption data and process information. The knowledge would compare with the expected energy consumption to adjust the process route in time and to meet energy-saving requirements of production. Through

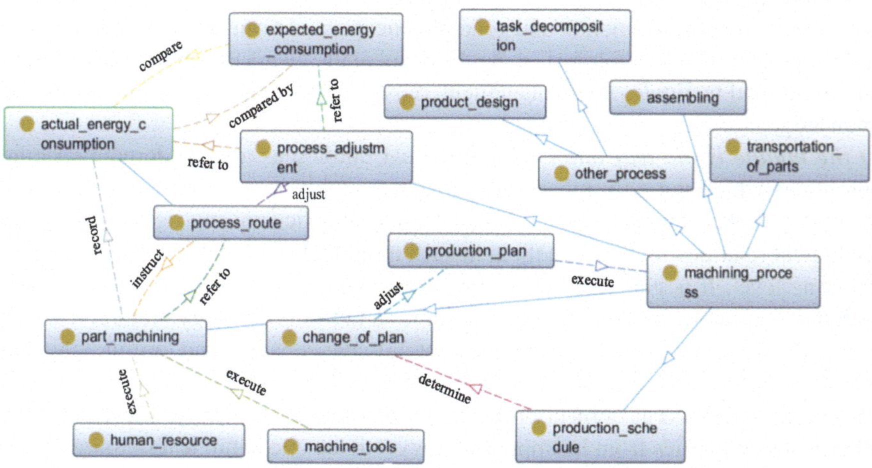

Fig. 3.12 Manufacturing-process-related knowledge ontology

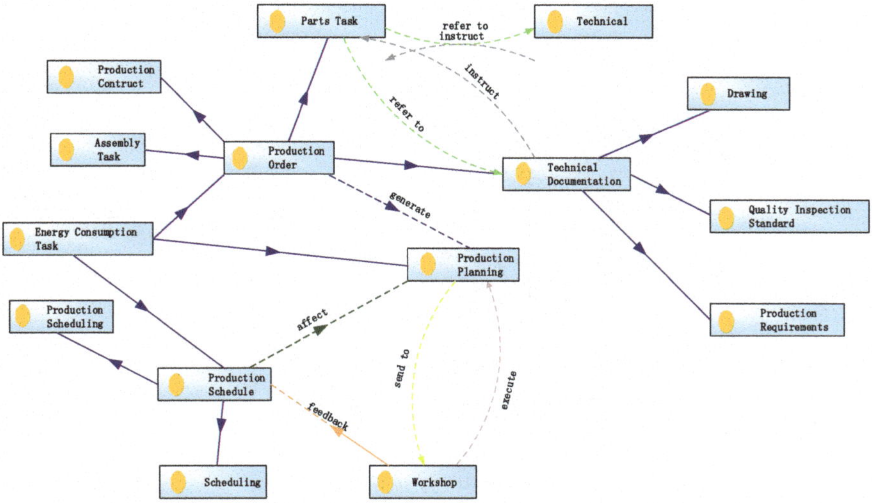

(a) Process energy ontology

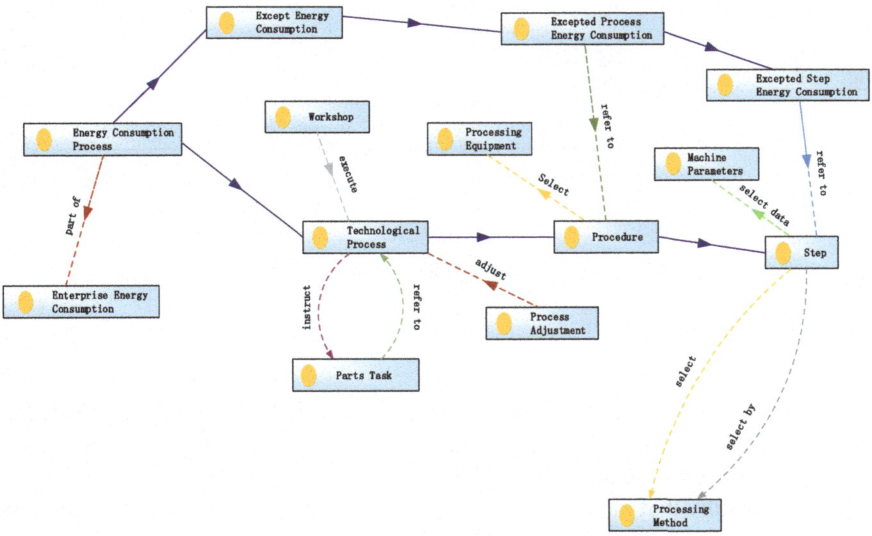

(b) Operating energy ontology

Fig. 3.13 Four class of energy knowledge ontologies

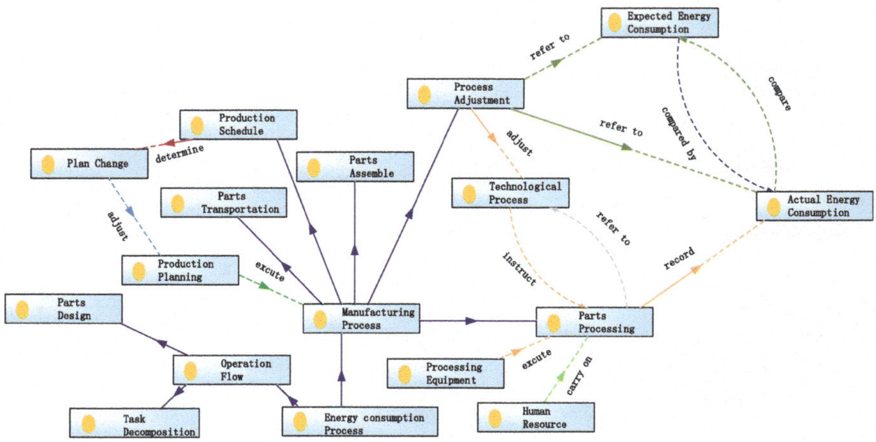

(c) Task energy ontology

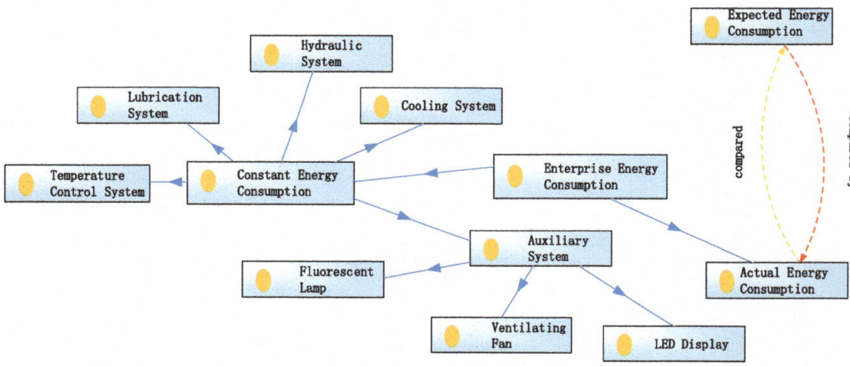

(d) Auxiliary energy ontology

Fig. 3.13 (continued)

the definition of related concepts in production, manufacturing process knowledge ontology ensures the unified semantic annotations of manufacturing data in real-time production and helps the decision-makers to reuse knowledge.

It is noteworthy that these five knowledge ontologies do not exist independently. These ontologies form a multi-granularity knowledge ontology network connected by various properties in leaf ontology layer. The unified definition and description of energy-related terms, conception and their relations could reduce the semantic ambiguity and facilitate the unified management and application of heterogeneous knowledge.

3.5.3 Description of Energy Consumption Knowledge Element

Knowledge element is the most basic knowledge unit that has semantic knowledge. According to the objects and functions of energy in discrete manufacturing, energy consumption knowledge can be divided into five types: resources knowledge, task knowledge, process route knowledge, additional energy consumption knowledge and manufacturing process knowledge. In this book, energy consumption element is formally described as:

$$ECKE = (Basic - Info, Content, Energy, Neighbor, Link) \tag{3.41}$$

(1) ***Basic-Info*** aims to represent basic information of energy consumption knowledge element: *Basic-Info* = (*ID, Name, Ontology, Category, Update-Time*). Here, *ID* is the only index of knowledge elements; *Name* is the name of this knowledge element; *Ontology* means the upper ontology; *Category* is the type of this knowledge element; *Update-Time* records the date that this knowledge element added or updated.
(2) ***Content*** is the detail of this knowledge element. Storage method depends on knowledge content because of different types of knowledge.
(3) ***Energy*** records the information related to energy consumption: *Energy* = (*Related, Manner*). The former records whether it is related to energy; the latter records the way this knowledge affects energy consumption.
(4) ***Neighbour*** is a set of knowledge elements that record all of neighbourhood knowledge elements of this knowledge.
(5) ***Link*** is the set of relationships between this knowledge element and its neighbours. Knowledge network inherits properties of upper ontology and uses object properties to describe relationships between knowledge elements.

The energy consumption-oriented knowledge and knowledge ontology turn the tacit knowledge in the enterprise into explicit knowledge. This knowledge is expressed in the form of words and numbers which are easy to exchange and share. The definition and description of the energy consumption knowledge provides the basic processing unit for a knowledge network and also ensures the unified form of various knowledge. As instances of ontologies, energy consumption knowledge element ensures that different personnel could understand knowledge through ontology semantic annotation.

3.6 Manufacturing Facility's Energy Consumption Knowledge Model Based on the Rough Set

Usually, the energy consumption model seeks to find the relationship between machining parameters and input power of machine tools. Balogun et al. [13] divided the electrical energy requirements of a machining process into four parts: basic state energy, ready state energy, cutting state energy and non-cutting state energy. Peng et al. [14] proposed three types of function blocks to model the energy consumption of CNC machine tools. Guo et al. [15] divided the workpiece process into three parts and predicted energy consumption through an operation-mode-based simulation approach.

Traditional approaches to the model energy consumption of CNC machine tools are mainly based on analytical models, which require clear knowledge about machine tools. However, this is challenging now due to complex construction and varying parameters of CNC machine tools. Hence, there is a need to develop a novel modelling approach that could avoid these problems. Kant and Sangwan [16] used an artificial neural network to predict the energy consumption. Garg et al. [17] proposed a multi-gene genetic programming approach combined with the complexity and orthogonal basis functions to model the energy consumption of the milling process.

Typically, an operator could make favourable decisions based on production experience obtained from observing inputs and outputs. This process is similar to the main idea of Rough Set (RS). RS is a new mathematical technique in knowledge discovery. It aims to find the relationship between different data, but does not require a classical mathematical description of this process.

In this section, a knowledge modelling method based on RS is presented to model the energy consumption of manufacturing equipment as CNC lathes and improves the modelling method appropriately according to the specific application. It includes a novel approach to discretize decision attributes, a method based on Boolean logic and attribute importance to discretize continuous condition attributes, a computing method based on information space to obtain output. As the basis of sustainable production, this study has a focus on energy consumption model of cutting period in CNC lathe.

3.6.1 Energy Consumption Model of CNC Machine Tools

CNC machine tools are widely used in today's society. In this case, an accurate energy consumption model is necessary for follow-up studies, like energy consumption optimization. Related researches [18] showed that the operational state of CNC machine tools has a bearing on energy cost. Typically, the process of CNC machine tools includes four periods: startup period, standby period, idle period and cutting period. Each period has its own characteristics, which are also shown in Fig. 3.14.

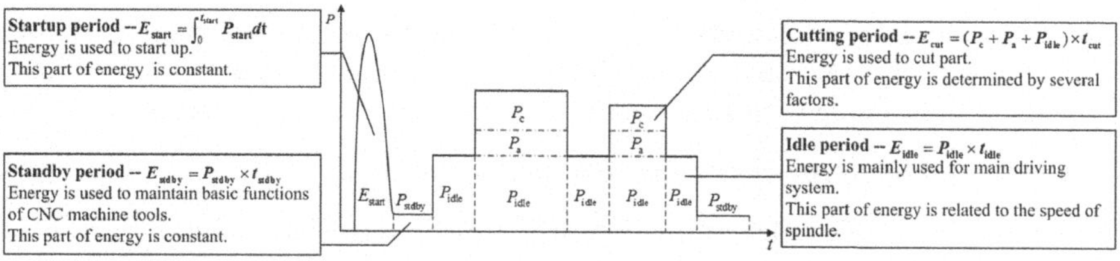

Fig. 3.14 Input power curve of CNC machine tools

The rough set is a soft computing approach that Pawlak Z presented in 1982 to deal with the uncertain and incomplete problem [19]. An information system in RS is a rule-based knowledge, which is expressed in a decision table. That is to say, it does not require a specific description of process mechanism, but consists of sets of If-Then rules instead [20].

(1) ***Decision table***

In RS, the information system is denoted as a nonempty finite set $S = (U, A, V, f)$. Here, $U = \{x_1, x_2, \ldots, x_n\}$ means a finite set of samples; $A = C \cup D$, $C \cap D = \varnothing$, C and D denote condition (C) and decision (D) attributes of samples, respectively; $V = \bigcup\limits_{a \in A} V_a$ is the set of these attributes' range; $f : U \times A \rightarrow V$ is an information function that gives every attribute a value.

(2) ***Indiscernibility relation***

According to RS, indiscernibility relation can be defined. Typically, the indiscernibility relation is:

$$\text{ind}(R) = \{(x, y) \in U \times U : f(x, a) = f(y, a), \forall a \in A\} \tag{3.42}$$

where (x, y) is a pair of samples and $f(x, a)$ denotes the value of attribute a for sample x. Generally, sample set U could be divided into several elementary sets, such as U/R. If $(x, y) \in \text{ind}(R)$, then it is hard to distinguish x, y according to R.

3.6.2 Rough-Set-Based Modelling Method

This section presents a novel RS-based modelling method, which is applied to the cutting period in this section.

(1) ***Data preprocessing***

Decision table built from original sample mostly has many problems, like data redundancy, data incompatibility and data incompleteness [21]. This step is aimed to clean the sample set and improve data quality.

In this book, the cutting period of CNC lathe is seen as an information system $S = (U, CUD, V, f)$, where $U = \{x_1, x_2, \cdots, x_n\}$, $C = \{v_c, f, a_p\}$, $D = \{P_i\}$. Table 3.3 is an example with 12 groups of data that have been preprocessed.

(2) ***Discretization of the continuous decision attribute***

Current studies provide little information on how to discretize the continuous decision attribute. Domain experts discretize continuous decision attribute in most studies [21]. Besides, it would be discretized into several equidistant intervals sometimes. However, the former method is subjective that may have potential problems; the latter method rarely considers data distribution. Therefore, a novel discretization method is urgently required. In this book, the input power of CNC lathe during the cutting period is chosen as a decision attribute, which is a continuous variable. Here presents a novel discretization method based on data distribution.

Algorithm 1 *Discretization of the decision attribute*

(a) *Set a precision value* Δ. *Turn decision attribute values* v_d *(input power) into intervals* $v_d \pm \Delta$.
(b) *If the length of overlapping portion between two intervals is greater than threshold value k, then combine these two intervals into one.*
(c) *If any two of these intervals have an overlapping portion that less than threshold value k, then partition the overlapping portion to the interval with less length.*
(d) *Divide all of the intervals into* $\text{round}(length/2\Delta)$ *equal parts, in which* $\text{round}(\bullet)$ *returns rounded result.*

Using the decision attribute value set in Table 3.1, $D = \{2497, 2505, 2693, 2725, 2911, 2949, 3081, 3083, 3184, 3234, 3272, 3479\}$, In

Table 3.3 Example of the information system

U	v_c(m/min)	f(mm/s)	a_p(mm)	P_i(W)
x_1	100	0.14	1.1	2497
x_2	110	0.11	1.1	2505
x_3	110	0.14	1.1	2693
x_4	110	0.12	1.3	2725
x_5	120	0.13	1.2	2911
x_6	130	0.11	1.2	2949
x_7	130	0.14	1.1	3081
x_8	120	0.17	1.1	3083
x_9	130	0.14	1.2	3184
x_{10}	120	0.16	1.3	3234
x_{11}	120	0.14	1.5	3272
x_{12}	130	0.15	1.4	3479

*These data have been collected from experiments in Sect. 3.4
* is used for annotation

this example, $\Delta = k = 50$. Then intervals formed from each decision attribute value are $[2447, 2547)$, $[2455, 2555)$, ..., $[3429, 3529)$, respectively. According to step (b) and step (c), the range of this decision attribute value set is divided into six parts: $[2447, 2555)$, $[2643, 2775)$, $2861, 2999)$,$[3031, 3133)$, $[3134, 3322)$, $[3429, 3529)$. Here, round $((3322-3134)/100) = 2$. So, interval $[3134, 3322)$ needs to be divided into two parts: $[3134, 3227)$, $[3228, 3322)$. 1, 2, 3, 4, 5, 6, 7 is used to represent these intervals, respectively. Hence, set $D^* = \{1, 1, 2, 2, 3, 3, 4, 4, 5, 6, 67\}$ is the decision attribute value set after distribution.

In this novel method, distribution quality is directly determined by precision value and the threshold value, which should be decided based on the actual situation.

(3) ***A new information system based on Boolean logic***

After the decision attribute being discretized, this section aims to propose a novel discretization method based on Boolean logic and attribute importance.

Nguyen and Skowron [22] introduced Boolean logic into condition attribute discretization and proposed an effective discretization method. This method builds a new information system $S^*= (U^*, C^* \cup D^*, V^*, f^*)$ according to the discernibility relation between two samples, whose decision attribute values are different. Besides, this new information system does not change the indiscernibility relation of the original system.

In this new information system, $U^* = \{(x_i, x_j) \in U \times U : d(x_i) \neq d(x_j)\}$, $C^* = \{c_i^a | i \in \{0, 1, \cdots, k_a\}, a \in C\}$, c_i^a is the cut point of attribute a,

$$f^*(u, c) = \begin{cases} 1 & \min[a(x_i), a(x_j)] < c < \max[a(x_i), a(x_j)] \\ 0 & \text{otherwise} \end{cases}.$$

(4) ***A novel discretization method based on Boolean logic and attribute importance.***

Attribute importance is a crucial concept in RS. The more important the attribute is, the more attention it should get. In RS, attribute importance describes classification capacity of this attribute.

Let C, D be condition attributes set and decision attributes set, respectively, $C' \subseteq C$. Then, the importance of attributes set C' is defined as:

$$\sigma_{CD}(C') = \gamma_C(D) - \gamma_{C-C'}(D) = \frac{|POS_C(D)|}{|U|} - \frac{|POS_{C-C'}(D)|}{|U|} \tag{3.43}$$

In particular, when $C' = \{a\}$, the importance of attribute $a \in C$ is

$$\sigma_{CD}(a) = \gamma_C(D) - \gamma_{C-\{a\}}(D) \tag{3.44}$$

In the information system shown in Table 3.1, the importance of v_c, f, a_p is 0.25, 0.33, 0.167, respectively.

In this book, a novel discretization method combined with Boolean logic and attribute importance is presented:

Algorithm 2 *Discretization method based on Boolean logic and attribute importance*

(a) *Calculate attribute importance according to* Eq. (3.3).
(b) *Sort condition attributes in ascending order based on attribute importance; sort condition attributes with the same importance in descending order according to the cut point numbers of these attributes.*
(c) *Build a new information system S^* based on Boolean logic.*
(d) *Decide the cut point will stay or remove in turn. Observe all rows marked with 1 in the corresponding column. If there is one row that only has one 1, namely only the corresponding column is 1 in this row and others are 0, then keep this cut point and delete the corresponding column, together with all rows marked with 1 in it; otherwise, remove this cut point and delete the corresponding column.*

According to Steps (a) and (b), the initial cut point set is:

$$CUT = C_{ap} \cup C_{vc} \cup C_f$$

where

$$C_{ap} = \{(a_p, 1.15), (a_p, 1.25), (a_p, 1.35), (a_p, 1.45)\}$$

$$C_{vc} = \{(v_c, 105), (v_c, 115), (v_c, 125)\}$$

$$C_f = \{(f, 0.115), (f, 0.125), (f, 0.135), (f, 0.145), (f, 0.155), (f, 0.165)\}$$

The new information system S^* based on Boolean logic is a 61×13 Boolean matrix. The part of it is shown in Table 3.4.

According to step (c), the final cut point is $CUT = \{(\mathrm{a_p}, 1.15), (\mathrm{v_c}, 105), (\mathrm{v_c}, 125), (f, 0.105), (f, 0.135), (f, 0.145)\}$. Set $\{0, 1\}$ is used to represent interval set of attribute a_p, namely $\{(-\infty, 1.15), [1.15, +\infty)\}$. For attribute v_c, set $\{0, 1, 2\}$ is used to represent interval set $\{(-\infty, 105), [105, 125), [125, +\infty)\}$. For attribute *f*, set $\{0, 1, 2, 3\}$ is used to represent interval set $\{(-\infty, 0.125), [0.125, 0.135), [0.135, 0.145), [0.145, +\infty)\}$. The discretized information system is shown in Table 3.5.

Classical discrete method neither takes attribute importance into consideration, nor obtains the minimal number of cut points. However, Algorithm 2 can find the most reasonable cut point set with fewest cut points.

So far, a feasible knowledge base has been obtained. In this knowledge base, each sample is seen as a rule. For a new sample, the decision attribute value could be predicted by matching every condition attribute.

(5) ***A novel output method of the knowledge base***

Table 3.4 Part of corresponding new information system

	1.15	1.25	1.35	1.45	105	115	125	0.115	0.125	0.135	0.145	0.55	0.165
(x_1, x_3)	0	0	0	0	1	0	0	0	0	0	0	0	0
(x_1, x_4)	1	1	0	0	1	0	0	0	1	1	0	0	0
(x_1, x_5)	1	0	0	0	1	1	0	0	0	1	0	0	0
(x_1, x_6)	1	0	0	0	1	1	1	1	1	1	0	0	0
(x_1, x_7)	0	0	0	0	1	1	1	0	0	0	0	0	0
(x_1, x_8)	0	0	0	0	1	1	0	0	0	0	1	1	1
(x_1, x_9)	1	0	0	0	1	1	1	0	0	0	0	0	0
(x_1, x_{10})	1	1	0	0	1	1	0	0	0	0	1	1	0
(x_1, x_{11})	1	1	1	1	1	1	0	0	0	0	0	0	0
(x_1, x_{12})	1	1	1	0	1	1	1	0	0	0	1	0	0
(x_2, x_{13})	0	0	0	0	0	0	0	1	1	1	0	0	0
(x_2, x_{14})	1	1	0	0	0	0	0	1	0	0	0	0	0

Table 3.5 Discretized information system

U	v_c(m/min)	f(mm/s)	a_p(mm)	P_i(W)
x_1	0	2	0	1
x_2	1	0	0	1
x_3	1	2	0	2
x_4	1	0	1	2
x_5	1	1	1	3
x_6	2	0	1	3
x_7	2	2	0	4
x_8	1	3	0	4
x_9	2	2	1	5
x_{10}	1	3	1	6
x_{11}	1	2	1	6
x_{12}	2	3	1	7

In the situation of continuous decision attribute, knowledge base only outputs an interval, not a value. Hence, a novel output method based on knowledge space is proposed.

When condition attributes all have an upper and lower limit, knowledge system S becomes an enclosure space with $|C|$-dimension, namely knowledge space. Usually, this knowledge space is divided into several subspaces by cut points, in which samples have the same decision attribute value. These subspaces are called basic hypercube.

Typically, each basic hypercube corresponds to an interval when the decision attribute of this knowledge space is continuous. As shown in Fig. 3.15, samples x_1 and x_2 are in the same hypercube, but the output of x_1 should be greater than x_2 when the corresponding decision attribute value is an interval.

Let the corresponding decision attribute value be $[v_i^d, v_{i+1}^d)$, then the output of x_1 could be calculated by Eq. (3.4)

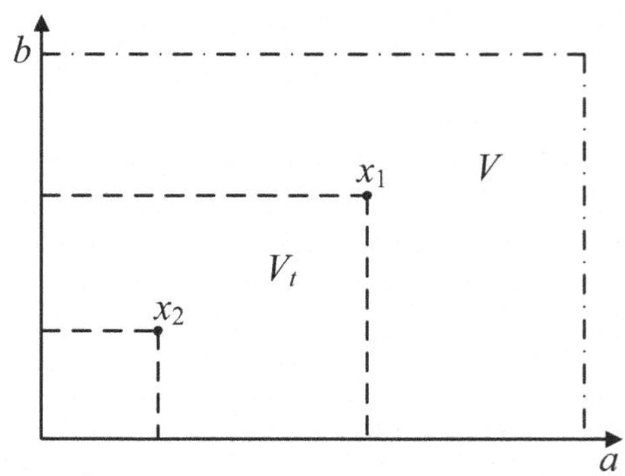

Fig. 3.15 Example of basic hypercube

$$v^d = \left(v_{i+1}^d - v_i^d\right)\frac{V_t}{V} + v_i^d = \left(1 - \frac{V_t}{V}\right)v_i^d + \frac{V_t}{V}v_{i+1}^d \quad (3.45)$$

where V is the space volume of this basic hypercube; V_t is the volume of the space surrounded by this sample (Fig. 3.15).

3.6.3 Experiment Analysis and Validation

In this book, CK60 NC lathe, whose numerical control system is Hua-zhong NC system, is taken as an example and related data are collected from the model presented by Zhou [23]. The workpiece material is 45# HRS (Hot-Rolled Steel). The diameter of the workpiece is 50 mm, and the cutting length is 100 mm.

Here, experiments are designed based on orthogonal experimental design, and related cutting parameters are shown in Table 3.6. Ten samples are recorded to build initial decision table.

In order to obtain a feasible knowledge base, continuous decision attribute is needed to be discretized first. The decision attribute value ranges from 1449 to 10,234. In this case, precision value $\Delta = 50$, threshold value $k = 50$. So, the decision attribute is divided into 65 intervals.

Next step is condition attribute discretization. The calculating result shows that the importance of three cutting parameters is all 1. This result is generated because of orthogonal design and small precision value. Actually, no cut point will be deleted according to the calculating result. Finally, the knowledge base is obtained.

Tables 3.7, 3.8 and 3.9 show the comparison of the RS-based model and real

Table 3.6 Values of cutting parameters

Parameters	Values
v_c(m/min)	70, 100, 130, 160, 190, 220, 250
f(mm/s)	0.1, 0.15, 0.2, 0.25, 0.3
a_p(mm)	0.5, 0.8, 1.1, 1.4, 1.7, 2.0

Table 3.7 Error analysis with different v_c

v_c(m/min)	RS(W)	Real Value(W)	Error(%)
75	3903	3005	3.39
105	3898	3923	0.64
135	4792	4815	0.48
165	5699	5687	0.21
195	6565	6544	0.32
225	7457	7388	0.93

* $f = 0.125\text{mm/s}$, $a_p = 1.6\text{mm} = 1.6$ mm
* is used for annotation

Table 3.8 Error analysis with different f

f(mm/s)	RS(W)	Real Value(W)	Error(%)
0.12	3201	3389	5.55
0.17	3798	3897	2.54
0.22	4275	4384	2.49
0.27	4794	4854	1.24

$^*v_c = 130\text{m/min}$, $a_p = 1.6\text{mm}$
* is used for annotation

Table 3.9 Error analysis with different a_p

a_p(mm)	RS (W)	Real value (W)	Error (%)
0.7	3308	3143	5.25
1	3794	3654	3.83
1.3	4271	4162	2.62
1.6	4791	4668	2.63
1.9	5307	5172	2.61

$^*v_c = 130\text{m/min}$, $f = 0.25\text{mm/s}$
* is used for annotation

value when cutting parameters change. The result indicates that this RS-based energy consumption model is precious enough to be applied to practical production.

The energy consumption model is the foundation of the following research, like energy saving. This RS-based energy consumption model is proposed under the trend of manufacturing information, which makes it possible to record and store production data. Knowledge base used in this model could be built from real production data and improved with the accumulation of data. Advantages of this RS-based model include simplifying computing process of output and providing enough precision. The only problem of this RS-based modelling approach is that it requires mass data to find the mapping relationship between different data. Besides, this, RS-based method solves the modelling problem of an information system whose input and output are both continuous. Hence, it is viable to be generalized to other fields.

3.7 CBR-Based Discrete Manufacturing System's Energy Consumption Knowledge Modelling

Due to the complexity and uncertainty of the discrete manufacturing system's energy consumption, it is not easy to construct an accurate and easy-to-calculate energy consumption model. There is a desperate demand for a data-driven, widely applicable method to predict energy consumption. Many scholars have constructed processing facility's energy consumption models from the perspective of data modelling. Based on massive observation data and SPSS software, Kara et al. [24] simulated an energy consumption experience model, which reveals that the material removing rate is in reverse proportion to the unit energy consumption. Garg et al. [25] proposed a multi-genetic inheritance planning method based on complexity and orthogonal basis function and applied it to the energy consumption modelling of the milling progress. Kant and Sangwan [26] combine the grey relational analysis and the response surface methodology, constructing a prognostic model of energy consumption and surface roughness, with technological parameters as variables. The manufacturing energy consumption has such a complex dynamic correlation with multiple processing variables that it creates a complex modelling issue with character and value variables combined.

Based on the working step-procedure-parts-product-integrated model proposed in Sect. 3.4, here we construct a discrete manufacturing system's energy consumption knowledge model, combining energy consumption data, technological information and scheduling information. Furthermore, taking into consideration the value and the character types of input variables, we use the power fluctuation degree to calculate the importance of properties. Through the examination method of layered examples, we guarantee the uniformity or relevance of the characters' input variables. Finally, we achieve energy consumption prediction through searching similar working steps and experiment simulation also testifies the effectiveness of the method proposed in this section.

3.7.1 Relevant Knowledge Integration Concerning Energy Consumption

About the massive data produced in the discrete manufacturing system, this section divides knowledge of energy consumption to three categories: technological knowledge, facility knowledge and parts knowledge, the knowledge integration framework of which is illustrated in Fig. 3.16.

In this framework, the present section treats technological knowledge as the interface for the entire system's knowledge description. After describing the scheduling plan, the technological agenda and the product information, we reflect such description in the knowledge of parts and facility, thereby describing and recording the actual processing progress from the perspectives of parts and facility, respectively, thus generating a knowledge case base applicable to energy consumption prediction and analysis. Meanwhile, we send the prediction and analysis results to the administrator, thereby achieving the effective adjustment of the scheduling scheme and the technological plan.

This figure also reflects the interactive relationship between these three sorts of knowledge. Technological knowledge is the basis and core of the entire knowledge system. After describing the scheduling scheme, technological plan and product information, it transmits such information to the facility's and the parts' knowledge. The latter two, based on the information guidance provided by the technological

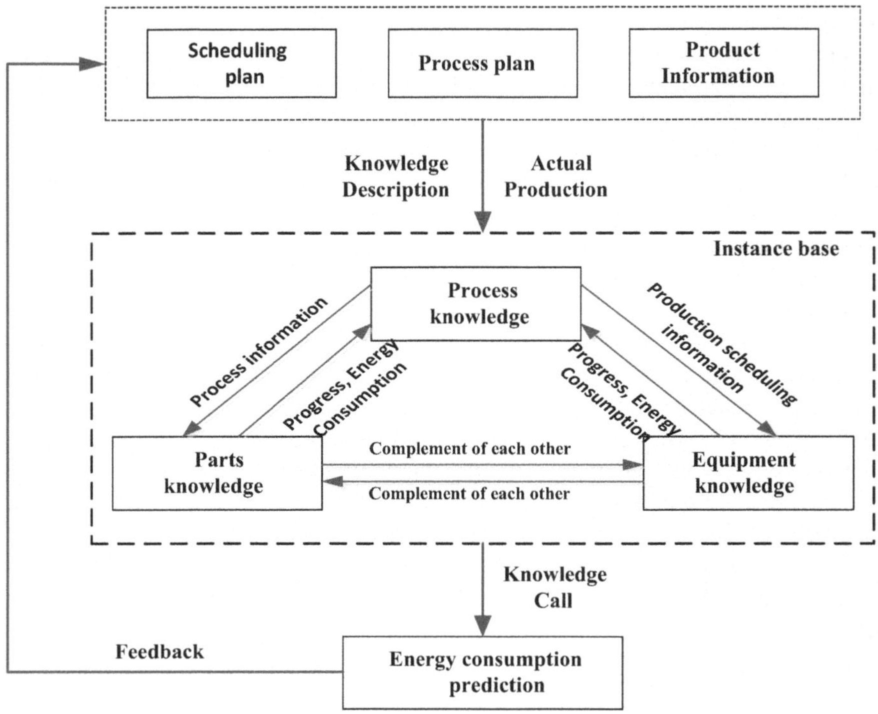

Fig. 3.16 Knowledge integration framework relevant to the discrete manufacturing system's energy consumption

knowledge, will supervise the actual production of products, keep real-time tracking and record of the production progress and energy consumption information, while sending statistics and responses of such information back to the technological knowledge, to be stored in the knowledge base as the enterprise's knowledge examples.

(1) **Technological knowledge relevant to energy consumption**

The discrete manufacturing system is essentially processing and assembling different components. Its energy is mostly consumed during the processing progress of these parts. Technological knowledge mainly describes the technological plan, which also combines product information, technological routes and scheming plan, to provide guidance and supplement for facility's and component's knowledge. Its descriptive formula is:

$$Route = \left(Route - ID, Product - Info, Part_1, Part_2 \ldots Part_p, \right. \\ \left. Assembly, Product - EC\right)$$

Route-ID represents the product's technological plan number;

Product-Info is the product's basic information. It is represented by a six-unit group: *Product-Info =(Product-ID, Batch, Order-Info, Product-Type, Specification, Remark).*

Product-ID is the product's number.;

Batch is the product's production batch;

Order-Info is the order information;

Product-Type is the product type;

Specification is the product's standards, including the product's actual parameters and processing requirements;

Remark is used for recording the additional requirements and conditions for processing products.

Part is the part's information, its descriptive formula is $Part_p$ = *(Part-Info,* $Process_1$, $Process_2$, … , $P\ rocess_w$, *Part-EC). Here,* $Process_w$ = *(Process-ID,* $Step_1$, $Step_2$, … , $Step_s$, *Machine-ID, Process-Stage, Process-EC),* Meanwhile, the working step uses a three-unit set to represent $Step_s$ = *(Method, Parameters, Step-EC), Process-Stage* is used to represent the processing stage of the procedure; *Part-EC* is the average energy consumption of the parts' production.

Assembly is the product's assembling information. *Assembly = (ASSY-Equip, ASSY-EC),* namely the assembly equipment and the energy consumption.

Product-EC is the average energy consumption for producing a product.

(2) **Knowledge relevant to the parts' energy consumption**

The parts' knowledge is used to describe the parts' basic information, processing schedule and energy consumption. Its descriptive formula is:

$$Part = (Part{-}ID, Part{-}Info, Completed{-}P, Incompleted - P,\\ Part{-}EC, Part{-}State, Location)$$

Part-ID represents the part's serial number. Its serial number is made up of two parts: the part's type number and code number. *Part-Info* is part's basic information. It can be represented by a four-unit group:

$$Part{-}Info = (Type, Basic{-}Tech, Material, Part{-}Stage).$$

In above, *Type* is the part's type; *Basic-Tech* refers to the basic craft used for processing parts; *Material* describes the materials used for parts; *Part-Stage* represents the processing stage before the part's being processed.

Completed-P is the assemblage of those parts with procedure finished;

Incompleted-P records the energy consumed by the current part's processing;

Part-EC is the assemblage of those parts without finished procedure;

Part-State refers to the current condition of the parts, *Part-State* ∈{ waiting, processing, finished}.

Location refers to the current location of the part. *Location*∈{ buffer zone code, equipment code, finished product code}. The part's location corresponds to its condition.

(3) **Facility Knowledge Relevant to Energy Consumption**

Facility knowledge is refined technological knowledge, which describes production progress from the perspective of the processing facility. The facility knowledge mainly records the facility's processing condition and its energy consumption changes. Its descriptive formula is:

$$\begin{aligned} Machine = (&Machine{-}ID, Completed{-}T, \\ &Incompleted{-}T, Machine{-}State, Function) \end{aligned}$$

Machine-ID represents the facility's code, including the facility's type number and code number.

Completed-T is processing facility's finished task set, including the part's code, procedure code, working step code, working step energy consumption and the start and finish time of the completed procedure.

Incompleted-T is the unfinished task set of the processing facility, including the part's code and corresponding procedure code of the unfinished tasks.

Machine-State records the facility's current operation state.

Function is the facility's functions, acting as a reference for scheduling improvement.

3.7.2 CBR-Based Energy Consumption Estimation

One advantage of CBR technology lies in its feature of incremental learning. Through the ever expansion and update of the example base, its performance keeps improving as well. When we use CBR technology to predict the energy consumption, essentially we are constructing a simulation relationship between the actual example and the object of prediction through analogical reasoning, based on which we calculate the prognostic object's energy consumption. The working step's energy consumption is the most basic energy-consuming unit of the entire manufacturing system. Accurate prediction of the working step level's effective processing energy consumption is the key to the analysis of the system's energy consumption.

When a new order arrives, first we ascertain the scheduling scheme and technological plan based on previous experience. Afterwards, in accordance with the scheduling scheme, technological plan and product information, we can filter such similar working steps from existing knowledge base via case retrieval and then predict the current product's processing energy consumption. If the prognostic result satisfies the processing demands, then we can put into actual production; otherwise, we will rearrange the scheduling scheme and technological plan and rerun the prediction. Once the production order is finished, we can run data preprocessing of the data produced by this order, and record it in the database. We update the property's importance for the data in the database and then update the knowledge base for later usage.

(1) Construction of the Example Base for Energy Consumption

After acquiring the massive data from actual production, we can use the energy consumption data to construct a knowledge base. There are various ways to represent knowledge, for instance, first-order predicate logic, production rules and the case-based knowledge [27]. Borrowing the basic concepts of the rough set, this section transforms knowledge into the form of the decision table. Then, we adopt the energy consumption as the decision attribute, the other variable as the conditional attribute. From the above discussions, we know that the energy consumption of the discrete manufacturing system's can be classified into four layers. Thus, the construction of a knowledge case base should also reflect the layers of energy consumption.

The energy consumption produced by the processing facility's actual operation, once combined with the relevant working step information and technological information in the technological knowledge, forms the "working step" knowledge, namely the working step entry in technological knowledge. Based on such working step knowledge, we calculate the energy consumption of each procedure and record it in the procedure knowledge. We then add all the procedures' consumed energy and get the parts level's energy consumption. Finally, we consider assembling energy consumption and calculate the average total energy consumption for producing a product.

1. ***Data Preprocessing***

Cases in the knowledge case base come directly from actual processing. Given the extreme repetitiveness of the batch production and technology of a product, there exists much redundant knowledge in the knowledge base. For the convenience of reusing this knowledge later and to reduce the search space in the knowledge base, it is necessary to deduce the redundancy. But before that, to ensure the model's accuracy, we need to determine whether we can correctly distinguish the decision attributes through the current conditional attributes. In this section, it means whether we have failed to take into consideration any additional, conditional attributes of great impact on energy consumption.

Algorithm 1 *Testing for the sanity of the conditional attributes*

Step 1: *Classification of the cases according to the conditional attributes;*
Step 2: *Calculation of the standard deviation of each category of decisional attributes, namely*

$$\sigma_i = \sqrt{\frac{1}{n_{cmp}} \sum_{j=1}^{n_{cmp}} \left| v_{d_{ij}} - \overline{v_{d_i}} \right|^2} \tag{3.46}$$

where n_{cmp} is the case number of the category i. $v_{d_{ij}}$ is the decisional attribution—i.e. energy consumption—of case no. j in the category i. $\overline{v_{d_i}}$ is the average energy consumption of category i.

Step 3: *We define as the mean value of σ_i. If is larger than the threshold value, it means that some important conditional attribute is unconsidered, and we need to redefine the make-up of the conditional attribute.*

Once precision is guaranteed, we can simplify the redundant knowledge of the same conditional attributes. We can take the average value of the energy consumption volume. With regard to the same product, the knowledge gained by the most recent processing can overwrite the knowledge recorded in the last processing, thus describe the changes of energy consumption performance caused by causes like abrasion during the continuous production.

2. ***Calculation of the Attributes' Importance***

During the discrete manufacturing system's production process, the processing energy consumption is affected by many factors. Among these factors, there are value-type variables, such as the processing parameter; and the character-type variables, such as the processing method, parts material and the processing equipment. Since here we choose the value-type variable of energy consumption as the decisional attribute, we propose a method to calculate the attributes' importance, thereby achieving calculation of the attributes' importance under different types of condition.

Algorithm 2 Calculating the Importance of the Conditional Attributes

Step 1: *Removing the attribute c to be measured, we reclassify cases in the case base according to the remaining conditional attributes;*
Step 2: *With regard to each category, we quantify the energy consumption's fluctuation;*

$$I_{ci} = \sqrt{\frac{1}{n_{\text{cond}}} \sum_{j=1}^{n_{\text{cond}}} \left| v_{d_{ij}} - \overline{v_{d_i}} \right|^2} \tag{3.47}$$

where n_{cond} is the case number of the category i. $v_{d_{ij}}$ is the decisional attribution—i.e. energy consumption—of case no. j in the category i. $\overline{v_{d_i}}$ is the average energy consumption of category i.

Step 3: *We solve the mean value of each category's energy consumption's fluctuation, which is exactly the importance of its conditional attribute.*

Algorithm 2 starts with the value type's decisional attribute. If the conditional attribute has a large impact on the result, the removal of the decisional attribute will lead to huge fluctuation of the energy consumption, for instance, that of the processing method; if not, there would be little fluctuation of the energy consumption, for instance of the parts' material. This algorithm disregards the difference in the types of conditional attributes, thereby enabling a direct comparison of the character variable and value variable under the same dimension. The drawback of this algorithm lies in its inapplicability to the case of the character-type decisional attribute.

3. ***Update of the Knowledge Base***

In real-life production, the discrete manufacturing system would accumulate massive knowledge of energy consumption, which greatly raises the time cost of knowledge search and knowledge repetitiveness. While data preprocessing has removed redundant knowledge cases, there still exists a massive quantity of highly similar knowledge cases in the case base, which contains quite similar energy consumption information as well. The case reapplication stage would lead to repetitive use of the same information, which affects prognostic precision. The purpose of updating the knowledge base is to remove highly repetitive knowledge, so that we may narrow down the search space and improve the efficiency of the energy consumption prediction, without affecting the prognostic precision.

Algorithm 3 *Knowledge Base Update*

Step 1: *We classify the knowledge cases in the knowledge base according to the character-type conditional attributes;*
Step 2: *Under the condition of the identical character-type conditional attributes, we undertake normalization of the processing parameters, and we calculate the similarity of the knowledge case* via *weighted Euclidean distance.*

$$d = \sqrt{\sum_{j=1}^{n_{num}} (\Delta x_j^2 \alpha_j)} \tag{3.48}$$

where n_{num} *is the number of processing parameters;* Δx_j *represents the difference after the processing parameters are normalized;* α_j *represents the deviation after the standardization of the processing parameters;* $\sum \alpha_j = 1$ *represents the attributional importance after the normalization, Formula (3.48) quantifies the difference between various cases, which intuitively compares the distance between the working steps of various cases.*

Step 3: *We set a threshold value* γ. *If the dissimilarity between the two cases is smaller than* γ, *then we delete one case; otherwise, we preserve both cases.*

4. ***Case Search***

The purpose of the case search is to find cases similar to the prognostic object from the massive cases in the case base. From the perspectives of both processing object and processing facility, this section uses the layered case search methodology to seek knowledge cases of identical or similar character-type conditional attributes. Then we propose a similarity quantified calculation method of each layer, to raise the precision of the case search.

(a) **Processing subject search**

In the discrete manufacturing system, a description of the processing progress normally reflects the arrangement of each procedure on the processing facility with three processing plans. The technological plan reflects the processing methods and parameters adopted in each procedure. Product information reflects the type and materials of the processing object. Such information has a direct influence on the processing energy consumption. A search of the processing object is to seek knowledge cases similar to the to-be-processed working step from the case base, based on the structural features and processing features of the processing objects. This ensures the realistic reference value of the sought knowledge case.

Features of the processing object include parts' material, processing method and geometrical features. As for the character-type variable, its relativity can be solved by using Formula (3.49):

$$s = \begin{cases} 1, \ The\ target\ part\ is\ the\ same\ as\ the\ example\ part \\ 0, \ Target\ parts\ are\ different\ from\ example\ parts \end{cases} \quad (3.49)$$

Thus, the parts' similarity is:

$$s_{obj} = \sum_{i=1}^{n_{obj}} (\beta_i s_i) \quad (3.50)$$

where n_{obj} is the number of the selected features. β_i is the weighting factor after normalising the attributes' importance of various features. Given the identical dimension of s_i, there is no need for normalising treatment. We select the case $s_{obj} = 1$ from the case base, so as to ensure the uniformity of the processing object's processing performance, and to erase the impact of character-type variable on the final result in energy consumption prediction.

(b) **Processing subject search**

Given the different processing scheme of one part, the processing facility and its energy consumption of each procedure will be different as well. Once we have made sure that the processing object is the same, we also need to take into consideration the impact of the processing facility.

First, we need to run quantization of the processing facility, namely to acquire the energy consumption affinity list of the processing facility. This affinity list quantifies the processing facility from the perspective of energy consumption, for the convenience of intuitive comparison and processing. Suppose the system has M number of the processing facility. We define a random facility M_{basic} as the criterion facility; we set this facility degree of similarity as $s_{basic} = 1$. Apart from this attribute of the processing facility, we classify the knowledge cases into C categories according to the other conditional attributes, with each category having M cases. In each category, we seek the energy consumption ratio between other facility M_l and the reference facility, namely $r_{cl} = P_l / P_{basic}$. Here, r_{cl} is the ratio of the facility M_l against M_{basic} in category $c(c \in C)$, with P_l and P_{basic} as their respective energy consumption.

Thus, the energy consumption similarity degree of the facility M_l, as opposed to the reference facility, is

$$s_{ml} = \frac{1}{C}\sum_{c=1}^{C} r_{cl} \tag{3.51}$$

As for each procedure, the chosen processing facility is not a fixed one. Through calculating the affinity value of the to-be-used processing facility and comparing it with the other processing facility, we can achieve a relatively accurate matching and retrieval of the processing facility. Compared with the traditional method of character matching, the advantage of this method lies in its intuitive comparison of the quantified value and the mutual conversion calculation between each facility's energy consumption.

For energy consumption prediction, first, we select processing cases identical with the to-be-used processing facility. In other words, only in the prerequisite that there is not enough searchable case, facility affinity degree—set at 1—would select similar facilities. It is because the facility's affinity degree deviates a lot when applied to the energy consumption prediction and can only be a supplementary option when there are not enough cases.

5. ***Case reapplication***

Case reapplication applies the sought cases into new issues through certain methods. In this section, it means predicting the processing energy consumption of the new working step. The above cases and predictive objects sought among the processing objects and processing facility have highly similar craft and processing methods. The sought cases eliminate the ill impact of character-type variable on the prediction results. Based on the processing parameters, this section selects knowledge cases similar to the prognostic objects from the above-sought results. Compared with the character variables, value variables can intuitively quantify the distance between different values.

Algorithm 4 *Working step energy consumption prediction based on cases*

Step 1: *Normalizing the processing parameters and calculating the case affinity with the Formula* (3.51)*:*

Step 2: *Form an affinity case set from cases of the least affinity and calculate the weight function of case no. k with the following formula:*

$$\omega_i = \frac{\max\{d_1, d_2, \ldots, d_k\} - d_i}{\sum_{i=1}^{k}(\max\{d_1, d_2, \ldots, d_k\} - d_i)} \tag{3.52}$$

Step 3: *The input power of the object working step is:*

$$v_d = \sum_{i=1}^{k} (\omega_i v_{d_i}) \tag{3.53}$$

Once we have ascertained the input power of the target working step, we can solve this working step's cutting energy consumption through its processing time. Once we have ascertained the energy consumption of each working step, we can deduce the energy consumption of the product at each level through Formulas (3.18)–(3.21).

3.7.3 Experiment Simulation Analysis

This section uses mechanical processing progress as an example to validate the effectiveness of the algorithm proposed in this chapter. During the mechanical processing progress, the character variables include parts material, processing method and cutters. Here, parts material reflects the cutting performance of the part itself; the processing method reflects the technological demands for processing the part. The cutter reflects the precision requirements for part processing. The chosen value variables are cutting speed v_c, feed volume f and depth of cut a_p.

We use the knowledge case base of mechanism model construction in four documents for simulation and analysis. Relevant parameters are displayed in Table 3.10. Through the input power model of Table 3.1, we construct a case base with the orthogonal experiment method. The knowledge base update threshold is set as 0.025; the cutting speed value is set as $\{70, 100, 130, 160, 190, 220, 250\}$. The feed volume value is set as $\{0.1, 0.15, 0.2, 0.25, 0.3\}$. The depth of cut value is set as $\{0.5, 0.8, 1.1, 1.4, 1.7, 2.0\}$. The orthogonal experiment method ensures that all cases are evenly distributed in the sample space to prevent uneven prediction precision caused by aggregated cases.

(1) **Influence of the number of similar cases on precision**

In Algorithm 4, the number of similar cases, i.e. k, directly influence the final result. If the value of k is too large, it will bring unnecessary computational complexity; if the value of k is too small, the prognostic result will be decided by the most similar case, disregarding information of other cases. To analyse the impact of k value on the prognostic precision, Table 3.11 chooses five target cases as prognostic targets, to analyse the prognostic deviation under different value. The results are shown in Fig. 3.17. From Fig. 3.17, we know that, as k value increases, the deviation decreases. However, as the value increases to a certain extent, the deviation alteration tends to stay stable. Thus, as precision is allowed, it is appropriate to choose $k = 12$. In real-life applications, we need to select the most appropriate number of similar cases through similar tests.

(2) **Validation of Energy Consumption's Estimated Precision**

In the case base constructed in Table 3.10, we calculate the importance of the three processing parameters with Algorithm 3.2, and the results are shown in Table 3.15.

Table 3.10 Experiment case relevant parameters

Source	Equipment in use and its processing method	Input power model	Notes
Document [28]	CK60 numerically controlled lathe(turney)	$P_i = 376.74 + 11.9v_c + 1.163 \times 10^{-4} \times v_c^2 + 102.96a_p^{0.982}v_c^{0.832}f^{0.856}$	Huazhong Numerically Controlled System equipped; workpiece being 45# hot-rolled steel of 50 mm diameter
Document [29]	numerically controlled lathe(turney)	$P_i = 40.6 + 1.445v_c - 2.7 \times 10^{-5} \times v_c^2 + 3.456a_pv_c^{0.6}f^{0.5}$	Workpiece being 45# hot-rolled steel of 50 mm diameter
Document [30]	Numerical control processing centre (milling)	$P_i = 90.115 + 24.7287v_c + 7.9157 \times 10^{-3} \times v_c^2 + 1.154a_pv_cf^{0.86}$	The cutter being high-speed vertical steel of $\varphi 60$ Milling cutter with 3 teeth
Document [31]	TC500R drilling and milling processing centre (milling)	$P_i = 448 + 2v_c + 8.23 \times 10^{-4} \times v_c^2 + 0.2v_cf + 6.185a_pv_c^{0.723}f^{0.86}$	The cutter being 3-teeth super-hard straight shank cutter of $\varphi 16$

Table 3.11 Target case data of model analysis

Target case	v_c(m/min)	f(mm/r)	a_p(mm)
1	120	0.15	1
2	160	0.2	0.6
3	200	0.25	1.5
4	220	0.22	1.7
5	150	0.16	1.8

We can see that cutting speed has the largest impact on input power, the impact of feed volume and depth of cut volume are relatively the same, which fits reality situation.

Tables 3.12, 3.13 and 3.14 reveals that as cutting speed, feed volume and depth of cut volume change, the deviation between the predicted energy consumption and the real energy consumption changes as well. From this, we can see that, no matter how

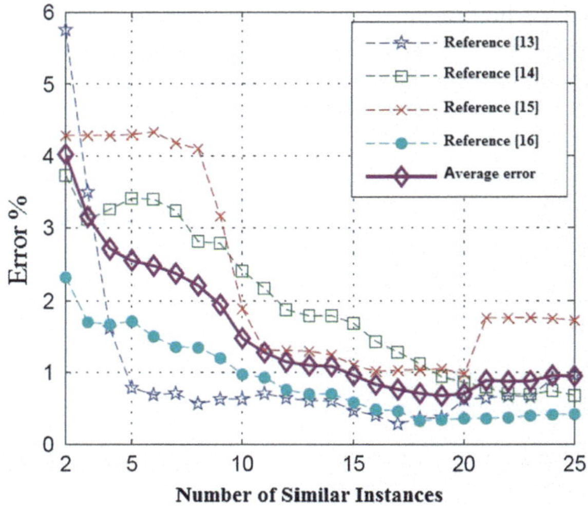

Fig. 3.17 Curve of how deviation changes with similar cases

parameters change, so long as the to-be-predicted processing parameters are within the sample space, we can get good prognostic results.

(3) **Comparison between this chapter's algorithm and algorithms**

To validate the performance of the algorithm proposed in this chapter, we compare it with the CBR algorithm of fusion fuzzy rough set proposed by Zhou et al. [36] and the CBR algorithm based on relevant factors proposed by Want et al. [37]. In our comparison, all three algorithms adopt the case base used in Sect. 3.7.2 and choose the five groups of target case in Table 3.12 as comparison objects, without losing generality. The algorithm in this chapter chooses $k = 12$. The parameters of the algorithm proposed by Zhou et al. are set as: $\lambda_{ci} = 2/3$, $\mathrm{SIM}_{yz}^{(i)} = 1 - 0.025i$, $b^{(i)} = 12$, $\omega = 6$. The algorithm of Wang et al. chooses $k = 12$.

Table 3.15 is the attribute importance of the three algorithms' calculation. This algorithm calculates the importance of each attribute directly according to the fluctuation of decisional attributes (input power). Zhou et al. calculated the categorizing capacity of the attributes with the fuzzy rough set method, thereby ascertaining the importance of the attribute. Wang et al. determined the impact of a conditional attribute on decisional attribute according to the relative factors. From Table 3.9, we can see that the calculation results of the three methods remain relatively the same, which means that all these three methods are relatively reasonable.

Table 3.16 is the prognostic deviation of these three algorithms on five groups of target cases. We can see that, among these three algorithms, ours has the strongest applicability and better result. Of course, the algorithm proposed by Zhou et al. is mainly applied to the soft measuring of the grinding process, and therefore of better effects and more robust in cases with missing attributes, whereas the algorithm of Wang et al. is mainly used for parameter refinement of expected output in polishing treatment.

Table 3.12 Deviation changes under different cutting speed

v_c(m/min)	Model in document [32]			Model in document [33]			Model in document [34]			Model in document [35]		
	$P_{predict}$	P_{real}	$E(\%)$	$P_{predict}$	P_{real}	$E(\%)$	$P_{predict}$	P_{real}	$E(\%)$	$P_{predict}$	P_{real}	$E(\%)$
75	3055.5	2969	2.91	174.74	183	4.51	2188.8	2029	7.88	664.67	674	1.38
105	3878.8	3876	0.07	225.59	234	3.59	2879.7	2829	1.79	749.25	759	1.28
135	4761.4	4757	0.09	275.76	284	2.90	3684.3	3644	1.11	834.16	844	1.17
165	5624.8	5618	0.12	325.35	334	2.59	4503.5	4473	0.68	919.05	929	1.07
195	6472.6	6465	0.12	374.31	383	2.27	5337.1	5316	0.40	1004.7	1015	1.02
225	7308.2	7299	0.13	422.75	431	1.91	6184.6	6173	0.19	1091.2	1101	0.89

* $f = 0.25\text{mm/r}$, $a_p = 1.5\text{mm}$, $k = 12$.
* is used for annotation

Table 3.13 Deviation changes under different feed volumes

f (mm/r)	Model in document [32]			Model in document [33]			Model in document [34]			Model in document [35]		
	$P_{predict}$	P_{real}	$E(\%)$	$P_{predict}$	P_{real}	$E(\%)$	$P_{predict}$	P_{real}	$E(\%)$	$P_{predict}$	P_{real}	$E(\%)$
0.12	3360.3	3358	0.07	263.01	261	0.77	3456.5	3475	0.53	782.03	781	0.13
0.17	3861.2	3856	0.13	267.00	268	0.37	3456.8	3488	0.89	798.70	800	0.16
0.22	4340.4	4333	0.17	272.54	273	0.17	3457.5	3500	1.21	817.36	819	0.20
0.27	4803.8	4794	0.20	276.74	278	0.45	3458.8	3512	1.51	834.94	837	0.25
0.12	3360.3	3358	0.07	263.01	261	0.77	3456.5	3475	0.53	782.03	781	0.13
0.17	3861.2	3856	0.13	267.00	268	0.37	3456.8	3488	0.89	798.70	800	0.16

$^{*}v_c = 130\text{m/min}, a_p = 1.5\text{mm}, k = 12.$

* is used for annotation

Table 3.14 Deviation changes under different depth of cut

a_p(mm/r)	Model in document [32]			Model in document [33]			Model in document [34]			Model in document [35]		
	$P_{predict}$	P_{real}	$E(\%)$	$P_{predict}$	P_{real}	$E(\%)$	$P_{predict}$	P_{real}	$E(\%)$	$P_{predict}$	P_{real}	$E(\%)$
0.7	3184.0	3196	0.38	251.46	250	0.58	3458.4	3470	0.33	777.82	776	0.23
1.0	3716.3	3729	0.34	260.51	260	0.20	3458.3	3484	0.74	796.32	796	0.04
1.3	4245.5	4239	0.15	270.00	270	0.00	3458.3	3498	1.14	816.62	816	0.08
1.6	4772.3	4787	0.31	279.46	279	0.16	3485.3	3511	1.50	836.74	836	0.09
1.9	5266.0	5313	0.88	285.29	289	1.28	3458.3	3525	1.89	852.56	856	0.40
0.7	3184.0	3196	0.38	251.46	250	0.58	3458.4	3470	0.33	777.82	776	0.23

*$v_c = 130\text{m/min}$, $f = 0.25\text{mm/r}$, $k = 12$.
* is used for annotation

Table 3.15 Comparison of the importance of these three algorithms' attributes

Processing parameters	Model in document [32]			Model in document [33]			Model in document [34]			Model in document [35]		
	Zhou	Wang	This Chapter	Zhou	Wang	This Chapter	Zhou	Wang	This Chapter	Zhou	Wang	This Chapter
v_c	0.5087	0.6214	0.4776	0.9768	0.9669	0.8030	1	0.9997	0.9757	0.8787	0.9374	0.7337
f	0.1889	0.1388	0.2258	0	0.0057	0.0617	0	0.0001	0.0104	0.0335	0.0239	0.1172
a_p	0.3023	0.2398	0.2967	0.0232	0.0274	0.1353	0	0.0002	0.0139	0.0139	0.0387	0.1491

Table 3.16 Comparison of the three algorithms' prognostic deviation

Target cases	Model in document [32] (%)			Model in document [33] (%)			Model in document [34] (%)			Model in document [35] (%)		
	Zhou	Wang	This chapter	Zhou	Wang	This chapter	Zhou	Wang	This chapter	Zhou	Wang	This chapter
1	10.67	0.60	0.73	0.42	7.12	3.60	7.81	8.42	2.13	4.03	3.46	1.27
2	10.43	1.37	0.14	2.75	0.16	0.27	0.35	0.06	0.11	0.51	0.53	0.38
3	0.26	0.50	0.40	16.37	4.82	2.29	0.71	5.23	0.10	8.16	2.96	1.03
4	4.33	0.32	0.32	13.11	0.02	0.06	1.66	0.02	1.41	6.85	0.11	0.03
5	3.90	1.29	0.67	7.82	5.40	3.16	5.44	6.80	2.80	0.12	2.88	1.11
Average deviation	5.915	0.8101	0.6504	8.094	3.506	1.877	3.194	4.106	1.310	3.934	1.989	0.7641

To compare the efficiency of these three algorithms, we record the average time of these algorithms, without losing generality. The algorithm time includes the entire progress of the CBR algorithm. The results display that the average time of Zhou et al.'s algorithm is 0.19 s, and the algorithm and the textual algorithm of Wang et al. are both 0.09 s. Thus, all three algorithms are pretty fast, with relatively the same efficiency.

3.8 Conclusion

Based on the analysis of the dynamic correlations between energy consumption, equipment status and process parameters in discrete manufacturing systems, a multi-source and multi-level integrated energy consumption model and an intelligent identification method of key parameters are proposed in this chapter. In order to compensate for the dynamic uncertainties not taken into account in the integrated model, a method of building energy consumption knowledge model is put forward. Here, a modular multi-granularity hierarchical representation model of energy consumption knowledge based on ontology is built by top-down method to realize knowledge integration. Then, the knowledge model of energy consumption based on case-based reasoning technology is established by integrating process information and production scheduling information. The validity and reliability of the models are verified by numerical simulations.

References

1. Dietmair A, Verl A (2009) A generic energy consumption model for decision making and energy efficiency optimization in manufacturing. Int J Sustainable Eng 2(2):123–133
2. Cai W, Liu F, Dinolov O, Xie J, Liu P (2018) Energy benchmarking rules in machining systems. Energy 142:258–263
3. Pellegrinelli S, Valente A, Tosatti LM (2012) An integrated setup planning and pallet configuration approach for highly automated production systems with energy modelling of manufacturing operations. Procedia CIRP 3:49–54
4. Gutowski T, Branham M, Dahumus J (2009) Thermodynamic analysis of resources used in manufacturing processes. Environ Sci Technol 43(5):1584–1590
5. Matibenga PT, Rajemi MF (2011) Calculation of optimum cutting parameters based on minimum energy footprint. CIRP Ann Manuf Technol 60(1):149–152
6. Mose C, Weinert N (2015) Process chain evaluation for an overall optimization of energy efficiency in manufacturing-the welding case. Robot Comput Integr Manuf 24:44–51
7. Bland M, Wheeler P, Clare J et al (2001) Comparison of calculated and measured losses in direct AC-AC converters. In: Proceeding of IEEE 32nd annual power electronics specialists conference, pp 1096–1101
8. Hu JH, Li JG, Zou JB et al (2009) Losses Calculation of IGBT Module and Heat Dissipation System Design of Inverters. Trans China Electr Soc 3:159–163
9. Yan JH, Li L (2013) Multi-objective optimization of milling parameters the trade-offs between energy, production rate and cutting quality. J Clean Prod 52(1):462–471

10. Hu T (2012) Energy consumption characteristics of feed drive system in CNC machine tools. Chongqing University, Chongqing
11. Commission E (2030) Climate and energy framework. https://ec.europa.eu/clima/policies/strategies/2030/index_en.htm
12. Tridech S, Cheng K (2011) Low carbon manufacturing: Characterization, theoretical models and implementation. Int J Manuf Res 6(2):110–121
13. Balogun VA, Mativenga PT (2013) Modelling of direct energy requirements in mechanical machining processes. Clean Prod 41(2):179–186
14. Peng T, Xu X, Wang L (2014) A novel energy demand modelling approach for CNC machining based on function blocks. J Manuf Syst 33(1):196–208
15. Guo Y, Duflflou JR, Qian J et al (2015) An operation-mode based simulation approach to enhance the energy conservation of machine tools. J Clean Prod 101:348–359
16. Kant G, Sangwan KS (2015) Predictive modelling for energy consumption in machining using artificial neural network. Procedia CIRP 37:205–210
17. Garg A, Lam JSL, Gao L (2015) Energy conservation in manufacturing operations: modelling the milling process by a new complexity-based evolutionary approach. Clean Product 108:34–45
18. Zhou L, Li J, Li F et al (2015) Energy consumption model and energy efficiency of machine tools: a comprehensive literature review. Clean Prod 112:3721–3734
19. Pawlak Z (1991) Rough sets: theoretical aspects of reasoning about data. Kluwer Academic Publishers, Dortrecht
20. Tan T (2003) Rough set-based process modeling, control and fault diagnosis. Zhejiang University, Zhejiang
21. Innovation T. Horizon 2020. https://ec.europa.eu/programmes/horizon2020/
22. Nguyen, S.H., Skowron, A (1995) Quantization of real value attributes-rough set and Boolean reasoning approach. In: Proceedings of the second joint conference on information sciences, pp: 34–37
23. Zhou Z, Zhang C, Xie Y et al (2015) Cutting parameters optimization for processing energy and efficiency in CNC lathe. Comput Integr Manuf Syst 21(09):2410–2418
24. Kara S, Li W (2011) Unit process energy consumption models for material removal processes. CIRP Ann Manuf Technol 60(1):37–40
25. Garg A, Lam JSL, Gao L (2015) Energy conservation in manufacturing operations: Modelling the milling process by a new complexity-based evolutionary approach. J Clean Prod 108:34–45
26. Kant G, Sangwan KS (2014) Prediction and optimization of machining parameters for minimizing power consumption and surface roughness in machining. J Clean Prod 83:151–164
27. Liu JJ (2012) Research on key technologies of knowledge integration and application supporting networked manufacturing for complex products. Nanjing: College of Mechanical and Electrical Engineering, Nanjing University of Aeronautics and Astronautics
28. Zhou ZH, Zhang CY, Xie Y et al (2015) Cutting parameters optimization for processing energy and efficiency in CNC lathe. Comput Integr Manuf Syst 21(9):2410–2418
29. Li CB, Cui LG, Liu F et al (2013) Multi-objective NC machining parameters optimization model for high efficiency and low carbon. J Mech Eng 49(9):87–96
30. Hu D, Zhang H, Yan W et al (2015) Method research on milling parameters of NC machining for energy consumption optimization[J]. Mach Des Manuf 8:110–113
31. Huang ZT, Yang J, Zhang CY et al (2016) Energy-oriented CNC milling process modelling and parameter optimization[J]. China Mech Eng 27(18):2524–2532
32. Li C-B, Li L-L, Tang Y et al (2019) A comprehensive approach to parameters optimization of energy-aware CNC milling. J Intell Manuf 30(1):123–138
33. Cai W, Liu F, Zhou XL (2016) Fine energy consumption allowance of workpieces in the mechanical manufacturing industry. Energy 114:623–633
34. Velchev S, Kolev I, Ivanov K (2014) Empirical models for specific energy consumption and optimization of cutting parameters for minimizing energy consumption during turning. J Clean Prod 80:139–149

35. Eisele C, Schrems S, Abele E (2011) Energy-efficient machine tools through simulation in the design process. Springer, New York
36. Zhou P, Lu SW, Chai T (2014) Data-driven soft-sensor modeling for product quality estimation using case-based reasoning and fuzzy-similarity rough sets. IEEE Trans Autom Sci Eng 11(4):992–1003
37. Wang G, Zhou X, Liu J et al (2017) Polishing process planning based on fuzzy theory and case-based reasoning. Int J of Adv Manuf Technol 90(1–4):907–915

Chapter 4
Constructing a Multi-layered Energy Efficiency Quantitative Analysis Index System

4.1 Introduction

The construction of an energy efficiency quantitative analysis index system is the precondition for the quantitative energy consumption analysis of the discrete manufacturing system. This energy efficiency quantitative analysis index system is directly related to the reliability and accuracy of the quantitative analysis results. A scientific energy efficiency quantitative analysis system can comprehensively cover all the quantitative analysis indexes which influence the quantitative analysis of the object's energy consumption level. It can truthfully reflect the connections between the energy efficiency quantitative indexes, while objectively reflecting the energy utilization level of the discrete manufacturing system. Therefore, it is extremely crucial to construct a reasonable energy efficiency analysis system of the discrete manufacturing system. At present, relevant research at home and abroad is about the current condition of the enterprises' energy consumption. Instead, there is an inadequate analysis of energy efficiency based on such a condition.

As for the normal method of constructing an energy efficiency quantitative analysis index system, the first step is to have the evaluators settle down the original energy efficiency index. The second step is to reduce the redundancy of the quantitative analysis index system via effective methods of index selection, thereby picking out the more representative indexes. The main index selection and optimization methods include hierarchical analysis [1], principal component analysis [2], the rough set method [3] etc. Taking into consideration the features of the discrete manufacturing system's energy consumption composition, this chapter sums up several indexes of the product, the facility, the system and the technology levels, thereby constructing a basic energy efficiency index system. We then use both the characteristic value method and GI method to filter and optimize the energy efficiency indexes, before finally constructing a three-layered index system of the discrete manufacturing system's energy efficiency quantitative analysis, including four level-1 indexes of the economic energy efficiency, product energy efficiency, facility energy efficiency and task energy efficiency, as well as ten level-2 indexes.

Y. Wang et al., *Quantitative Analysis and Optimal Control of Energy Efficiency in Discrete Manufacturing System*, https://doi.org/10.1007/978-981-15-4462-0_4

4.2 Principles in Constructing an Energy Efficiency Quantitative Analysis Index System

The principles in constructing an energy efficiency quantitative analysis index system mean that such a system should be flexibly applicable with regard to the manufacturing characteristics and energy utilization of the discrete manufacturing system, based on references to the construction principles of the energy efficiency evaluation system, energy administration system, discrete manufacturing index system and the service industry's manufacturing index system. The energy efficiency quantitative analysis system is the basis for conducting a quantitative analysis of the discrete manufacturing system. The energy efficiency quantitative analysis index system is directly related to the reliability and accuracy of the quantitative analysis results.

There are various and wide-ranging factors affecting the discrete manufacturing system's energy efficiency quantitative analysis. Therefore, we should choose the energy efficiency index scientifically and cautiously, to construct a neat, precise, and objective index system. When selecting the energy efficiency quantitative analysis index, we should follow these principles:

(1) **Comprehensiveness and systematicness**

The proposed energy efficiency indexes should cover all the energy consumption facilities and sections; they should fully reflect the energy efficiency level and energy utilization status of the discrete manufacturing system, neither adding nor lacking any information. The energy efficiency index system is a systematic description of the discrete manufacturing system, and it has to be objective, systematic and comprehensive.

(2) **Scientificity and independence**

The energy efficiency quantitative analysis index system thus constructed should be based on scientific evidence and be capable of truly and objectively reflecting the energy efficiency level of the discrete manufacturing system. All the quantitative analysis indexes should be formulated according to the goals of the quantitative analysis. Every level of these indexes should remain independent from each other's influence.

(3) **Representativeness and process-orientedness**

The energy efficiency quantitative analysis indexes should cover not only the major factors representative of the discrete manufacturing system, but also other typical potential factors which reflect the chief energy efficiency of the enterprise. The indexes chosen should include not only the energy efficiency indexes which reflect the results, such as the energy consumption index and energy consumption added value per ten thousand yuan, but also the energy efficiency index which reflect the middle process, such as the technological energy efficiency index and the production materials scheduling energy efficiency index.

(4) **Feasibility and operability**

In the energy efficiency quantitative analysis index system, we can choose both the quantitative index and the qualitative index. The former should be easy to collect and be acquired via a certain route, or should have an explicit method of calculation. The qualitative index should have an explicit concept, with which the experts can reach an evaluation result. The selection of these indexes should be feasible and compatible with the actual operation of the enterprise.

4.3 Constructing a Primary Energy Efficiency Quantitative Analysis Index System

With regard to the main energy consumption condition, economic information, workshop information, main electric consumption facility, manufacturing technology information and the manufacturing materials' scheduling information, we analyse the impact of these factors on the enterprise's energy efficiency, filtering the typical and effective energy efficiency indexes, with which we construct a three-layered index system of the discrete manufacturing system's energy efficiency quantitative analysis.

In accordance with the layer characteristics of the discrete manufacturing system, namely the four levels (economy, product, facility and technology) of the discrete manufacturing system, we establish four level-1 index, namely that of the economic energy efficiency, product energy efficiency and the task flow energy efficiency, with each level-1 index containing several level-2 indexes. Here, we introduce each energy efficiency index.

4.3.1 Economic Energy Efficiency Index

The economic energy efficiency index can reflect the economic development, the revenue output value and the industrial consumption of the discrete manufacturing system. The economic index contains the following level-2 indexes:

(1) ***The Industrial Total Output Value*** means the currency representation of the industrial products' total amount produced during by a factory during the period of the report, including the current session's production the product's value, the external processing income, self-made semi-finished product and the difference between the beginning and the end values of the manufacturing process.
(2) ***Product Energy Consumption per Ten Thousand Yuan*** means the electric energy consumed by the enterprise for each ten thousand yuan value of industrial output, which reflects the energy consumption level of the enterprise under a certain industrial output value.

(3) ***Energy Consumption per Ten Thousand Yuan Added Value*** means the electric energy consumed by the enterprise for the additional industrial output value per ten thousand yuan, which reflects the additional energy consumption level of the enterprise under a certain industrial output value.
(4) ***Energy Resilience Factor*** reflects the change of one relevant variable caused by the change of another variable; here it means the impact of energy consumption on the industry's economic growth.

4.3.2 Product Energy Efficiency Index

The product energy efficiency index can reflect the amount of energy consumed and saved by the discrete manufacturing system for producing each unit of products. The product energy consumption index includes these level-2 indexes:

(1) ***Comprehensive Energy Consumption per Unit Product*** is the comprehensive energy consumed by an enterprise for producing each unit of products.
(2) ***Energy Savings per Unit Product*** refers to the difference between the energy savings of an enterprise for producing each unit of products before and after the optimization of the production materials scheduling or that of the technology level.
(3) ***Product Energy Utilization Level:*** Compared with the advanced level of product energy utilization level at home and abroad, if the unit product energy consumption reaches 110% of the top level or below, then we evaluate the energy utilization level as "Excellent"; if the unit product energy consumption is 110–120% of the top level, then we evaluate it as "Good"; if the unit product energy consumption is 120–130% of the top level, then we evaluate it as "Fair"; and if the unit product energy consumption exceeds 130% of the top level, then we regard it as "Poor".

4.3.3 Facility Energy Efficiency Index

The facility energy efficiency index reflects the processing facility's performance, technology level and energy conversion efficiency of the discrete manufacturing system. The facility energy efficiency index includes the following level-2 indexes:

(1) ***Processing Facility Energy Efficiency:*** Refers to the energy utilization ratio of the facility during a given period. Energy utilization ratio = cutting energy consumption/total energy consumption. Cutting energy consumption refers to the energy consumed when the facility is cutting. Before calculation, we need to determine the operation status of the facility, namely whether it is in the "off", "standby", "idle" or the "cutting" state.

(2) ***Energy Transmission Efficiency:*** During the enterprises' manufacturing and processing, the percentage of the actual energy amount used in the transmission can reflect the transmission level of the energy (including coal, electricity, water, natural gas and oil).
(3) ***Energy Processing Conversion Efficiency:*** In a given period when the energy is being converted and processed, the percentage of the amount of the products produced against the energy input amount during the processing and conversion. It reflects whether the energy processing conversion device and the processing technology are advanced or backward, as well as whether the administration level is high or low.
(4) ***Overall Facility Efficiently:*** Refers to the availability ratio, quality index and performance of the overall facility. Due to reasons such as the enterprises' management problems, shortage of raw materials or equipment failure, the operation statue of each facility varies—some may be idle, or temporarily go out of service.
(5) ***Motor Deterioration Ratio:*** Refers to the ratio of the electricity consumed by the motor of the processing facility to the total electric power consumption of the entire enterprise. It reflects the utilization ratio and the overall deterioration degree of the enterprise's motor energy consumption.

4.3.4 Task Flow Energy Efficiency Index

The task flow energy efficiency index reflects the craft production level and the production resources scheduling degree of the discrete manufacturing system. It is an important part of the energy efficiency quantitative analysis, and also an important section of energy conservation optimization. The task flow energy efficiency index includes the following level-2 indexes:

(1) ***Processing Technique Energy Efficiency:*** It reflects the energy efficiency level and the administration level of the enterprise's production process. Normally, the ratio of the facility's consumed energy to the total energy consumption during the processing flow reflects the magnitude of the index.
(2) ***Processing Resources Scheduling Energy Efficiency:*** it reflects the energy efficiency level and the scheduling level during the workshop's processing and scheduling. Mainly, it uses the percentage of the energy consumption difference before and after the processing resources scheduling compared with the energy consumption before the scheduling, to reflect the magnitude of the index.

With regard to the actual conditions of the discrete manufacturing system, and taking into consideration all the major factors affecting the discrete manufacturing system's energy efficiency, we ascertain the major indexes affecting the discrete manufacturing system's energy efficiency, and construct a primordial energy efficiency quantitative analysis index system, shown in Fig. 4.1.

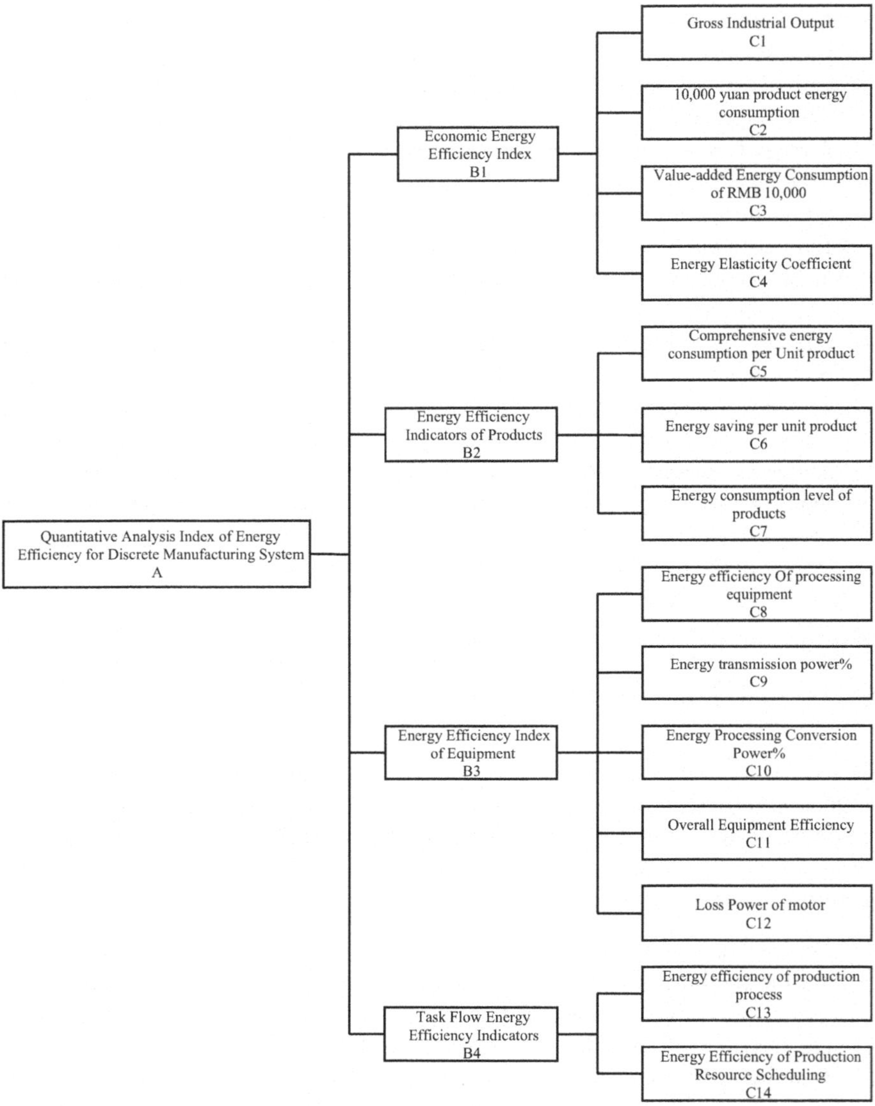

Fig. 4.1 Primordial index system of the discrete manufacturing system's energy efficiency quantitative analysis

4.4 Index Filtering Method Based on Feature Values and G1 Method

The feature value method arrives at the judgement matrix by comparing the indexes at the same level. If the matrix fits the consistency test condition, then the largest the feature value is set as the weight value for all the indexes [4]. G1 method is proposed by Professor Guo Yajun of the Northeastern University [5]. It is similar to the feature value method, but it does not require a consistency test. Therefore, this chapter combines the feature value method and the G1 method in filtering and optimising the indexes. When the judgement matrix fulfils the consistency test, we adopt the feature value method for filtering; when the judgement matrix does not fit the consistency test condition, we adopt the G1 method for filtering. Such a combination overcomes the shortcomings of the feature value method, rendering the selection and optimization of the indexes more scientifically reliable. The following are the specific steps:

Step 1 We compare the energy efficiency quantitative analysis indexes at the same level. The judgement matrix Q is as follows:

$$Q = \begin{bmatrix} \frac{w_1}{w_1} & \frac{w_1}{w_2} & \cdots & \frac{w_1}{w_m} \\ \frac{w_2}{w_1} & \frac{w_2}{w_2} & \cdots & \frac{w_2}{w_m} \\ \cdots & \cdots & & \cdots \\ \frac{w_m}{w_1} & \frac{w_m}{w_1} & \cdots & \frac{w_m}{w_1} \end{bmatrix} = Q(q_{ij})\ i, j = 1, 2, \ldots, m \tag{4.1}$$

where q_{ij} represents, in the energy efficiency quantitative analysis index set C, the relative importance of index c_i to index c_j. The rules for scoring the importance of the judgement matrix are shown in Table 4.1.

Step 2 Deriving the feature vector corresponding to the largest feature value w_i.

$$p_i = \left(\prod_{j=1}^{m} q_{ij} \right)^{1/m}, \quad i = 1, 2, \ldots m \tag{4.2}$$

$$w_i = p_j / \sum_{k=1}^{m} p_k, \quad j = 1, 2, \ldots, m \tag{4.3}$$

Step 3 Deriving the largest feature value $\lambda_{\max}$.

Table 4.1 Rules for scoring the importance of the judgement matrix

a_{ij}	Degree of importance
6	Extremely important
5	Strongly important
4	Obviously important
3	Relatively important
2	Somewhat important
1	Equally important

$$\lambda\text{max} = \frac{1}{m}\sum_{i=1}^{m}\frac{\sum_{j=1}^{m} q_{ij}w_j}{w_i} \tag{4.4}$$

Step 4 Deriving the consistency index value $C.I.$.

$$C.I.{=}\frac{\lambda\text{max} - m}{m - 1} \tag{4.5}$$

Step 5 Deriving $C.R.$, for judging the consistency of the matrix.

$$C.R. = \frac{C.I.}{R.I.} \tag{4.6}$$

where $R.I.$ is the average random consistency index at the same order in Table 4.2. Its value is derived from multiple repetitions of the random judgement matrix value root, and subsequent calculation of the arithmetic mean value. The following are the average random consistency index after 1000 times repeated calculations of the judgement index order 1–8.

If $C.R. < 0.1$, then the judgement index passes the consistency test.

Step 6 If Q passes the consistency test, then the weight function of the index is exactly the feature vector. Otherwise, we derive the index weight function with the method G1, entering Steps 7 and 8.

Step 7 We rank index c_i in the order of importance. If the importance of c_i is no smaller than c_j, then we record it as $c_i \geq c_j$. Similarly, if the ratio between expert quantitative analysis index c_{k-1} and the importance degree of c_k is f_k [6].

Table 4.2 Consistency index value corresponding to a different order

Order number	$R.I.$
1	0
2	0
3	0.52
4	0.89
5	1.12
6	1.26
7	1.36
8	1.41

$$f_k = w_{k-1}/w_k \tag{4.7}$$

$f_k = 1$, then it means that c_{k-1} is equally important as c_k. If $f_k = 1.2$, it means that c_{k-1} is slightly important to c_k. If $f_k = 1.4$, then c_{k-1} is obviously important to c_k. If $f_k = 1.6$, then c_{k-1} is strongly important to c_k. If $f_k = 1.8$, then c_{k-1} is extremely important to c_k. If $f_k = 1.1, 1.3, 1.5, 1.7$, the degree of importance corresponds to the medium condition of the above two neighbouring judgements.

Step 8 Calculating the index weight w.

$$w_m = \left(1 + \sum_{k=2}^{m} \prod_{i=k}^{m} f_i\right)^{-1} \tag{4.8}$$

$$w_{k-1} = f_k w_k \quad k = m, m-1, \ldots, 3, 2 \tag{4.9}$$

The procedure of filtering the energy efficiency quantitative analysis indexes runs as Fig. 4.2.

4.5 Case Analysis of Index Filtering

4.5.1 The Procedure for Filtering the Indexes

Based on the primordial energy efficiency quantitative analysis index system, the task flow energy efficiency index includes the following three level-2 index which requires no filtering. The level-1 indexes which need to be filtered are economic energy efficiency index, product energy efficiency index and the facility energy efficiency index.

(1) The economic energy efficiency index includes the four indexes of total industrial output value, energy consumption per ten thousand yuan, energy consumption per ten thousand yuan additional value, and the energy elastic coefficient. Based on their degrees of importance, we derive the matrix Q_1 as

$$Q_1 = \begin{bmatrix} 1 & 1/4 & 1/5 & 1 \\ 4 & 1 & 2 & 4 \\ 5 & 1/2 & 1 & 5 \\ 1 & 1/4 & 1/5 & 1 \end{bmatrix}$$

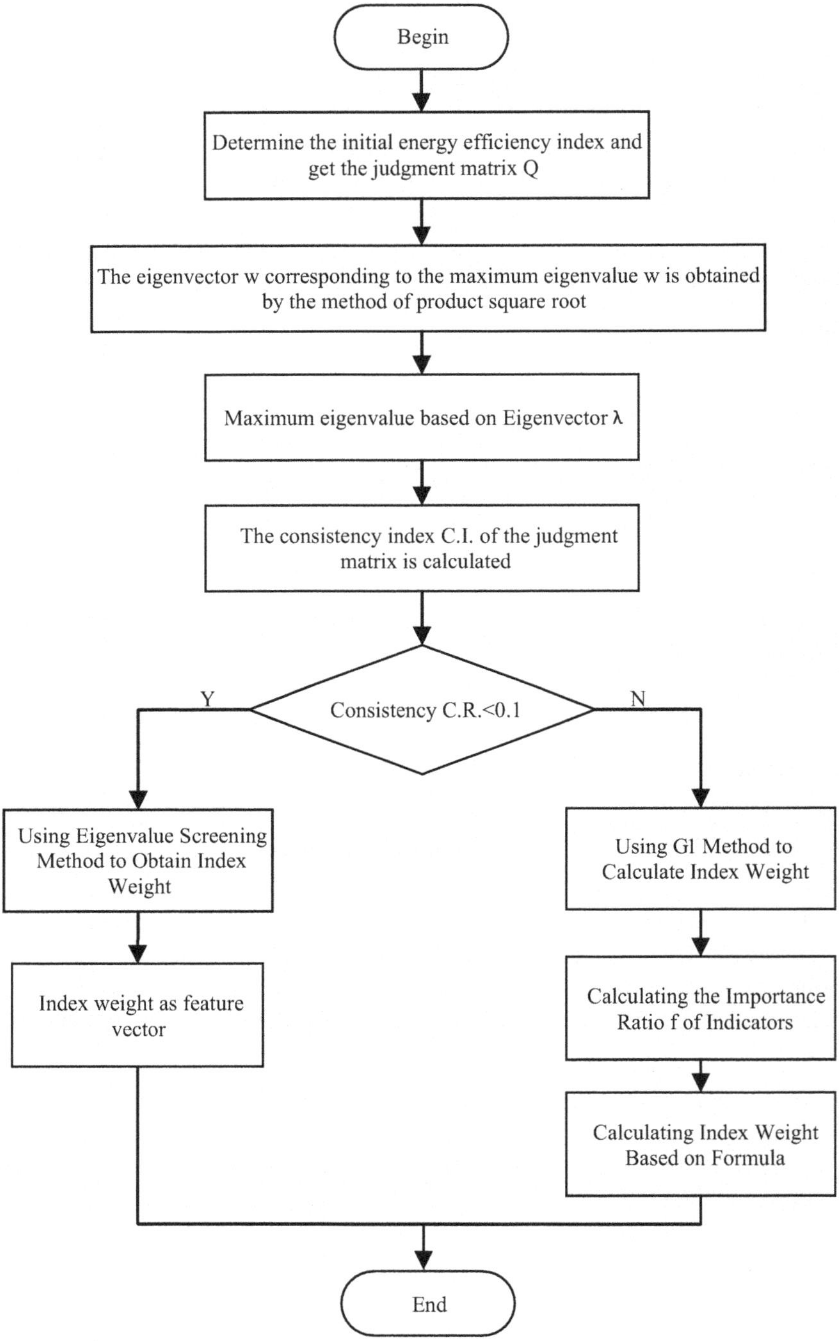

Fig. 4.2 Procedure of filtering the energy efficiency quantitative analysis indexes

Calculation based on Formulas (4.2) and (4.3) leads to the feature vector.

$$w_1 = \begin{bmatrix} 0.091 \ 0.457 \ 0.362 \ 0.090 \end{bmatrix}$$

Calculation based on Formula (4.4) leads to the largest feature value $\lambda_{\max}$=4.106.

Calculation based on Formula (4.5) leads to the consistency index $C.I. = 0.036$.

Calculation based on Formula (4.6) $R.I. = 0.89$ leads to $C.R. = 0.04$. Given $C.R. < 0.1$, we decide that matrix Q_1 satisfies the consistency test. Then, the index weight is the vector value of its feature. Analysis: both the total industrial output value and the energy elastic coefficient weight are below 0.1. Hence, they are weak weight indexes which can be deleted.

(1) The product energy efficiency index includes energy consumption per unit product, energy savings per unit product and the product energy utilization level. Based on their degrees of importance, we derive the judgement matrix Q_2 as

$$Q_2 = \begin{bmatrix} 1 & 1 & 1/2 \\ 1 & 1 & 3 \\ 2 & 1/3 & 1 \end{bmatrix}$$

Calculation based on Formulas (4.2) and (4.3) leads to the feature vector:

$$w_2 = \begin{bmatrix} 0.255 \ 0.464 \ 0.281 \end{bmatrix}$$

Calculation based on Formula (4.4) leads to the largest feature value $\lambda_{\max}$=3.367.

Calculation based on Formula (4.5) leads to the consistency index $C.I. = 0.184$.

Calculation based on Formula (4.6) $R.I. = 0.52$ leads to $C.R. = 0.346$; since $R.I. > 0.1$, we decide that matrix Q_2 does not satisfy consistency. Thus, we adopt the G1 method to calculate the weight of the energy efficiency index. The precedence relation of these three indexes is $c_2 > c_3 > c_1$. For the convenience of calculation, the precedence relation can be written as $c_1^{'} > c_2^{'} > c_3^{'}$.

Based on Formula (4.7), we get $f_2 = 1.4$, $f_3 = 1.2$.

Calculation based on Formulas (4.8) and (4.9) leads to the weight value:

$$w_2^{'} = \begin{bmatrix} 0.258 \ 0.434 \ 0.310 \end{bmatrix}$$

All the index value weights are larger than 0.1, thus none of these three indexes shall be deleted.

(3) The facility energy efficiency index includes the processing facility's energy efficiency, energy transmission efficiency, energy processing conversion efficiency, the overall facility efficiency and the motor's deterioration efficiency. Based on their degrees of importance, we derive the judgement matrix Q_3 as:

$$Q_3 = \begin{bmatrix} 1 & 1/2 & 1 & 3 & 4 \\ 2 & 1 & 1/2 & 4 & 5 \\ 1 & 2 & 1 & 4 & 3 \\ 1/3 & 1/4 & 1/4 & 1 & 2 \\ 1/4 & 1/5 & 1/3 & 1/2 & 1 \end{bmatrix}$$

Calculation based on Formulas (4.2) and (4.3) leads to the feature vector:

$$w_3 = \begin{bmatrix} 0.234 & 0.298 & 0.309 & 0.087 & 0.072 \end{bmatrix}$$

Calculation based on Formula (4.4) leads to the largest feature value λ_{max}=5.238.
Calculation based on Formula (4.5) leads to the consistency index $C.I. = 0.059$.
Calculation based on Formula (4.6), $R.I. = 1.12$ leads to $C.R. = 0.0526$; given $R.I. < 0.1$, we decide that matrix Q_3 satisfies consistency. Once the consistency test is passed, then the feature vector is the index weight. Analysis: both the total facility efficiency and the motor's deterioration efficiency weight are below 0.1. Hence, they are weak weight indexes which can be deleted.

4.5.2 *The Energy Efficiency Quantitative Analysis System After the Filtering*

The energy efficiency quantitative analysis index system of the discrete manufacturing system derived after the filtering is shown in Fig. 4.3.

In the energy efficiency index system, the computation expressions of each index are:

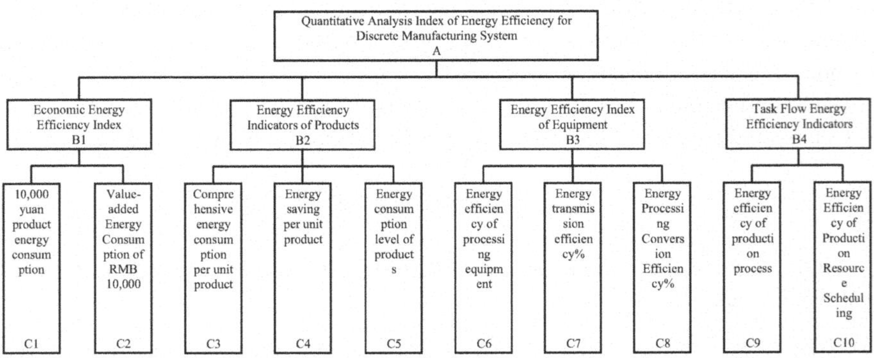

Fig. 4.3 Energy efficiency quantitative analysis system after the filtering

(1) **Energy consumption per ten thousand yuan product**

$$C_1 : C_1 = \frac{S_1}{M_1}.$$

where S_1 represents the total amount of energy consumption, with 1 ton of standard coal as the unit; M_1 represents the industry added value, its unit being ten thousand yuan.

(2) **Energy consumption per ten thousand yuan added value**

$$C_2 : C_2 = \frac{S_2}{M_2}.$$

where S_2 represents the total added value of the energy consumption, its unit being 1 ton standard coal; M_2 represents the industrial added value, its unit being ten thousand yuan.

(3) **Comprehensive energy consumption per unit product**

$$C_3 : C_3 = \frac{S_1}{N}.$$

where S_1 represents the total amount of energy consumption, with 1 ton of standard coal as the unit; N represents the production output, with each product as a unit.

(4) **Amount of energy saving per unit product**

$$C_4 : C_4 = C_3 - C_3^{'}.$$

where C_3 represents the total primary amount of energy saving per unit product, its unit being 1 ton standard coal; $C_3^{'}$ is the total amount of energy consumption per unit product after the energy-saving optimization, with 1 ton standard coal as its unit.

(5) **The product energy utilization level** C_5:

When $\frac{C_3}{C_0} \leq 110\%$, C_5 is excellent; when $110\% \leq \frac{C_3}{C_0} \leq 120\%$, C_5 is good;

When $120\% \leq \frac{C_3}{C_0} \leq 130\%$, C_5 is fair; and when $\frac{C_3}{C_0} \geq 130\%$, C_5 is poor.

where C_3 represents the total amount of energy consumption per unit product, its unit being 1 ton standard coal; C_0 is the comprehensive energy consumption per unit among the advanced-level products in China, with 1 ton standard coal as its unit.

(6) **Processing Facility's Energy Efficiency**

$$C_6 : C_6 = \frac{S_3}{S_4} \times 100\%.$$

where S_3 represents the total amount of energy consumption when the processing facility is cutting, its unit being 1 ton standard coal; S_4 is the total amount of energy consumption of the processing facility throughout the entire production process, with 1 ton standard coal as its unit.

(7) **Energy transmission efficiency**

$$C_7 : C_7 = \frac{S_5}{S_6} \times 100\%.$$

where S_5 represents the total amount after the energy has been transmitted, with 1 ton of standard coal as the unit; S_6 represents the total energy input amount, with each product as a unit.

(8) **Energy conversion processing efficiency**

$$C_8 : C_8 = \frac{S_7}{S_8} \times 100\%.$$

where S_7 represents the total amount of energy processing and conversion output, its unit being 1 ton standard coal; S_8 is the total amount of energy processing and energy input, with 1 ton standard coal as its unit.

(9) **Processing technique energy efficiency**

$$C_9 : C_9 = \frac{S_9}{C_{10}} \times 100\%.$$

where S_9 represents the total amount of the energy consumption per unit product's technological process, its unit being 1 ton standard coal; C_{10} is the comprehensive energy consumption per unit product's entire production cycle, with 1 ton standard coal as its unit.

(10) **Production resources scheduling energy efficient**

$$C_{11} : C_{11} = \frac{C_{12} - C_{12}^{'}}{C_{12}} \times 100\%.$$

where C_{12} represents the total amount of energy consumption per unit product prior to the production resources scheduling, its unit being 1 ton standard coal; $C_{12}^{'}$ is the total amount of energy consumption per unit product after the application of the production resources scheduling scheme, with 1 ton standard coal as its unit.

4.6 Conclusion

This chapter investigates the energy consumption condition of the discrete manufacturing system. Based on the energy consumption features of this system, we list the main factors for energy efficiency quantitative analysis. And with the precondition of obeying the principles set by the energy efficiency quantitative analysis system, we construct a primary energy efficiency quantitative analysis system which comprises four level-1 indexes and several level-2 indexes. Then, we screen out the important energy efficiency indexes by combining the feature value method and the G1 method. Finally, we settle down on a three-layered index system for the discrete manufacturing system's energy efficiency quantitative analysis, which serves as an excellent foundation for quantitative analysis of the discrete manufacturing system's energy efficiency.

References

1. Xu H, Luo C, Liu ZG (2007) Study on the selection method of evaluation indexes by AHP. China Offshore Oil and Gas 19(6):415–418
2. Zhang H, Zhao QH (2013) An economic indicator screening method based on fundamental principle of principal components analysis. J Shandong Univ Finance 2:52–61
3. Li YY (2009) Research on the construction of indicator system and comprehensive evaluation method based on rough set, Master Thesis. Wuhan University of Technology, Wuhan
4. Zhao GT, Ren TG (1995) Characteristic value method and its application. J Univ Sci Technol Beijing (1): 27–30
5. Wang XJ, Guo YJ (2004) Aggregate analysis of group decision making based on G1-method. Chin J Manag Sci, pp 14–16
6. Huang S, Jiang YC, Chen X et al (2010) Eigenvalue method and G1 method based index screening and optimization in energy efficiency assessment. Microcomput Inf 26(36):24–26

Chapter 5
The Quantitative Analysis of Energy Efficiency Based on Rough Set Theory and AHM

5.1 Introduction

After building the discrete manufacturing logistics system and quantitative analysis system of energy efficiency, we need to use a scientifically efficient method to solve the energy efficiency problem. In the quantitative analysis system, the indexes can show one aspect of the energy efficiency of discrete manufacturing system, but cannot analyze them comprehensively. Therefore, we should choose a highly efficient method to quantify energy efficiency. Such method should be able to covert the many quantitative analysis indexes into the index in the quantitative analysis system, which is the result of energy efficiency quantification, providing objective quantitative analysis for the energy efficiency of the discrete manufacturing system. Scholars from all over the world have used various methods for the energy efficiency quantification in the discrete manufacturing system, which includes attribute hierarchical model (AHM) [1], Fuzzy [2], entropy [3], Grey [4], AHM-GF [5] and so on.

This chapter focuses on laying out a combinational method for energy efficiency quantification, which in the phase of weight determination uses rough set-AHM method, and fully considers the objective and subjective side of the standard weight value. If we use only the calculation method that is based on the objective data, though it can to some extent mitigate the influence of expert knowledge and experience, it is also subject to the influence of the sample chosen and the number of samples used. Especially, when the data of the sample is not comprehensive enough, the weight obtained will deviate gravely away from reality. In the other segments of the quantitative analysis of energy efficiency, grey and fuzzy are used, which overcomes the grey and uncertainty in the quantitative analysis of energy efficiency. At last, through the analysis of real examples and simulation comparison, the chapter proves that this combinational method is doable and stable.

Y. Wang et al., *Quantitative Analysis and Optimal Control of Energy Efficiency in Discrete Manufacturing System*, https://doi.org/10.1007/978-981-15-4462-0_5

5.2 Combinational Method for Quantitative Analysis of Energy Efficiency

5.2.1 *The Weight Determination Method Based on Rough Set Theory*

Rough set theory was brought up by a Polish scientist Palawk, which is used to deal with systems that are inaccurate and not complete. There are two advantages to use rough set theory. First, when dealing with data, this theory is not influenced by outside factors and does not need any prior knowledge. It can use the data provided to analyze. Second, through attribute reduction, it can deduct redundant information. The attribute significance and weight determination method are both important research content in the rough set theory. The weight determination method that is based on rough set theory is below:

(1) In the decision table $S = (U, C, D, V, F)$, the decision attribute set $D(U/D = \{D_1, D_2 \ldots D_k\})$ relative to the conditional attribute set $C(U/C = \{C_1, C_2 \ldots C_m\})$ conditional information entropy is defined as [6–8]

$$I(D|C) = \sum_{i=1}^{m} \frac{|C|^2}{|U|^2} \sum_{j=1}^{k} \frac{|D_j \cap C_i|}{|C_i|} \left(1 - \frac{|D_j \cap C_i|}{|C_i|}\right) \tag{5.1}$$

where U is object collection, C is condition attribute set, D is decision attribute set, $C \cap D = \emptyset$, $D \neq \emptyset$, V is attribute value set, F represents an information function; it represents the attribute values of each object in the universe on the corresponding attributes.

(2) In the decision table $S = (U, C, D, V, F)$, $\forall c \in C$, the weight of condition attributes (indexes) C is defined as

$$\mathrm{Sig}(c) = I(D|C - \{c\}) - I(D|C) \tag{5.2}$$

where $I(D|\{C\})$ means the weight significance of condition attribute C in the system, $Sig(c)$ means the weight significance of condition attribute C in the decision table; this weight significance is relative to the condition attribute set.

(3) In the decision table $S = (U, C, D, V, F)$, $\forall c \in C$, the weight of condition attributes (indexes) [9]

$$W_{Ai}(c) = \frac{\mathrm{Sig}(c) + \mathrm{I}(D|\{c\})}{\sum\limits_{a \in C} \{\mathrm{Sig}(a) + \mathrm{I}(D|\{a\}\}} \tag{5.3}$$

when $Sig(c)$ and $I(D|\{C\})$ are combined to calculate the weight, not only the weight significance of condition attribute C and the weight significance of the attribute itself are looked at, but also the weight assignment of every attribute is more reasonable, the value is not 0.

5.2.2 The Weight Determination Method Based on AHM

AHM and AHP are very similar to each other. Compared to AHP, AHM is easier to conduct. It does not require eigenvector, and it does not require consistency checking. The weight calculation based on AHM is below:

Step 1: Calculate the relative significance of the same layer index, establishing a judgement matrix:

$$A = \{a_{ij}\}, \text{ in this } a_{ij} = 1/a_{ji}, \quad a_{ii} = 1$$

Step 2: Use formula to convert $A = \{a_{ij}\}$ into measurement matrix

$$\mu = \begin{cases} \frac{\beta k}{\beta k+1} & a_{ij} = k \\ \frac{1}{\beta k+1} & a_{ij} = \frac{1}{k} \\ 0.5 & a_{ij} = 1, \quad i \neq j \\ 0 & a_{ij} = 1, \quad i = j \end{cases} \tag{5.4}$$

where k takes a positive integer and is greater than 1,

Step 3: Calculate single-layer index weight, to obtain the set of weights of each index relative to its upper index.

$$W = [w_1, w_2 \ldots w_{10}], \quad w_i = \frac{2}{n(n-1)} \sum_{j=1}^{n} \mu_{ij}, \quad i = 1, 2, \ldots, n \tag{5.5}$$

$$\sum_{i=1}^{n} w_i = 1, 0 \leq w_i \leq 1 \quad n = 10 \tag{5.6}$$

Step 4: Calculate the combinational weight of the base-layer indexes:

$$w_j = w_i * w_{ij} \tag{5.7}$$

where w_j means the combinational weight the item *j* sub-index relative to the total index, w_i means the combinational weight of the item *i* sub-index, w_{ij} means the weight of item *j* sub-index relative to *i* sub-index. Item ***i*** sub-index is located on the layer above the item *j* sub-index. Combinational weight is mainly used to analyze the significance of each index, not used for the calculation afterwards.

5.2.3 The Computation Function of the Synthetic Weight of Quantitative Analysis Indexes

Use the rough set theory and AHM method to obtain the weight of both objective and subjective indexes. Rough set theory can be used to deal with uncertain, inaccurate data, which overcomes the influence of the subjective experience of the expert. AHM theory, on the other hand, can fully use the experience of the expert. Therefore, the combinational usage of both of these two methods can result in a final set of weight indexes for quantitative analysis of energy efficiency. The formula to calculate the synthetic weight of quantitative analysis of energy efficiency is as below:

$$W = uw_{Ai} + (1 - \mu)w_{Bi} \tag{5.8}$$

where w_{Ai} is objective weight value, w_{Bi} is subjective weight value, the value of μ depends on the specific circumstances. When the decision leans towards expert knowledge, $\mu \in [0, 0.5]$, when the decision leans towards objective data, $\mu \in [0.5, 1]$. The final weight resulted is the synthetic weight based on the subjective and the objective.

5.3 The Method to Analyze Multi-layer Energy Efficiency Quantification

5.3.1 The Non-dimensionalization Treatment of Quantitative Indexes

For quantitative indexes, because the units of measurement and magnitude of each index are different, the original data indexes need to go through non-dimensionalization treatment, to reduce the interference of random factors. Set the original data value of number k sub-index as C_k^i, $C_i(k) \in (0, 1)$, then we need to apply non-dimensionalization treatment based on Formula (3.9).

$$C_i(k) = \frac{c_k^i - \min c_k^i}{\max c_k^i - \min c_k^i} \quad i = 1, 2 \ldots \xi, \quad k = 1, 2 \ldots \varsigma \tag{5.9}$$

where ς is the number of decision index, ξ is the number of alternative options.

5.3.2 The Quantitative Analysis of Qualitative Indexes

In the system of the quantitative analysis of energy efficiency, there is a qualitative index component. Qualitative index means that which cannot go through the quantitative analysis process solely based on data, but needs to be described and analyzed objectively, to reflect the result of quantitative analysis. For qualitative indexes, we need to convert qualitative indexes into quantitative indexes, using methods such as hierarchical scoring, expectation description and critical event method. Hierarchical scoring can use data to clearly define the subject of quantitative analysis, dividing the subject into several layers. This method limits the subjectivity of the individual conducting quantitative analysis and makes the quantitative analysis more objective and the result of quantitative analysis more reliable. This chapter uses hierarchical scoring method and gives each layer a score, layers of "excellent, good, moderate and bad" correspond respectively to the scores of "4, 3, 4, 1".

5.3.3 The Fuzzy Quantitative Analysis for a Single Index

Apply the method of Fuzzy into the quantitative analysis, consider comprehensively all the indexes that influence the system of quantitative analysis, based on the significance of each index to get the result of their quantitative analysis, apply quantitative treatment to the indexes of qualitative and quantitative analysis, deal satisfactorily with problems such as multiple indexes, subjective judgement and fuzzy in the system of the quantitative analysis of energy efficiency. Start from a single index, to ascertain the degree of membership for the subject of quantitative analysis relative to the set indexes of quantitative analysis. The fuzzy mapping from U to $F(V)$:

$$f\colon U \to F(V), \quad \forall ui \in U, \quad ui| \to f(ui) = \frac{ri,1}{c_1} + \frac{ri,2}{c_2} + \cdots + \frac{ri,m}{c_m} \tag{5.10}$$

where $r_{i,m}$ means the degree of membership of u_i to c_j.

The methods to determine membership function are: Functional reasoning method, binary comparative ranking method, fuzzy statistics method, three-part method and fuzzy distribution method, etc. [10]. This article uses membership function model as below [11, 12].

When the smaller the value of the index, the better it is, Membership function model is as below:

$$\begin{aligned} u_1(x) &= \begin{cases} 1 & x \le a_1 \\ (x - a_2)/(a_1 - a_2) & a_1 < x \le a_2 \\ 0 & x > a_2 \end{cases} \\ u_2(x) &= \begin{cases} 0 & x \le a_1 \; or \; x \ge a_3 \\ 0 & x_1 < x < a_2 \\ (x - a_1)(a_2 - a_3) & a_2 < x < a_3 \end{cases} \\ u_3(x) &= \begin{cases} 0 & x \le a_2 \\ (x - a_2)/(a_3 - a_2) & a_2 < x \le a_3 \\ 1 & x > a_3 \end{cases} \end{aligned} \tag{5.11}$$

The bigger the value of the index, the better it is. The membership function model is as below:

$$\begin{aligned} u_1(x) &= \begin{cases} 0 & x \le a_2 \\ (x - a_2)/(a_1 - a_2) & a_2 < x \le a_1 \\ 1 & x > a_1 \end{cases} \\ u_2(x) &= \begin{cases} 0 & x \le a_3 or \; x \ge a_1 \\ (x - a_3)/(a_2 - a_3) & a_3 < x \le a_2 \\ (x - a_3)(a_2 - a_3) & a_2 < x < a_1 \end{cases} \\ u_3(x) &= \begin{cases} 1 & x \le a_3 \\ (x - a_2)/(a_3 - a_2) & a_3 < x \le a_2 \\ 0 & x > a_2 \end{cases} \end{aligned} \tag{5.12}$$

For the index whose value is within a range, its membership function model is as below:

$$u_1(x) = \begin{cases} 1 & a_{11} \le x \le a_{12} \\ (x - a_{11})(a_{11} - a_{21}) & a_{21} \le x < a_{11} \\ (x - a_{22})(a_{12} - a_{22}) & a_{12} < x < a_{22} \\ 0 & x < a_{21}\ or\ x > a_{22} \end{cases}$$

$$u_2(x) = \begin{cases} 0 & x \le a_{31}\ or\ x \ge a_{32}\ or\ a_{11} \le x \le a_{12} \\ (x - a_{31})(a_{21} - a_{31}) & a_{31} < x < a_{11} \\ (x - a_{11})/(a_{21} - a_{11}) & a_{21} < x < a_{11} \\ (x - a_{12})/(a_{22} - a_{12}) & a_{12} \le x \le a_{22} \\ (x - a_{32})/(a_{22} - a_{32})\ a_{22} \le x < a_{32} & \end{cases} \tag{5.13}$$

$$u_3(x) = \begin{cases} 1 & x \le a_{31}\ or\ x \ge a_{32} \\ (x - a_{11})(a_{11} - a_{21}) & a_{31} < x < a_{21} \\ (x - a_{22})(a_{12} - a_{22}) & a_{22} < x < a_{32} \\ 0 & a_{21} \le x \le a_{22} \end{cases}$$

where u_1, u_2, u_3 are the degrees of membership when the single index is good, moderate or bad, and u_1, u_2, u_3 fulfil the below relation:

$$u_1 + u_2 + u_3 = 1 \tag{5.14}$$

Use $f(u)$ we can get the set of the quantitative analysis of single indexes:

$$R_i = \left(r_{i,1}, r_{i,2}, \ldots r_{i,m}\right) \tag{5.15}$$

Based on this set, we can calculate the result of quantitative analysis of qualitative energy efficiency indexes.

5.3.4 The Synthetic Qualitative Analysis of Multi-layer Grey

Grey is a method to evaluate the degree of association between each index, which mainly takes advantage of the degrees of similarity of dissimilarity between the trend of each index, and is therefore also called "grey relational analysis". The basic concept of relational analysis based on the similarity of the geometric shape of sequence curve to judge whether the correlation and relevance belong to the geometric process category [13]. The closer the sequence curve gets to the corresponding sequence, the greater the correlation degree is.

The steps for first-order grey synthesis method is:

Step 1: Calculate the optimal energy efficiency index set:

$$C^* = \left[c_1^*\ c_2^*\ c_m^* \right] \tag{5.16}$$

Step 2: Calculate the primitive quantitative analysis matrix [14]:

$$D = \begin{bmatrix} c_1^* & c_2^* & \cdots & c_m^* \\ c_1^1 & c_2^1 & \cdots & c_m^1 \\ \cdots & \cdots & \cdots & \cdots \\ c_1^n & c_2^n & \cdots & c_m^n \end{bmatrix} \tag{5.17}$$

where m is the number of decision indexes, n is the number of alternative options, c_k^* is the best value of the item k sub-index, c_k^i is the original value of item k sub-index in the number i option.

Step 3: Calculate minimum difference between poles:

$$\text{TOW}_{\text{min}} = \max_i \max_k \left| c_k^\star - c_k^i \right| \tag{5.18}$$

Calculate the maximum difference between poles:

$$\text{TOW}_{\text{max}} = \max_i \max_k \left| c_k^\star - c_k^i \right| \tag{5.19}$$

Step 4: Calculate the grey relational coefficiency:

$$\mathrm{L}_i^k = \frac{\text{TOWmin} + \rho \text{TOW}_{\text{max}}}{\left| c_k^* - c_k^i \right| + \rho \text{TOWmax}} \tag{5.20}$$

Step 5: Calculate the quantitative analysis matrix:

$$R = \begin{bmatrix} L_1(1) & L_2(1) & \cdots & L_n(1) \\ L_1(2) & L_2(2) & \cdots & L_n(2) \\ \cdots & \cdots & \cdots & \cdots \\ L_1(m) & L_2(m) & \cdots & L_n(m) \end{bmatrix} \tag{5.21}$$

Step 6: The result of the grey relational analysis is:

$$J = W \times R \tag{5.22}$$

where W is weight matrix, R is quantitative analysis matrix.

If the index system for quantitative analysis of energy efficiency has indexes consist of n layers, then we need to conduct grey synthetic quantitative analysis based on n layer. The set for a single index quantitative analysis is $J = W \times R$. In this, j_i is a quantitative analysis index for a single index c_i. When the index has n layers, and each layer has multiple indexes, first we conduct single index fuzzy quantitative analysis for the base layer (a.k.a level n), then conduct single-layer fuzzy quantitative analysis, then conduct grey synthetic quantitative analysis from layer n to layer $n - 1$, so on so forth, then conduct grey synthetic quantitative analysis for the first layer, so as to get the result for the system of quantitative analysis of system energy efficiency.

5.3.5 *The Flow Chart for Multi-layer Energy Efficiency Combinational Quantitative Analysis*

The method for combinational quantitative analysis is shown Fig. 5.1, which include five steps: ascertain weight, dimensionalization of quantitative indexes, quantitative analysis of qualitative indexes, fuzzy quantitative analysis of single index, grey synthetic quantitative analysis.

5.4 Model Demonstration

The machine tool industry is an important component of the discrete manufacturing industry, is the foundation and the heart of the manufacturing industry, is the industry that plays a leading role in the process of industrialization, and occupies an important position in the national economy. From a statistical point of view, in the world machine tool industry, Japan is the number one production country, occupying 22% of the world machine tool production, and Germany ranks the second, occupying 21%, which is followed by America, 13%, Italy, 11%, Switzerland, 6% and China 3%. China occupies an important place in the world machine tool manufacturing industry and has created enormous output value, but the energy efficiency rate is low, and there is a lot of pressure for energy conservation. Efficient qualitative analysis for energy efficiency and energy conservation analysis are important tasks in the discrete manufacturing industry.

Below I choose A and B two machine tool manufacturing companies as examples to conduct combinational qualitative analysis. The data collected through surveys are listed in Table 5.1

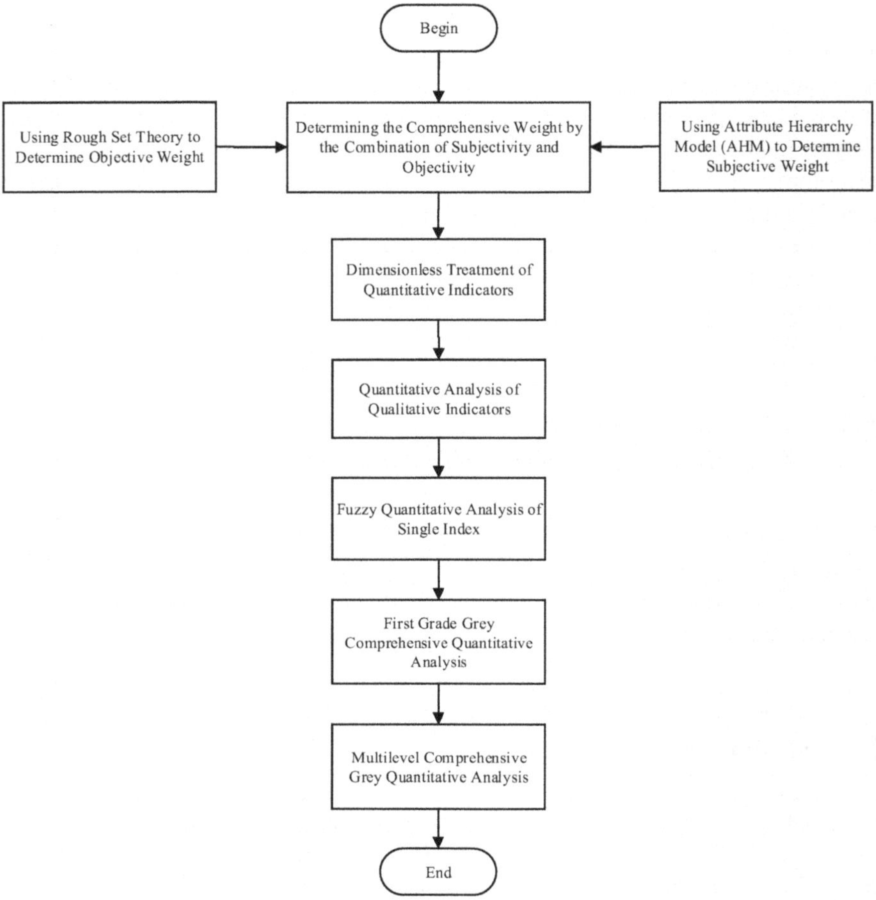

Fig. 5.1 Main procedures of quantitative analysis method for energy efficiency

5.4.1 Determine the Weight Set of Indexes

(1) ***Calculate the eight based on Rough Set Theory***

Calculate the significance of each index based on Formulas (5.1) and (5.2):
First index:

$$\mathrm{sig}(B_1) = \frac{8}{32} \quad \mathrm{sig}(B_2) = \frac{7}{32} \quad \mathrm{sig}(B_3) = \frac{8}{32} \quad \mathrm{sig}(B_4) = \frac{9}{32}$$

Secondary index:

$$\mathrm{sig}(C_1) = 0.618 \quad \mathrm{sig}\ (C_2) = 0.382 \quad \mathrm{sig}(C_3) = 0.323 \quad \mathrm{sig}(C_4) = 0.31$$
$$\mathrm{sig}(C_5) = 0.367 \quad \mathrm{sig}(C_6) = 0.439 \quad \mathrm{sig}(C_7) = 0.412 \quad \mathrm{sig}(C_8) = 0.149$$

Table 5.1 The value of each secondary index

Second-level index	*A*	*B*
Energy consumption per 10,000 Yuan product (per ton standard coal/¥10,000)	2.5	3.2
Value-added energy consumption per 10,000 Yuan Product (per on standard coal/¥10,000)	3.7	5.1
Single product combinational energy consumption (per ton standard coal)	89	92
Single product energy conservation (per ton standard coal)	4.9	2.6
The energy consumption level of products	Good	Moderate
The energy efficiency of processing equipment	0.4	0.5
Energy transmission efficiency (%)	67	55
Energy processing conversion efficiency (%)	60	65
The energy efficiency of the production process	0.5	0.8
The energy efficiency of production resource scheduling	0.6	0.6

Note The energy consumption unit is per ton standard coal, 10,000 kWh = 3.27 tons of standard coal

$$\text{sig}(C_9) = 0.34 \quad \text{sig}(C_{10}) = 0.66$$

Calculate the weight based on Formula (5.3):

$$w = [0.26\ 0.21\ 0.241\ 0.289]$$
$$w_1 = [0.63\ 0.37] \quad w_2 = [0.30\ 0.33\ 0.37] \quad w_3 = [0.452\ 0.41\ 0.138]$$
$$w_4 = [0.52\ 0.48]$$

Calculate combinational weight based on Formulas (5.2) and (5.3):

$$w(C_1) = \frac{3}{25} \quad w(C_2) = \frac{3}{25} \quad w(C_3) = \frac{1}{25} \quad w(C_4) = \frac{2}{25} \quad w(C_5) = \frac{3}{25}$$
$$w(C_6) = \frac{3}{25} \quad w(C_7) = \frac{3}{25} \quad w(C_8) = \frac{1}{25} \quad w(C_9) = \frac{3}{25}$$
$$w(C_{10}) = \frac{3}{25}$$

(2) ***Calculate the weight based on AHM***

The weight for single-layer Index:

$$w = [0.251\ 0.263\ 0.244\ 0.242]$$
$$w_1 = [0.6\ 0.4] \quad w_2 = [0.251\ 0.372\ 0.377] \quad w_3 = [0.491\ 0.302\ 0.207]$$
$$w_4 = [0.392\ 0.608]$$

Combinational weight:

Table 5.2 The weight for each first-level index

First-level index	Objective weight	Subjective weight	Synthetic weight
B_1	0.260	0.251	0.257
B_2	0.210	0.263	0.230
B_3	0.241	0.244	0.242
B_4	0.289	0.242	0.271

$w_1' = [0.128\ 0.14]$ $w_2' = [0.076\ 0.083\ 0.092]$ $w_3' = [0.117\ 0.113\ 0.046]$
$w_4' = [0.107\ 0.129]$

(3) ***calculate the final weight***

Use respectively rough set theory and AHM to get the weight for objective and subjective indexes for quantitative analysis of energy efficiency. Based on Formula (5.8) $W = uw_{Ai} + (1-\mu)w_{Bi}$, set $\mu = 0.62$, calculate combinational weight, the result leans towards objective weight, synthetic weight is shown in Tables 5.2 and 5.3:

The weight for the first-level index:

$$W = [0.257\ 0.230\ 0.242\ 0.271]$$

The weight for second-level index:

$$W_1 = [0.62\ 0.318]\ \ W_2 = [0.281\ 0.346\ 0.373]$$
$$W_3 = [0.467\ 0.369\ 0.164]\ \ W_4 = [0.471\ 0.529]$$

Table 5.3 The weight for each second-level index

Second-level index	Objective weight	Subjective weight	Synthetic weight
C_1	0.63	0.6	0.62
C_2	0.37	0.4	0.38
C_3	0.30	0.251	0.281
C_4	0.33	0.372	0.346
C_5	0.37	0.377	0.373
C_6	0.452	0.491	0.467
C_7	0.41	0.402	0.369
C_8	0.138	0.207	0.164
C_9	0.52	0.392	0.471
C_{10}	0.48	0.608	0.529

5.4.2 The Quantitative Analysis of Single Index

(1) ***The non-dimensionalization of quantitative index***

Based on Formula (5.9), the non-dimensionalization of the quantitative index in machine tool manufacturing company A is:

$$C = [0.36\ 0.13\ 0.77\ 0.85\ 0.66\ 0.5\ 0.75\ 0.5\ 0.25\ 1]$$

The non-dimensionalization of the quantitative index in machine tool manufacturing company B is:

$$C = [0.58\ 0.64\ 0.62\ 0.036\ 0.33\ 1\ 0.24\ 0.9\ 1\ 0.5]$$

(2) ***Determine the value of the qualitative index***

The energy efficiency levels for the product of A and B machine tool manufacturing company are "good" and "moderate". Based on hierarchical scoring method, the corresponding scores are 3 and 2.

(3) ***Fuzzy analysis of single index***

Use Formula (5.10) to ascertain the degree of membership of each quantitative index and qualitative index:

$$R_i = [0.5\ 0.6\ 0.6\ 0.7\ 0.6\ 0.7\ 0.4\ 0.5\ 0.4\ 0.5]$$

5.4.3 Grey Synthetic Quantitative Analysis

(1) ***First-level grey synthetic quantitative analysis***

Get the optimal index set C^*, and undergo the quantitative treatment of qualitative indexes and the non-dimensionalization of quantitative indexes:

$$C^* = [0\ 0\ 0\ 1\ 1\ 1\ 1\ 1\ 1\ 1]$$

Based on Formulas (5.18) and (5.19):

$$\text{TOW}_{\text{min}} = 0, \quad \text{TOW}_{\text{max}} = 0.964$$

Based on Formula (5.20), (5.21), we set $\rho = 0.5$:

$$L_1(1) = 0.07 \quad L_1(2) = 0.79 \quad L_1(3) = 0.38 \quad L_1(4) = 0.76 \quad L_1(5) = 0.59$$

$L_1(6) = 0.49 \quad L_1(7) = 0.66 \quad L_1(8) = 0.49 \quad L_1(9) = 0.39 \quad L_1(10) = 1$

$L_2(1) = 0.45 \quad L_2(2) = 0.43 \quad L_2(3) = 0.44 \quad L_2(4) = 0.33 \quad L_2(5) = 0.43$
$L_2(6) = 1 \quad L_2(7) = 0.39 \quad L_2(8) = 0.83 \quad L_2(9) = 1 \quad L_2(10) = 0.49$

$$R = \begin{bmatrix} 0.57\ 0.79\ 0.38\ 0.76\ 0.59\ 0.49\ 0.66\ 0.49\ 0.39\ 1 \\ 0.45\ 0.43\ 0.44\ 0.33\ 0.43\ 1\ 0.39\ 0.83\ 1\ 0.49 \end{bmatrix}^{\mathrm{T}}$$

Based on Formula (5.22):

$$J_1 = [0.629\ 0.38] \times \begin{bmatrix} 0.57\ 0.45 \\ 0.79\ 0.43 \end{bmatrix} = [0.654\ 0.442]$$

$$J_2 = [0.281\ 0.346\ 0.373] \times \begin{bmatrix} 0.38\ 0.44 \\ 0.76\ 0.33 \\ 0.59\ 0.43 \end{bmatrix} = [0.59\ 0.398]$$

$$J_3 = [0.467\ 0.369\ 0.164] \times \begin{bmatrix} 0.49\ 1 \\ 0.66\ 0.39 \\ 0.49\ 0.83 \end{bmatrix} = [0.553\ 0.747]$$

$$J_4 = [0.47\ 0.529] \times \begin{bmatrix} 0.39\ 1 \\ 1\ 0.49 \end{bmatrix} = [0.713\ 0.73]$$

(2) ***Second level grey synthetic quantitative analysis***

$$R = [J_1, J_2, J_3, J_4]^T$$

Based on formula $J = W \times R$, we get the final result:

$$J = [0.63, 0.58]$$

5.4.4 *The Analysis of the Result of the Quantitative Analysis of Energy Efficiency*

Using combinational quantitative analysis of energy efficiency to conduct quantitative analysis for a discrete manufacturing company, we get the result of quantitative analysis of energy efficiency for A machine tool manufacturing company is 0.63, the result of quantitative analysis of energy efficiency for A machine tool manufacturing company is 0.58. The synthetic quantitative analysis of energy efficiency for A machine tool manufacturing company is good, the product energy efficiency and equipment energy efficiency are moderate and the rest of first-level indexes are good. The synthetic quantitative analysis of energy efficiency for B machine tool manufacturing company is moderate, the product energy efficiency and equipment energy efficiency are moderate and the rest of first-level indexes are good. A machine tool manufacturing company does better than B machine tool manufacturing company in economy energy efficiency index, product energy efficiency index, and equipment energy efficiency index sections; B machine tool manufacturing company should improve primarily in product energy efficiency index and equipment energy efficiency index sections.

Based on the weight calculation of each index, among the four indexes, task flow energy efficiency indexes occupy the weight of 0.217 and are the most important index among first-level indexes. Task flow energy efficiency indexes include production process energy efficiency index, production resource energy efficiency index. Therefore, production process energy efficiency index and production resource energy efficiency index have a significant influence upon the result of the quantitative analysis of energy efficiency. The result of the first-level task flow energy efficiency for A machine tool production company is 0.713, which is good; among second-level indexes, the result of quantitative analysis of production resource energy efficiency is good. These mean that the production resource scheduling of this company is good. The result of the first-level task flow energy efficiency for B machine tool production company is 0.73, which is good. The result of the quantitative analysis of economy energy efficiency index is 0.442; the result of the quantitative analysis of product energy efficiency is 0.398. Both of these two indexes belong to moderate, which reflects the fact that the economy energy efficiency index and product energy efficiency index for this company are low and need to be improved. Among second-level indexes, the results of the quantitative analysis of processing equipment energy efficiency and production process energy efficiency are good, the result of the quantitative analysis of single product is moderate, which means that the company should pay attention to energy saving and emission reduction and improve the utilization rate of resources.

Choose six sets of data, we will get six plans for quantitative analysis $\{y^1, y^2, y^3, y^4, y^5, y^6\}$, each plan has ten indexes $\{C_1, C_2, C_3, C_4, C_5, C_6, C_7, C_8, C_9, C_{10}\}$. If after the non-dimensionalization treatment: the optimal index set is $\{1, 1, 1, 1, 1, 1, 1, 1, 1, 1\}$, the worst index set is $\{0, 0, 0, 0, 0, 0, 0, 0, 0, 0\}$. The value of each index is shown in Table 5.4:

The C_3 in Plan Four is the global minimum. To compare the sensitivity level of the grey fuzzy energy efficiency evaluation method with that of the combinational energy efficiency method for extreme value, we adjust C_3 from 0.58 to 0.3, then to 0.1, the result of quantitative analysis is shown in Figs. 5.2 and 5.3. According to the result of the simulation, when changing the global extreme value, apart from Plan Four, the results of all the other plans using combinational quantitative analysis of energy efficiency do not change significantly, but are close to what they were

Table 5.4 The values of each plan for quantitative analysis

Plan	c_1	c_2	c_3	c_4	c_5	c_6	c_7	c_8	c_9	c_{10}
One	0.7	0.74	0.81	0.9	1	0.66	1. 0.68	0.59	0.75	0.6
Two	0.88	0.8	1	0.72	0.6	0.8	0.85	0.69	0.71	0.59
Three	0.6	0.69	0.59	0.7	0.75	0.9	0.66	0.74	0.65	0.7
Four	0.72	0.87	0.58	0.61	0.63	0.64	1	0.64	0.65	0.9
Five	0.9	0.64	0.94	0.82	0.8	0.72	0.65	0.86	0.59	0.67
Six	0.66	0.65	1	0.8	0.91	0.77	0.7	0.82	0.9	0.61

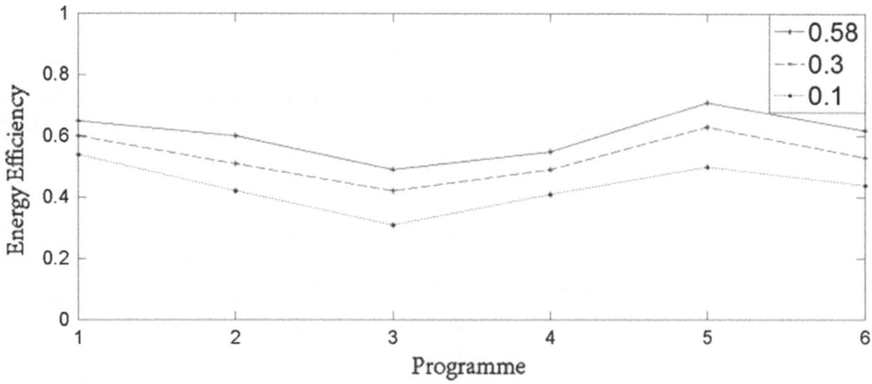

Fig. 5.2 The result of grey fuzzy quantitative analysis of energy efficiency method

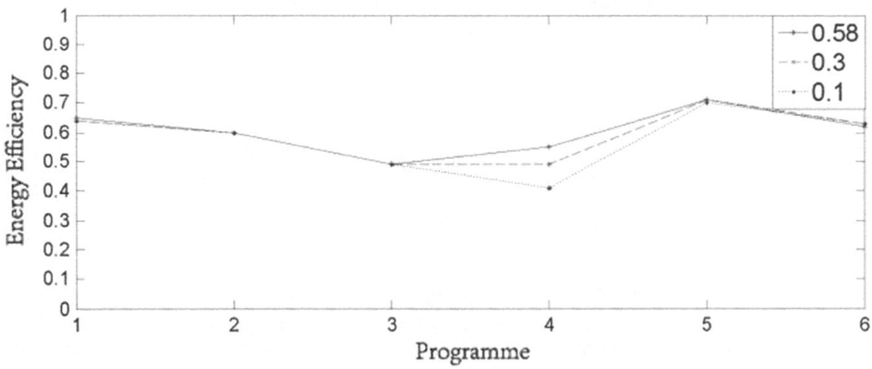

Fig. 5.3 The result of combinational quantitative analysis of energy efficiency as introduced in the chapter

before, but the fluctuation of traditional grey fuzzy method is obvious. Therefore, the combinational quantitative analysis method is more stable.

When using grey fuzzy method, if the minimum value is adjusted from 0.58 to 0.3, the sensitivity level of the result of the quantitative analysis in each plan to extreme value is about 25%; when the minimum value is adjusted to 0.1, the sensitivity level is about 29%. The sensitivity level of the combinational method introduced in this chapter is only about 5%, which means that it is insensitive to the change of extreme value, and this also reflects the influence of the value of indexes on the result of quantitative analysis. Therefore, the evaluation result ensures the stability of quantitative analysis, which is scientific and doable.

5.5 Conclusion

This chapter focuses on the quantitative analysis of energy efficiency in a discrete manufacturing system and proposes a combinational quantitative analysis method for energy efficiency. In the procedure of weight determination, it uses rough set theory and attributes hierarchical model combination, which overcomes the objectivity and subjectivity in the process of quantitative analysis of energy efficiency. In other procedures, it uses grey fuzzy calculation method, which overcomes the uncertainty and ambiguity in the quantitative analysis of energy efficiency. After ascertaining the quantitative analysis method, the chapter goes on to illustrate the method through an example. It chooses two machine tool manufacturing companies and investigates into the operational condition of the companies to get the values of energy efficiency indexes. The chapter applies the combinational method to machine tool manufacturing industry and through calculation gets the results of the quantitative analysis of energy efficiency for the two machine tool manufacturing companies, analyzing the results suitably and providing efficient energy conservation advice.

References

1. Han Y, Geng Z, Liu Q (2014) Energy efficiency evaluation based on data envelopment analysis integrated analytic hierarchy process in ethylene production. Chin J Chem Eng 22(11):1279–1284
2. Rad MB, Moghadam MP, Sheikh-El-Eslami MK (2007) Fuzzy evaluation of energy efficiency improvement impact on load shape. In: Power Tech(PT). IEEE, Lausanne, pp 1429–1434
3. Cao Z, Ma L, Wang N et al (2011) An entropy-based evaluation method of maintenance support system. In: Reliability, maintainability and safety international conference (ICRMS). IEEE, Guiyang. pp 842–848
4. Yuan S, Wang J, Dai YR (2009) Comprehensive industry energy efficiency evaluation based on combination evaluation method (AHM-GF). Manuf Autom 31(3):20–23
5. Pandey RK, Panda SS (2017) Optimization of bone drilling parameters using grey-based fuzzy algorithm. Measur 47(1):386–392
6. Tan ZF (2012) Research on method of attribute weight based on rough sets theory. Master thesis. Nanning Normal University, Gui Lin
7. Pei D, Xu ZB (2007) Transformation of rough set models. Knowl Based Syst 20(8):745–751
8. Mi JS, Leung Y, Zhao HY et al (2008) Generalized fuzzy rough sets determined by a triangular norm. Inf Sci 178(16):3203–3213
9. Jiang CH, He TR, Qu YH et al (2008) The method of ascertaining rule weight based on rough sets theory. J Gansu Lianhe Univ Nat Sci 22(6):35–37
10. Lin JL, Lin CL (2005) The use of grey-fuzzy logic for the optimization of the manufacturing process. J Mater Process Technol 160(1):9–14
11. Chakraborty MK (2014) Membership function based rough set. Int J Approx Reason 55(1):402–411
12. Pedrycz W, Vukovich G (2002) On elicitation of membership functions. IEEE Trans Syst Man Cybern Part A Syst Hum 32:761–767
13. Li J, Yu W (2013) An improved grey incidence analysis method based on TOPSIS thinking and its application on the project evaluation. Math Pract Theory 43(8):76–81
14. Li Y, Yan H, Yan W et al (2014) A framework for characterising energy consumption of machining manufacturing systems. Int J Prod Res 52(2):314–325

Chapter 6
Diagnosis of the Manufacturing Energy Consumption Bottleneck in a Complex Environment

6.1 Introduction

Using the energy efficiency quantitative analysis method to evaluate the energy efficiency of the discrete manufacturing system's job shop can fully reflect the workshop's energy efficiency level. But since there are so many departments and multiple types of equipment involved with the discrete manufacturing system's energy consumption, and since the energy consumption of various procedures varies a lot, such complexity, diversity and evolvability of the discrete manufacturing energy consumption system result in the difficulty of optimizing the energy efficiency of the manufacturing workshop. The most important factor affecting the optimized performance of energy efficiency is the system's energy consumption bottleneck. Therefore, analysis and identification of the energy consumption bottleneck are the primary task for optimizing the discrete manufacturing energy efficiency.

There are diversified definitions of a job shop bottleneck. For instance, document [1] treats the facility's utilization ratio as the bottleneck. Document [2] and document [3] define the bottleneck as the maximum completion time. Document [4] defines the bottleneck as the equipment load. So far there has not been much research on treating the discrete workshop energy consumption as the bottleneck. And this chapter studies precisely on the discrete workshop energy consumption as the bottleneck. Our research identifies where the energy consumption bottleneck is and treats it as the basis of the energy efficiency optimization. On the other hand, the common workshop bottleneck identification methods do not apply to the complex environment of the discrete manufacturing workshop. For instance, document [5] proposes the section TOPSIS multi-attribute identification method to identify the bottleneck machine, solving the problem that the feature attribute of the identification machine can hardly be represented by a definite value during the bottleneck identification. But this method is not applicable in complex processing environment (such as discrete manufacturing). Document [6] proposes a multiple bottleneck dynamic prognostic method based on two bottleneck degrees, which achieves continuous and accurate prediction of the system's multiple bottlenecks. However, such a method requires much historical data

Y. Wang et al., *Quantitative Analysis and Optimal Control of Energy Efficiency in Discrete Manufacturing System*, https://doi.org/10.1007/978-981-15-4462-0_6

and is hardly operable. To achieve energy consumption bottleneck identification in the complex environment of the discrete manufacturing system, this chapter adopts the fresh-new perspective of complex network and treats the discrete manufacturing workshop as the research object, combined with the complexity of the discrete manufacturing system. Thus, it constructs an energy consumption network model of the discrete manufacturing process. Taking into account features of the network and the workshop's energy consumption, we propose an energy consumption bottleneck identification method based on the complex network's discrete manufacturing process, thereby solving the problem of energy consumption bottleneck identification during the manufacturing progress in a complex environment.

6.2 Energy Consumption Model of the Discrete Manufacturing System

6.2.1 Problem Description

In the discrete manufacturing workshop, it is planned that n types of parts shall undergo processing operations at m processing equipment. The component $S_i(i = 1, 2, \ldots, n)$ is processed by multiple procedures according to the given operational path. The processing rules of the discrete manufacturing workshop are as follows: component S_i goes through H_i steps of the procedure and finishes the entire technological processing procedure. Each procedure is confined by a technological order of processing. The processing equipment needed for each procedure has been ascertained, and so are the standard processing time and preparation time spent on the equipment in each procedure. Here, T_{ijr} represents the standard processing time of component no. i, i.e.S_i, during procedure no. j at equipment no. r, and T'_{ijr} represents its processing preparation time. The discrete manufacturing process involves multiple aspects of energy consumption, and two major aspects of energy consumption are that which occur during the equipment processing and that during the route transportation of components between different procedures. The difference between the processing tasks and the components results in the various degrees of difference between $C_r(r = 1, 2, \ldots, m)$, i.e. each equipment's energy consumption during the technological process. Additionally, after the procedure H_i is completed, the transportation equipment would send the components to the equipment for the next processing procedure H_{i+1}, during which process energy consumption of various procedure path,$W_{st}(s, t = 1, 2, \ldots, m)$, would occur.

6.2.2 Constructing the Discrete Manufacturing System's Energy Consumption Network Model

We regard each processing equipment in the discrete manufacturing energy consumption system (such as procedures of polish, bore and mill processing) as the nodes in the network. Procedures of different processing tasks often fight for the same processing equipment, and the relationship between the procedure and the equipment is shown in Fig. 6.1. According to the technological process, equipments are allocated to the corresponding component processing route. Rescheduling of the procedures results in connection and cooperation among the equipments. We treat these connections as the edges in the energy consumption network, and as the technological procedure is directed, so are these edges. Then we construct a directed network of the discrete manufacturing process vie linking the equipment nodes and the equipment edges, shown in Fig. 6.2.

The entire workshop processing progress is accompanied by all sorts of energy consumption. Here, we pay special attention to two major energy consumption in the discrete workshop: the route energy consumption and equipment energy consumption. We take these two energy consumption sources into the discrete manufacturing network, with equipment nodes corresponding to the equipment energy consumption and link between the nodes corresponding to the route energy consumption, and then construct an energy consumption network model of the discrete manufacturing progress as shown in Fig. 6.3.

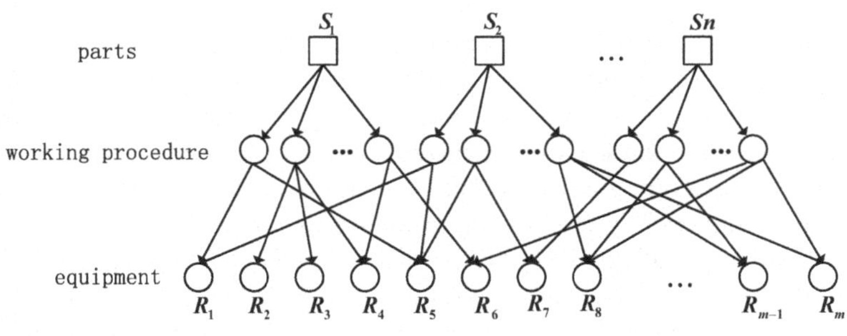

Fig. 6.1 Relationship between procedures and equipment

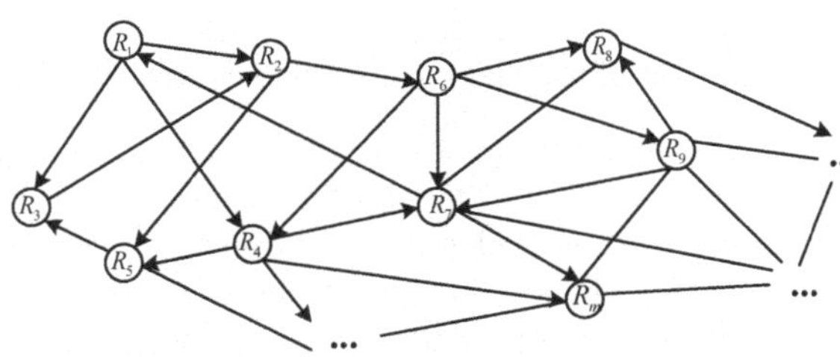

Fig. 6.2 Directed network of the discrete manufacturing progress

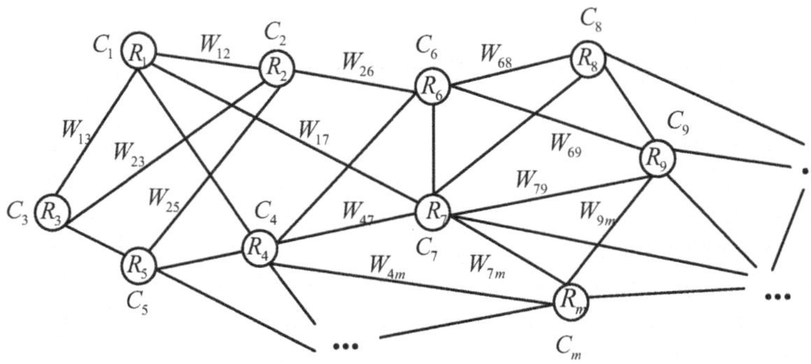

Fig. 6.3 Energy consumption network model of the discrete manufacturing progress

6.3 Energy Consumption Bottleneck Identification of the Discrete Manufacturing System Based on Complex Network

6.3.1 Definitions of the Energy Consumption Bottleneck and the Network Energy Consumption Feature

1. **Definition of the Discrete Manufacturing Progress's Energy Consumption Bottleneck**

Here is how we define the energy consumption bottleneck of the discrete manufacturing system: in the discrete manufacturing system, the node which plays a leading role in the trend of the entire workshop's processing progress energy consumption could also lead the direction and scope of the energy consumption's route of transmission in the discrete manufacturing progress. The existence of such a bottleneck maximizes the energy consumption of the discrete manufacturing workshop, and it is the key factor restraining energy efficiency optimization. Whether a node qualifies for being a node is determined by the quantified result called node bottleneck degree. Bottleneck degree is defined as the capability of the node to become a bottleneck in the energy consumption network. The bottleneck degree of a node is dynamic—a feature which holds direct influence over the discrete manufacturing energy consumption network.

2. **Definition of the Network Energy Consumption Feature**

Factors influencing the trend of the discrete manufacturing system's energy consumption include structure of the discrete manufacturing energy consumption network, the communication mechanism of the energy consumption network, features of the network nodes and node-relevant features. All these aspects affect the emergence of node bottleneck. Based on the above perspectives, we select these node identification indexes of the energy consumption network bottleneck: the number of network nodes, betweenness, network efficiency, route energy consumption and node energy consumption. Following are the specific definitions:

Definition 6.1 Node Degree

$$A_i = \sum_{j \in \tau(i)} a_{ij} \tag{6.1}$$

where A_i represents the node degree; a_{ij} represents whether node i and node j are connected, namely whether there exists any mobilization of the component procedure between these two equipment nodes. If there is indeed component procedure mobilization between node d and node j, then $a_{ij} = 1$; otherwise, $a_{ij} = 0$. As an important feature of network nodes, node degree is the sum of the times that the component is mobilized into and out of this node. The higher the degree, the more processing tasks that the node is undertaking. The higher the utilization ratio and the degree of importance, the bigger proportion it occupies in energy consumption.

Definition 6.2 Betweenness

$$B_r = \sum_{s \neq r \neq t} \frac{\lambda_{st}(r)}{\lambda_{st}} \tag{6.2}$$

where B_r represents the betweenness of node r, λ_{st} represents the number of all the shortest routes between node s and node t. $\lambda_{st}(r)$ represents the number of all shortest routes between node s and node t which pass node r. The higher the betweenness of the equipment node, the higher the possibility for the component processing to pass this node. Quantification of this index represents the centrality of the node in the network, emphasizes the relevance between one node and its neighbouring node while influencing the direction of the discrete manufacturing workshop's energy consumption propagation.

Definition 6.3 Network Efficiency In the discrete energy consumption network, node efficiency directly and effectively represents the mutual influence between nodes. Efficiency e_{st} refers to the reciprocal of the distance between node s and node t, i.e. the reciprocal of the number of shortest edges between nodes:

$$e_{st} = \frac{1}{d_{st}} \tag{6.3}$$

where d_{st} represents the number of the edges which pass through node s and node t. If $s = t$, there is no shortest route between the edges, then the efficiency is 0. If the two nodes are neighbouring nodes to each other, namely $d_{st} = 1$, then the efficiency is 1, which means the maximum impact between the two nodes. Therefore, the network efficiency matrix of a discrete manufacturing energy consumption network with m nodes is:

$$E = \begin{bmatrix} e_{11} & e_{12} & \dots & e_{1m} \\ e_{21} & e_{22} & \dots & e_{2m} \\ \dots & \dots & & \dots \\ e_{m1} & e_{m2} & \dots & e_{mm} \end{bmatrix}$$

Definition 6.4 Route Energy Consumption The components to be processed by the discrete manufacturing workshop are very complex, since they are large in number and have various forms. Nor can the transportation energy consumption caused by mobilization of components between two procedures be ignored, since it is a major component of the energy consumption during the discrete manufacturing progress and a breakthrough point for energy conservation. And since the transportation equipment varies according to the different components, the transportation energy consumption thus generated varies greatly as well. Therefore, in the energy consumption network of the discrete manufacturing progress, the route energy consumption between nodes is an important index measuring the level of the network energy consumption.

Different components are equipped with different transportation tools. The route energy consumption between node s and node t is:

$$W_{st} = \sum_{i=1}^{N_{\text{kind}}} a_{sti} T_{sti} P_{sti} \tag{6.4}$$

where W_{st} represents the route energy consumption between node s and node t within a given period. N_{kind} represents the kinds of components. a_{sti} represents whether there is the component mobilization of the component i between node s and node t: if there is, then $a_{sti} = 1$; if not, then it is 0. P_{sti} represents the different transportation equipment power equipped for component mobilization. T_{sti} represents the operation time of the transportation equipment for transporting i between node s and node t. m The route energy consumption matrix of each node's discrete manufacturing energy consumption network is:

$$W_{\text{path}} = \begin{bmatrix} W_{11} & W_{12} & \dots & W_{1m} \\ W_{21} & W_{22} & \dots & W_{2m} \\ \dots & \dots & & \dots \\ W_{m1} & W_{m2} & \dots & W_{mm} \end{bmatrix}$$

Definition 6.5 Node Energy Consumption Node energy consumption C_r is a feature of the node itself, which poses direct influence over the node's energy consumption function in the entire discrete manufacturing energy consumption network. The cutting time (processing time) and idle time (preparation time in wait) vary with each component. And each equipment has its unique cutting power and idle power. Therefore, the energy consumption of the equipment node is an important factor causing the discrete manufacturing energy consumption network's bottleneck.

$$C_r = \sum_{i=1}^{N} u_i \left(\sum_{j=1}^{H_i} \beta_{ijr} T_{ijr} P_{n\text{cut}} + \sum_{j=1}^{H_i} \beta_{ijr} T'_{ijr} P'_{n\text{idle}} \right) \tag{6.5}$$

where node energy consumption C_r refers to the energy consumed by the equipment node r in a given period. u_i refers to the number of component i. H_i represents the number of procedures of component i. β_{ijr} represents the equipment coefficient. If for component i, its procedure no. j can be finished at the equipment node r, then β_{ijr} is 1, otherwise 0. T_{ijr} And T'_{ijr} represent, respectively, the cutting time and idle time of component i at procedure no. j. $P_{n\text{cut}}$ and $P'_{n\text{idle}}$ represent, respectively, the cutting power and idle power of equipment node r.

6.3.2 Analysis of Discrete Manufacturing Progress's Energy Consumption Bottleneck Degree

In the energy consumption network of the discrete manufacturing progress, there are two major factors determining the size of an equipment's node bottleneck degree: (1) the nature of the node itself, including the route energy consumption and the equipment node energy consumption; and (2) the degree of relevance between one node and the other nodes in the network, namely the degree of other nodes' contribution to this node's bottleneck degree. In the discrete manufacturing workshop, the existence of a complete technological route ensures not only the close connection between neighbouring equipment nodes, but also the mutual influence between non-adjacent nodes in the network. Such a degree of relevance influences the emergence of the bottleneck. And to overcome the shortcoming of exclusive reliance on a neighbouring node in the other bottleneck identification methods, we adopt the node energy consumption and the node-relevant route energy consumption as the reference values contributing to the network bottleneck degree. We solve the degree of each node's reliance upon the other nodes based on the network efficiency matrix, thereby establishing the node's bottleneck degree in a more scientific and effective manner.

1. **The Node Route Energy Consumption's Degree of Importance Based on Efficiency Matrix**

Route energy consumption is an important energy consumption source in the discrete energy consumption network. When each node receives the component mobilized from the last procedure, and when the component is mobilized to the next procedure, route energy consumption occurs. If the network route energy consumption matrix W_{path} takes into consideration all the route energy consumption of the nodes, then the route energy consumption caused by node r is:

$$W_r = \sum_{i=1}^{m} W_{ri} \tag{6.6}$$

Based on the degree of reliance of the node-caused route energy consumption upon other nodes, namely the other nodes' contribution, we can calculate through the network energy efficiency matrix:

$$S_{\text{path}} = \begin{bmatrix} s_1 \\ s_2 \\ \dots \\ s_m \end{bmatrix} = E \times W_r = \begin{bmatrix} e_{11} & e_{12} & \dots & e_{1m} \\ e_{21} & e_{22} & \dots & e_{2m} \\ \dots & \dots & \dots & \dots \\ e_{m1} & e_{m2} & \dots & e_{mm} \end{bmatrix} \begin{bmatrix} W_1 \\ W_2 \\ \dots \\ W_m \end{bmatrix}$$

The importance of the route energy consumption of node $a_j^{\min}$ is:

$$s_r = \sum_{s \neq t}^{m} (e_{st} W_r) \tag{6.7}$$

2. **The Degree of the Node Energy Consumption's Importance based on the Efficiency Matrix**

Different types of equipment are chosen according to the degree of the node equipment's energy consumption importance, the primary energy consumed in the discrete workshop and the technological route. Hence, the entire discrete manufacturing network energy consumption alters greatly. According to the network efficiency and node energy consumption formula, we solve the equipment node's processing energy consumption degree of importance:

$$S_{\text{node}} = \begin{bmatrix} s'_1 \\ s'_2 \\ \dots \\ s'_m \end{bmatrix} = E \times C_r = \begin{bmatrix} e_{11} & e_{12} & \cdots & e_{1m} \\ e_{21} & e_{22} & \cdots & e_{2m} \\ \dots & \dots & & \dots \\ e_{m1} & e_{m2} & \cdots & e_{mm} \end{bmatrix} \begin{bmatrix} C_1 \\ C_2 \\ \dots \\ C_m \end{bmatrix} \tag{6.8}$$

The importance degree of the node energy consumption of node r is:

$$s'_r = \sum_{s \neq t}^{m} (e_{st} C_r) \tag{6.9}$$

3. **The Network Node's Energy Consumption Bottleneck Degree**

We take into consideration the network node degree, betweenness and other nodes' contribution in the energy consumption network of the discrete manufacturing progress and solve the node bottleneck degree as:

$$V_r = A_r B_r (\alpha_1 s_r + \alpha_2 s'_r) \tag{6.10}$$

where α_1 and α_2 are respectively the weight functions of route energy consumption and node energy consumption. Using the expert method, and in accordance with their importance in the workshop energy consumption, we assign values to the two weight functions. The assigned values of α_1 and α_2 are respective 0.3 and 0.7. We run normalization processing and obtain the bottleneck degree value of node no. r:

$$V'_r = \frac{V_r}{\sum_{k=1}^{m} V_k} = \frac{A_r B_r \left(\alpha_1 \sum_{s \neq t}^{m} e_{st} W_r + \alpha_2 \sum_{s \neq t}^{m} e_{st} C_r \right)}{\sum_{k=1}^{m} \left(A_k B_k \left(\alpha_1 \sum_{n \neq k}^{m} e_{kn} W_n + \alpha_2 \sum_{n \neq k}^{m} e_{kn} C_n \right) \right)} \tag{6.11}$$

6.3.3 Steps of Analyzing the Energy Consumption Bottleneck

Before using the complex network method to identify the discrete manufacturing system's job shop bottleneck, first, we need to process the raw production and energy consumption data acquired from the workshop and construct the workshop's energy consumption network model. Based on this model, we then run bottleneck identification and analysis in these following steps:

1. Convert the workshop raw production and energy consumption relevant data into a manufacturing progress's energy consumption network model
    - **Step 1**: Ascertain the time zone for acquiring the discrete workshop's data
    - **Step 2**: Ascertain the procedure mobilization between all the equipment nodes and the equipment, namely linking the nodes into edges
    - **Step 3**: Add the node energy consumption and node route energy consumption factor
    - **Step 4**: Convert the energy consumption nodes and the energy consumption edges into the discrete workshop's energy consumption network model
    - **Step 5**: Calculate the energy consumption network's feature data, which provides the basis for the bottleneck calculation and analysis.
2. Calculate the node's degree according to Formula 6.1
3. Calculate the node's betweenness according to Formula 6.2
4. Calculate the network efficiency matrix between every two nodes
5. Calculate the route energy consumption matrix between the nodes
6. Calculate the node's equipment energy consumption according to Formula 6.3
7. Solve each node's bottleneck degree according to the normalization formula and identify the bottleneck node.

6.3.4 Case Analysis and Validation

This chapter selects one machine tool manufacturing enterprise's production workshop as the object of analysis. This workshop holds 28 processing equipment $R_i(i = 1, 2, 3 \ldots, 28)$, which processes four types of components: S_1, S_2, S_3, and S_4. We acquire multiple layers of raw production and energy consumption data from this workshop's manufacturing execution system (EMS). Table 6.1 lists the processing time of each component at each equipment T_{ijr} and its preparation time T'_{ijr}, with minute as the temporal unit. Table 6.2 shows each equipment's cutting power P and idle power P', the power unit being KWH. Table 6.1 shows the number of four types of components, S_1, S_2, S_3 and S_4, processed each day (Table 6.3).

To validate the effectiveness and precision of the discrete manufacturing progress's bottleneck analysis method, which is based on the complex network, we undertake a simulation comparison analysis with the bottleneck closed-loop prognostic method [7]. The experiment data sets "day" as its unit. Both methods predict the same group of data for six continuous days, disregarding factors such as the workshop disturbance. The final simulation and comparison experiment is shown in Fig. 6.4a–f, each corresponding to six days' prognostic results.

From these figures, we can see that both methods have a similar prediction of the first ten types of equipment' bottleneck degree in six continuous days. They are identical in the prediction of the first major bottleneck, both having five days as R_3 and the precision rate at 83.3%. But there is a slight deviation concerning the prediction of the secondary bottleneck. The complex network-based discrete manufacturing progress's energy consumption bottleneck identification method has a precision rate of 83.3%, while that of the bottleneck closed-loop prognostic method is 66.7%. The method proposed in this chapter improves the drawback of the closed-loop prognostic method—i.e. its reliance upon the nearby environment and hence enhances the

Table 6.1 Component processing and preparation time

Component	R_1 T_{ijr}, T'_{ijr}	R_2 T_{ijr}, T'_{ijr}	R_3 T_{ijr}, T'_{ijr}	R_4 T_{ijr}, T'_{ijr}	…	R_{27} T_{ijr}, T'_{ijr}	R_{28} T_{ijr}, T'_{ijr}
S_1	/	/	25.2	/	…	28.4	/
S_2	22.2	18.3	/	16.1	…	/	30.5
S_3	20.2	/	24.3	/	…	27.4	/
S_4	16.3	22.2	/	28.4	…	/	/

Table 6.2 Equipment power

Power	R_1	R_2	R_3	R_4	…	R_{27}	R_{28}
P	2.2	2.5	3.1	2.6	…	3.6	3.5
P'	0.32	0.41	0.33	0.25	…	0.32	0.30

Table 6.3 Component processing number

Component	Day 1	Day 3	Day 4	Day 5	Day 6	Day 7	Day 8
S_1	10	10	8	15	15	10	10
S_2	18	20	16	12	25	20	14
S_3	12	22	18	10	16	15	20
S_4	15	20	22	18	25	18	15

prognostic rate of bottleneck prediction. The above proves that the complex network-based discrete manufacturing progress energy consumption bottleneck identification method is effective and precise at the same time.

6.4 Bottleneck Node Identification Method Based on the Structural Analysis Model

The existing bottleneck node identification method mainly focuses on one aspect during the processing progress. While the above methods have enhanced the system's energy efficiency level, but it is a limited improvement since only some, but not all aspects of the bottleneck are identified. As Fig. 6.5 shows, this chapter applies the Interpretative Structural Modelling Method (ISM) to the study of manufacturing system energy efficiency bottleneck index node identification problems from a comprehensive perspective. Based on the energy consumption-oriented discrete manufacturing evaluation index system proposed in Chap. 5, this chapter adopts the partial correlation coefficient analysis method to expel influence of other index nodes on the two node-relevant coefficients and runs a quantitative analysis of the correlative relationship between each index node. As for the problem of ascertaining adjacent matrix, we adopt the similarity level of the substance–relation–substance collection contained in the two index nodes as a threshold for defining the relation of influence. Based on the relation of importance between the relevant coefficients, the structural analysis model analyzes the level of importance between index nodes and the deep-level factors affecting the manufacturing system's energy efficiency level. This model uncovers the bottleneck node of the system's energy efficiency during the production progress, which serves as a reference to the system's energy efficiency optimization.

6.4.1 Method of Analyzing the Energy Efficiency Bottleneck

The Partial Relevant Analysis Method

The simple relevant analysis and the partial relevant analysis are two commonly used

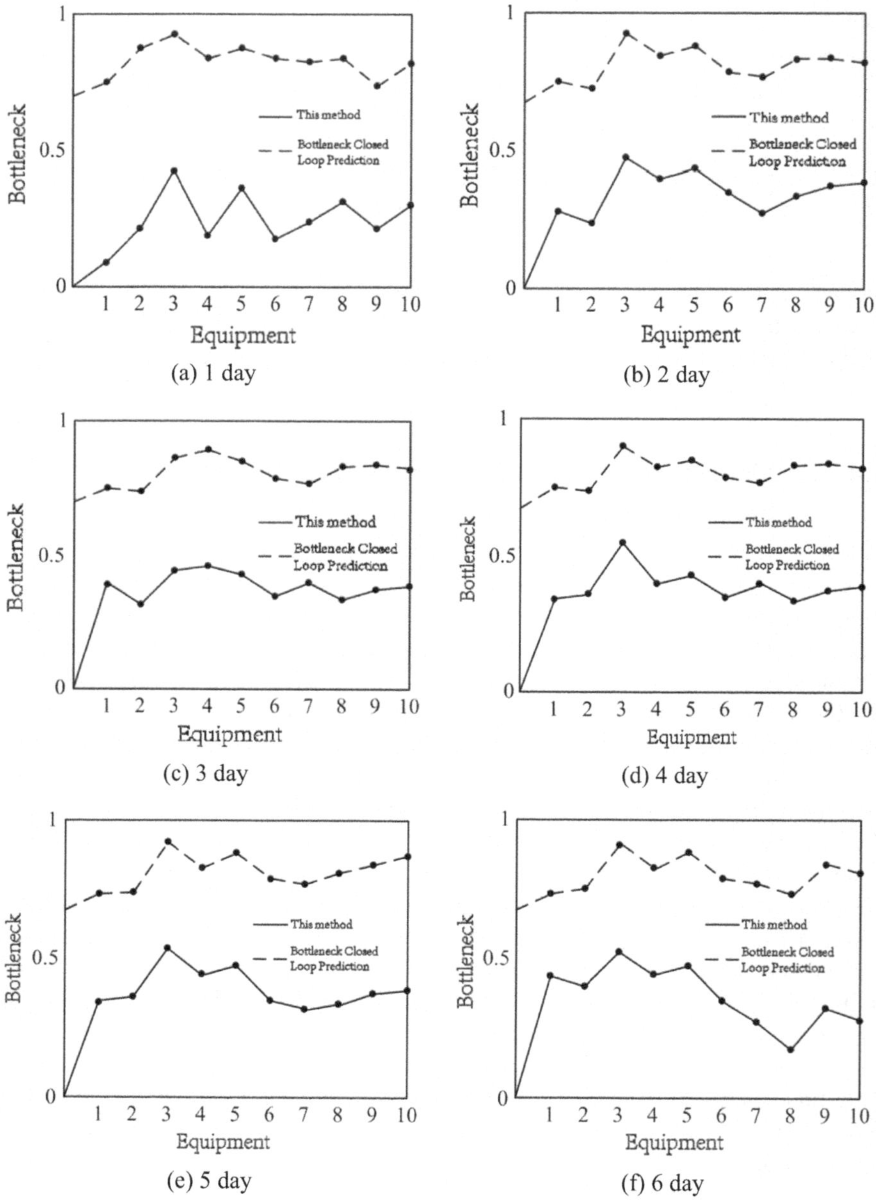

Fig. 6.4 **a** 1 day, **b** 2 day, **c** 3 day, **d** 4 day, **e** 5 day, **f** 6 day

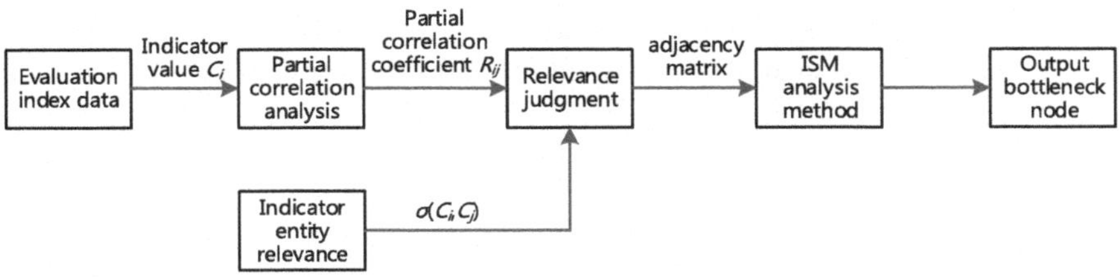

Fig. 6.5 Energy efficiency bottleneck identification

correlation analysis methods. Both modes use the quantitative analysis formula to describe the correlation between two variables. However, the simple relevant analysis only measures the historical values of the two variables when acquiring related coefficients, without excluding the potential impact of other variables upon the two's correlation. When analyzing the correlation between the two variables, the partial relevant analysis method takes into account the other variables. It also measures all the variables' historical values when acquiring partial relevant coefficients. Through controlling other variables, it truthfully reflects the closeness of the mutual relation between the two variables, to make its analytic results more precise and reliable [8]. Thus, this chapter adopts the partial relevant coefficient method to analyses the correlation between the two variables, with its size value chosen within $[-1, +1]$.

There are three ways to obtaining the partial relevant coefficients. (1) The iterative method: the partial relevant coefficient of order n can be seen as the relevant coefficient of two variables excluding the impact of the other n variables. The simple relevant coefficient is the partial relevant coefficient of order 0. The partial relevant coefficient of any order n can be obtained by using three $n-1$ orders' partial relevant coefficients. (2) The linear regression formula [9]: suppose Z represents all the other variables apart from variable X and Y, then R_x is the residual error acquired after linear regression analysis of X and Y, and R_y is the residual error acquired after linear regression analysis of Y and Z. The simple relevant coefficient between R_x and R_y is the partial relevant coefficient between variables X and Y. (3) Relevant matrix inverse: first we acquire the relevant coefficient between two random variables, constructing a relevant coefficient matrix including all variables. Then we can obtain the partial relevant coefficient between the variables via methods such as the inverse calculation of the matrix. Compared with first two methods, the third method requires less calculation. So without affecting the judgement results, this chapter adopts the third method in calculation and acquires the two variables' partial relevant coefficient via the relevant matrix inverse method. Its basic principles are:

1. Suppose $x_i(i = 1, 2, 3, \cdots, m)$ is value no. i of variable x, then the partial coefficient of variables x and y are:

$$r_{xy} = \frac{\sum_{i=1}^{M} (x_i - \bar{x})(y_i - \bar{y})}{\sqrt{\sum_{i=1}^{M} (x_i - \bar{x})^2}\sqrt{\sum_{i=1}^{M} (y_i - \bar{y})^2}} \tag{6.12}$$

where $\bar{x}$ is the average value of x, while $\bar{y}$ is the average value of y.

2. The relevant coefficient matrix r of all the variables can be solved by Formula 4.1

$$r = \begin{bmatrix} r_{1,1} & r_{1,2} & \cdots & r_{1,m-1} & r_{1,m} \\ r_{2,1} & r_{2,2} & \cdots & r_{2,m-1} & r_{2,m} \\ \cdots & \cdots & \cdots & \cdots & \cdots \\ r_{m,1} & r_{m,2} & \cdots & r_{m,m-1} & r_{m,m} \end{bmatrix}_{m*m} \tag{6.13}$$

3. Calculate the inverse matrix of r

$$D = r^{-1} = \begin{bmatrix} d_{1,1} & d_{1,2} & \cdots & d_{1,m-1} & d_{1,m} \\ d_{2,1} & d_{2,2} & \cdots & d_{2,m-1} & d_{2,m} \\ \cdots & \cdots & \cdots & \cdots & \cdots \\ d_{m,1} & d_{m,2} & \cdots & d_{m,m-1} & d_{m,m} \end{bmatrix}_{m*m} \tag{6.14}$$

D as the inverse matrix of relevant matrices;

4. Acquire the partial relevant coefficient between variables x and y from the following formula

$$R_{ij} = -\frac{d_{ij}}{\sqrt{d_{ii} * d_{jj}}} \tag{6.15}$$

where $j, i = 1, 2, \ldots, m$.

5. Once the relevant coefficients have been obtained, due to the impact of noise and sample data, the size of the coefficient cannot directly display whether there is relevance between the two. Therefore, we need to set a threshold value θ to ascertain whether the two nodes are relevant. The threshold value is decided by the substantial correlation level between the two index nodes. Following is the method:

$$\sigma(C_i, C_j) = \frac{U_{Ci} \cap U_{Cj}}{U_{Ci} \cup U_{Ci}} = \frac{U(C_i, C_j)}{U(C_i, C_j) + U(\overline{C_i}, C_j) + U(C_i, \overline{C_j})} \tag{6.16}$$

where U_{Ci} represents the substance–relation–substance set contained within node C. $U(C_i, C_j)$ represents the intersection of the two indexes' corresponding sets. We use the similarity degree as the threshold of relevance. If the partial relevant coefficient is within the threshold range, then the two nodes are relevant, otherwise non-relevant.

6.4.2 Analytical Structural Analysis Model

ISM splits the entire system into multiple sub-components. Aided by theoretical knowledge, progress experience and smart facility, it finally constructs a layered progressive relation model for analyzing the relevance between the system's components. ISM is an important method in systematic engineering analysis. It connects the conceptual analysis model with the quantitative analysis model. It is used extensively in scenarios hardly applicable to direct quantitative analysis. Following are its steps of building a model:

1. Construct correlative model describe whether one variable in the system set is directly related to the other variables. The relation between these two variables can be represented in value as

$$k_{ij} = \begin{cases} 1 \; x_i \text{ have an impact on } y_j \\ 0 \; x_i \text{ have no impact on or equal } y_j \end{cases} \tag{6.17}$$

In this chapter, we use the threshold value σ to decide whether two indexes are related, namely the relevance of the two nodes when $|R_{ij}| > \sigma$. When $R_{ij} > \sigma$ is a positive correlation: the adjacent matrix element of variable x_i opposed to variable x is 1, while the element value of x_j opposed to the adjacent matrix of x_i is 0. When $R_{ij} < \sigma$ is a negative correlation, the adjacent matrix element of variable x_i opposed to variable x_j is 0, while the element value of x_j opposed to the adjacent matrix of x_i is 1. Following is the adjacent matrix:

2. Acquire the adjacent matrix. Based on the above direct binary relation, the corresponding relation between the elements can construct a matrix of order n, and we call this matrix the adjacent matrix.

$$K = \left(k_{ij}\right)_{m*m} = \begin{bmatrix} k_{1,1} & k_{1,2} & \cdots & k_{1,m-1} & k_{1,m} \\ k_{2,1} & k_{2,2} & \cdots & k_{2,m-1} & k_{2,m} \\ \cdots & \cdots & \cdots & \cdots & \cdots \\ k_{m,1} & k_{m,2} & \cdots & k_{m,m-1} & k_{m,m} \end{bmatrix}_{m*m} \tag{6.18}$$

3. Calculate the accessibility matrix In matrix K, the value of element k_{ij} can be ascertained by Formula 6.17. Through relevant connection, variables without direct relation can generate indirect relations. Through accessibility matrix M, we can uncover the indirect correlative relation between elements, and the calculation formula is shown in 6.18:

$$(K + I) \neq (K + I)^2 \neq \cdots \neq (K + I)^m = (K + I)^{m+1} = M \tag{6.19}$$

Firstly, we sum up the adjacent matrix K and the unit matrix I. According to the Operational Rule of Boolean Algebra, the sun of these two then squares itself. When the obtained matrix no longer changes, the self square result is the accessibility matrix M.

4. Define the layered structure find the accessibility matrix $R(C_i)$ and the anterior cause set $A(C_i)$ of all the elements. If element C_i can fulfil the condition $R(C_i) \cap A(C_i) = P(C_i)$, then element C_i is the analytic model's highest level element, with all elements fulfilling such a condition at the same level. Once we have obtained the top-level elements, we delete all the other elements on the same row and column of these elements. We treat the matrix left behind as the new accessibility matrix and restart calculation of the second-level elements. By repeating the above steps, we can get all the elements of the next level, until the analysis ends.
5. Generate the layered structure diagram. According to the relevance between the elements, we draw a directed connection diagram. The connection diagram can clearly describe the layered structure between the system's elements as well as the transmission route of information, thereby assisting the administrators in analyzing the factors of influence.

Through the ISM analytic algorithm, we can analyze the relevance between each index node in the entire judgement index system and the importance of the index nodes. The layered analytic structure of the index nodes thus constructed can identify the node of biggest influence over the other nodes in the entire index system, which is essentially the bottleneck node of the entire system.

6.4.3 Case Analysis and Validation Analysis

This chapter uses the component processing index data of one machine tool factoring in Wuxi between October 2017 and November 2017 as the sample. We analyze the bottleneck affecting the energy efficiency level during the product's processing progress, thus providing an effective reference for the system's energy efficiency optimization and saving the energy cost. Firstly, we conduct partial relevant calculation of the energy consumption sample data according to Formulas 6.12–6.15. The partial relevant coefficient matrix is as follows:

$$\begin{bmatrix} -1.000 & 0.138 & -0.0490 & 0.039 & -0.070 & -0.052 & 0.072 & -0.643 & 0.056 & 0.052 & -0.091 \\ 0.138 & -1.000 & -0.125 & -0.088 & -0.153 & -0.097 & -0.197 & -0.086 & 0.617 & -0.514 & 0.152 \\ -0.049 & -0.125 & -1.000 & -0.192 & -0.057 & -0.203 & -0.579 & 0.059 & -0.058 & 0.648 & 0.103 \\ 0.039 & -0.088 & -0.192 & -1.000 & 0.167 & -0.095 & -0.122 & 0.516 & -0.229 & -0.018 & -0.021 \\ -0.070 & -0.153 & -0.057 & 0.167 & -1.000 & 0.792 & -0.215 & -0.558 & -0.053 & -0.072 & 0.173 \\ -0.052 & -0.097 & -0.203 & -0.095 & 0.792 & -1.000 & 0.093 & 0.042 & -0.179 & 0.177 & 0.153 \\ 0.072 & -0.197 & -0.579 & -0.122 & -0.215 & 0.093 & -1.000 & -0.084 & 0.087 & -0.105 & -0.026 \\ -0.643 & -0.086 & 0.059 & 0.516 & -0.558 & 0.042 & -0.084 & -1.000 & -0.158 & 0.130 & -0.765 \\ 0.056 & 0.617 & -0.058 & -0.229 & -0.053 & -0.179 & 0.087 & -0.158 & -1.000 & 0.143 & 0.182 \\ 0.052 & -0.514 & 0.648 & -0.018 & -0.072 & 0.177 & -0.105 & 0.130 & 0.143 & -1.000 & -0.081 \\ -0.091 & 0.152 & 0.103 & -0.021 & 0.173 & 0.153 & -0.026 & -0.765 & 0.182 & -0.081 & -1.000 \end{bmatrix}$$

According to Formula 6.16, we calculate the threshold value σ between the index nodes, and acquire the following adjacent matrix:

$$K = \begin{bmatrix} 0&0&0&0&0&0&0&1&0&0&0 \\ 0&0&0&0&0&0&0&0&0&1&0 \\ 0&0&0&0&0&0&1&0&0&0&0 \\ 0&0&0&0&0&0&0&0&0&0&0 \\ 0&0&0&0&0&0&0&1&0&0&0 \\ 0&0&0&0&1&0&0&0&0&0&0 \\ 0&0&0&0&0&0&0&0&0&0&0 \\ 0&0&0&1&0&0&0&0&0&0&1 \\ 0&1&0&0&0&0&0&0&0&0&0 \\ 0&0&1&0&0&0&0&0&0&0&0 \\ 0&0&0&0&0&0&0&0&0&0&0 \end{bmatrix}$$

Once we have obtained the adjacent matrix K, we further calculate the matrix sum of K and I and calculate the self-square of this sum, until Formula 6.19 is established and we obtain the accessibility matrix M. Elements in the accessibility matrix describe whether there is any correlation between the nodes. If $m_{ij} = 1$, then there is an accessible route between C_i and C_j, namely there is an impact relation between C_i and C_j.

$$M = \begin{bmatrix} 1&0&0&1&0&0&0&1&0&0&1 \\ 0&1&1&0&0&0&1&0&0&1&0 \\ 0&0&1&0&0&0&1&0&0&0&0 \\ 0&0&0&1&0&0&0&0&0&0&0 \\ 0&0&0&1&1&0&0&1&0&0&1 \\ 0&0&0&1&1&1&0&1&0&0&1 \\ 0&0&0&0&0&0&1&0&0&0&0 \\ 0&0&0&1&0&0&0&1&0&0&1 \\ 0&1&1&0&0&0&1&0&1&1&0 \\ 0&0&1&0&0&0&1&0&0&1&0 \\ 0&0&0&0&0&0&0&0&0&0&1 \end{bmatrix}$$

We disintegrate the accessibility matrix M and sort out respectively node C_i's relevant accessibility set $R(C_i)$ and its anterior cause set $A(C_i)$. The accessibility set $R(C_i)$ is the collection of nodes corresponding to all the value 1 rows corresponding to node C_i in the matrix M, described as $R(C_i) = \{C_j | r_{ij} = 1, j = 1, 2, \ldots, 11\}$. The anterior cause set $A(C_i)$ is composed of index nodes corresponding to the value 1 rows corresponding to node C_j in the accessibility matrix M, described as $A(C_j) = \{C_i | r_{ij} = 1, i = 1, 2, \ldots, 11\}$. In the analytic structure model, all the elements at the level n index nodes' element set L_n fulfil the condition $R(C_i) \cap A(C_j) = R(C_i)$. Then, from M, we delete the rows and columns corresponding to nodes in L_n till the end. The accessibility set and the anterior cause set corresponding to each element are shown in Table 6.4.

Based on the results in Table 6.5, the layered structural relation of the entire manufacturing system's energy consumption-related index is shown in Fig. 6.6.

Based on the mutual influence between the nodes in the above figure, we classify the entire evaluation system into three categories: basic nodes, middle nodes and deep nodes. Basic nodes are three evaluation points closest to the entire energy efficiency level's root node route, and they are direct factors affecting the whole energy efficiency level. Normally, they are the three biggest weight functions during quantitative analysis. Deep nodes mean those index nodes unaffected by other nodes.

Table 6.4 Analysis of energy efficiency factors

C_i	$R(C_i)$	$A(C_j)$	$R(C_j) \cap A(C_j)$
C_1	C_1, C_4, C_8, C_{11}	C_1	C_1
C_2	C_2, C_3, C_7, C_{10}	C_2, C_9	C_2
C_3	C_3, C_7	C_2, C_3, C_9, C_{10}	C_3
C_4	C_4	C_1, C_4, C_5, C_6, C_8	C_4
C_5	C_4, C_5, C_8, C_{11}	C_5, C_6	C_5
C_6	$C_4, C_5, C_6, C_8, C_{11}$	C_6	C_6
C_7	C_7	$C_2, C_3, C_7, C_9, C_{10}$	C_7
C_8	C_4, C_8, C_{11}	C_1, C_5, C_6, C_8	C_8
C_9	$C_2, C_3, C_7, C_9, C_{10}$	C_9	C_9
C_{10}	C_3, C_7, C_{10}	C_2, C_9, C_{10}	C_{10}
C_{11}	C_{11}	$C_1, C_5, C_6, C_8, C_{11}$	C_{11}

Table 6.5 Elements in L_n

Level L_n	Node C_i
L_1	C_4, C_7, C_{11}
L_2	C_3, C_8
L_3	C_1, C_5, C_{10}
L_4	C_2, C_6
L_5	C_9

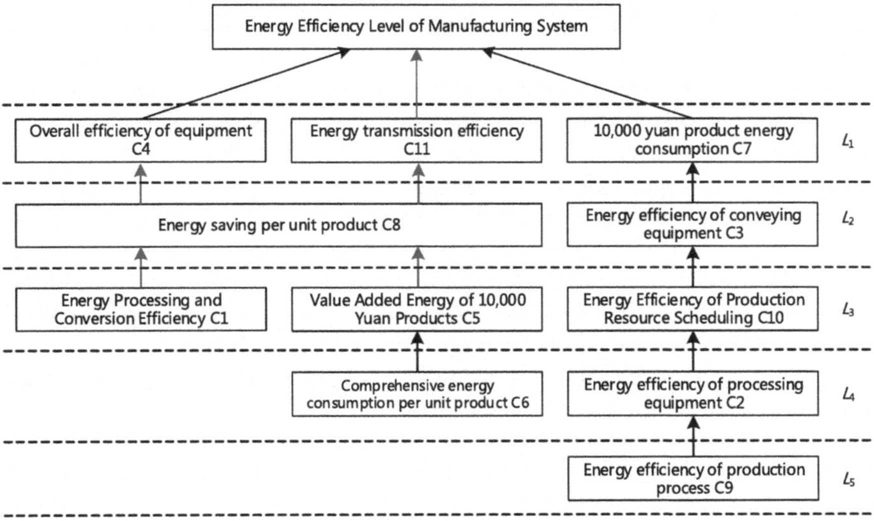

Fig. 6.6 Index analytical structural model

The remaining nodes are all categorized as middle nodes. The middle nodes are both under the influence of other nodes while posing influence over them as well. Classification of the nodes is shown in the following table.

In Table 6.6, the basic nodes are the equipment overall efficiency C_4, energy transmission efficiency C_{11} and the energy consumption C_7 per ten thousand Yuan product. These three nodes hold direct influence over the overall evaluation results and correspond respectively to the three index nodes of the biggest weight functions in the evaluation progress. The middle nodes influence other nodes and are influenced in return. If we conduct improvement analysis of only these nodes, then the overall energy efficiency would have limited improvement. Therefore, middle nodes cannot act as the bottleneck of the evaluation system. Energy processing conversion efficiency C_1, general energy consumption per unit product C_6, and the production technological energy efficiency C_9: these three are deep nodes unaffected by any

Table 6.6 Classification of index nodes

Classification	Nodes
Basic nodes	Equipment overall efficiency C_4; Energy transmission efficiency C_{11}; Energy consumption per ten thousand yuan product C_7
Middle nodes	Unit product energy savings C_8; Transmission equipment energy efficiency C_3; Energy consumption per ten thousand yuan product added value C_5; Production resources scheduling energy efficiency C_{10}; Processing equipment energy efficiency C_2
Deep nodes	Energy processing conversion efficiency C_1; General energy consumption per unit product C_6; Production technological energy efficiency C_9

other nodes. Among these three deep influence factors, the production technological energy efficiency node has the longest influence route and widest influence range. It has the biggest influence on the manufacturing system's energy efficiency level. Therefore, the production technology energy efficiency node can be regarded as the system's bottleneck node. If we optimize the production technology, then we can enhance the system's general level at maximum.

This chapter adopts the Tecnomatrix Plant Simulation 14.2 software to simulate the large product production workshop of a machine tool corporation in Wuxi [10, 11]. We select the processing progress of 8 machines ($M_1 \sim M_8$) in a typical manufacturing workshop as the source of simulation. The specific energy consumption condition of each processing procedure is written into the EnergyData table. The specific processing time of the machine tool's each processing procedure is written into the ProTable sheet.

First, we divide the entire simulation experiment into two parts. The first part is experiment data acquisition, which mainly analyzes the partial relevant coefficients between different nodes. To improve the reliability and dependability of the analysis results, we have to acquire energy consumption sample data under different processing conditions. Therefore, this chapter defines the simulation length as the time necessary for completing three contract tasks. The entire progress produces 18 kinds of components in total. The processing procedure for each component varies in kind and the equipment for conducting the processing operation differs as well. The second part is the validation stage. Based on the energy consumption data obtained in the first part, we use the method proposed in this chapter to analyse the bottleneck nodes of the energy consumption index system. We then optimize the bottleneck nodes and re-run the simulation test to validate the correctness of the analyzed results. Following are the detailed analyses.

Based on the energy consumption data of the simulation validation model during the simulation production progress, we can get the analytic structure of the entire evaluation index nodes, shown in Fig. 6.7, as well as classification of the nodes, shown in Fig. 6.8.

The deep node is the production technology energy efficiency node C_9. The production technology energy efficiency represents the proportion of the technological process's energy consumption per unit product in the total energy consumption. It represents the connection between the technological process factor in the manufacturing processing progress and the manufacturing system's energy consumption.

With regard to this issue, this chapter adopts the particle swarm optimization to optimize the current manufacturing technological parameters, ascertaining the new technological parameters before reapplying them to the above simulation system. The changes after the system node optimization are shown in Fig. 6.9.

Here, time 1–time 12 are the time points before the change; time 13–time 20 are the time points after the change. In Table 6.7, nodes with alteration value larger than 0.1 are all located between the C_9 node and the comprehensive energy efficiency level node. It means that optimization of node C_9 can improve the index values of most nodes, with all three basic nodes improved as well, leading to the simultaneous

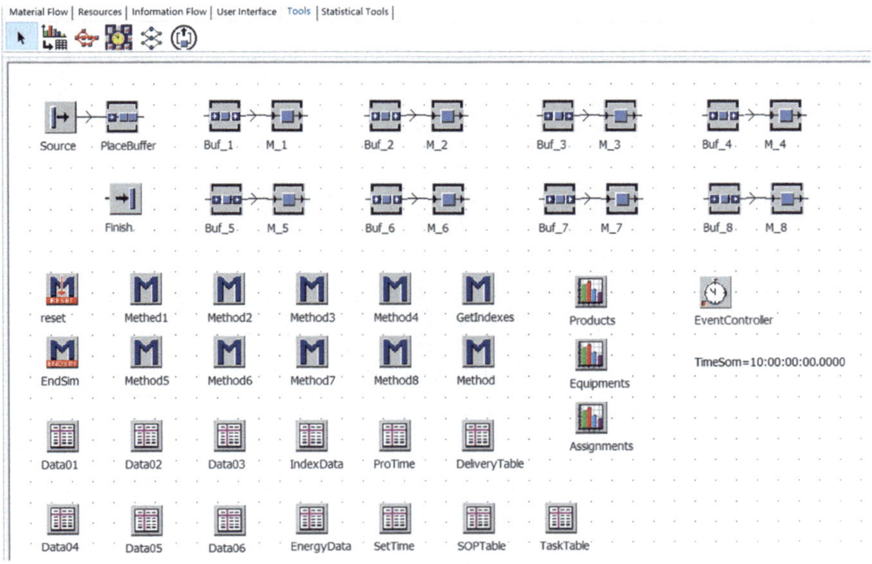

Fig. 6.7 Simulation model construction

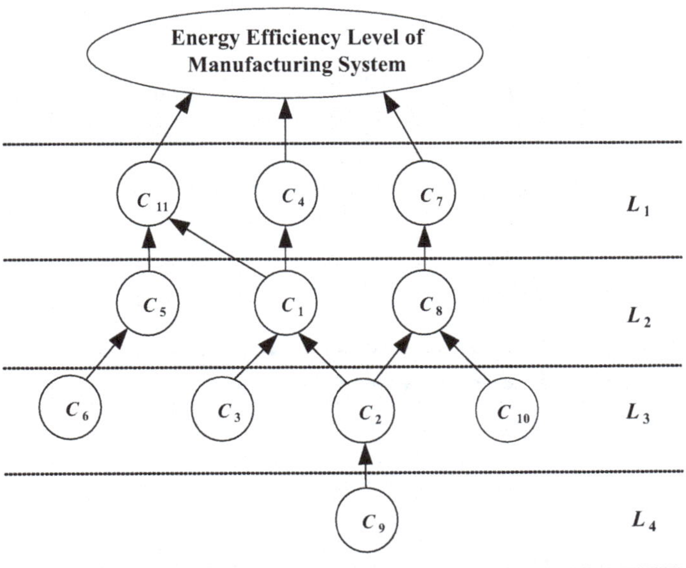

Fig. 6.8 Simulation system node ISM structure

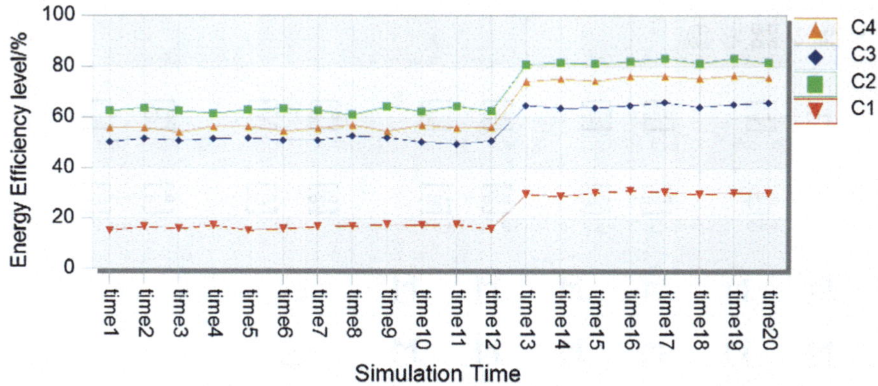

(a) Equipment level's energy efficiency node alteration

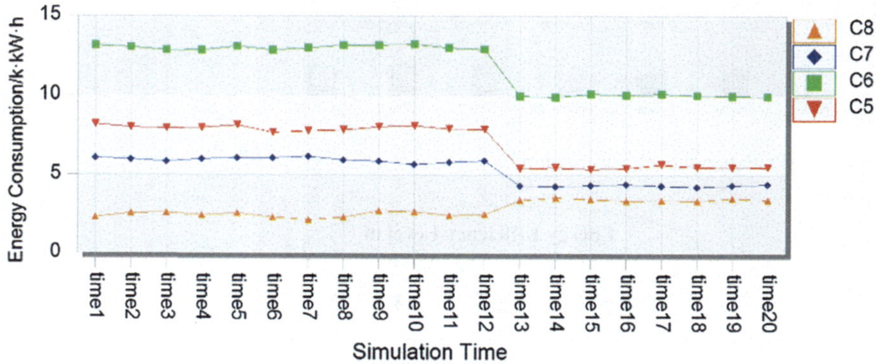

(b) Product level's energy efficiency node alteration

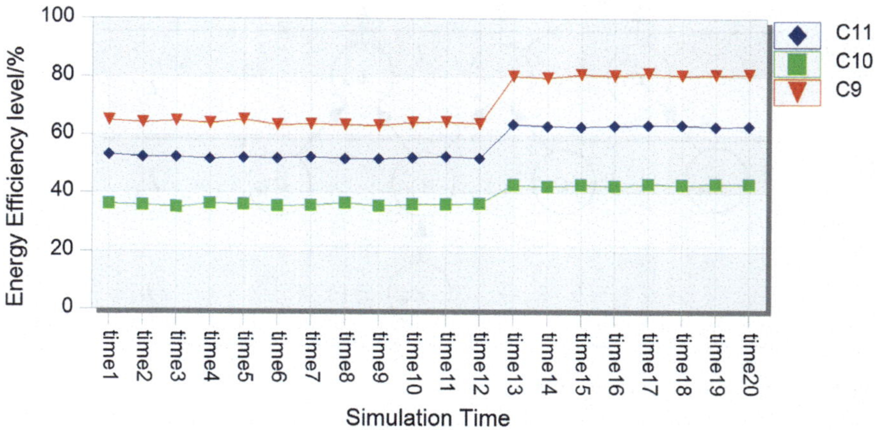

(c) Task level's energy efficiency node alteration

Fig. 6.9 **a** Equipment level's energy efficiency node alteration, **b** product level's energy efficiency node alteration, **c** task-level energy efficiency node alteration

Table 6.7 Classification of the simulation system nodes

Classification	Nodes
Basic nodes	Equipment overall efficiency C_4; Energy transmission efficiency C_{11}; Energy consumption per ten thousand yuan product C_7
Middle nodes	Unit product energy savings C_8; Transmission equipment energy efficiency C_3; Energy consumption per ten thousand yuan product added value C_5; Production resources scheduling energy efficiency C_{10}; Processing equipment energy efficiency C_2; Energy processing conversion efficiency C_1; Overall energy consumption per unit product C_6
Deep nodes	Production technology energy efficiency C_9

improvement of the system's overall energy efficiency level. Therefore, the manufacturing system energy efficiency bottleneck node identification method based on the analytic structure model proposed in this chapter is an effective one.

6.5 Conclusion

With regard to the complexity of the discrete manufacturing system's energy consumption, this chapter first proposes a bottleneck identification method based on complex network. This accurate quantitative evaluation and prediction of the bottleneck takes full consideration of the correlation between the nodes. It combines the node route energy consumption and the node equipment energy consumption and utilizes the network efficiency matrix to solve the node bottleneck degree, thereby overcoming the drawback of traditional algorithms which rely upon adjacent nodes. Furthermore, this chapter proposes a manufacturing system energy efficiency bottleneck identification method which combines the partial relevant coefficients with the ISM model. Based on real-life production index data, we obtain the partial relevant coefficient matrix between each index, and then decide whether the two factors of influence are related according to the threshold value, based upon which we construct the adjacent matrix. Then we construct a structural analytic model of the entire production evaluation nodes and obtain the system energy efficiency bottleneck through analysis. The methods proposed in this chapter not only solve the identification problem of the discrete energy consumption bottleneck. It can also be applied to solve energy consumption bottleneck identification problems in the other complex fields and industries.

References

1. Lu JS, Jing F (2014) Job shop bottleneck control research based on improved filtered-beam-search algorithm. J Zhejiang Univ Technol 25(14):263–273

2. Yang K, Li Y, Xiong Y et al (2015) Knowledge driven product innovation design based on complex network. Comput Integr Manufact Syst 21(9):2258–2269
3. He WJ, Lu JS, Li XL (2013) Research on production scheduling based on the production of logistics bottleneck. Light Ind Mach 31(1):101–110
4. Yan Z, Hanyu G, Yugeng X (2007) Modified bottleneck-based heuristic for large-scale job-shop scheduling problems with a single bottleneck. J Syst Eng Electron 18(3):556–565
5. Liu Z, Jiang Z, Gong B (2014) Dynamic prediction method of multi-bottleneck in manufacturing shop based on two bottleneck degrees. Zhongguo Jixie Gongcheng/China Mech Eng 25(14):1022–1031
6. Deng W, Jian-Sha LU, Weng YW (2013) Scheduling research of work-shop with bottleneck in cycle-time. J Mech Electr Eng 30(3):373–383
7. Liu Z, Tang J (2012) Based on bottleneck polymorphism of production logistics bottleneck closed-loop prediction method. Compu Integr Manufact Syst 12(11):2554–08
8. Geng ZQ, Bai J, Jiang DY et al (2018) Energy structure analysis and energy saving of complex chemical industries: a novel fuzzy interpretative structural model. Appl Therm Eng 2018(142):433–443
9. Wang ZS, Luo YN, Li AH et al (2017) The effect analysis of correlation between variables on the partial coefficient. J Railway Eng Soc 34(2):37–42
10. Pan CR, Wu NQ, Huang XJ (2012) EM-plant-based parameterization virtual cluster tool. Syst Eng-Theory Pract 32(8):1831–1840
11. Peng WM, Zhao XC (2004) Application of EM-plant in simulation to material flow operation. J Wuhan Univ Technol (Transp Sci Eng) 28(4):597–599

Chapter 7
The Energy Efficiency Quantitative Analysis Based on the Principal Component Analysis

7.1 Introduction

Based on the now-established energy efficiency quantitative analysis system of the discrete manufacturing system, we propose a scientific and effective method to solve the problems of energy consumption quantitative analysis. Each index in the quantitative analysis system can reflect a certain aspect of the discrete manufacturing system's energy efficiency, but it fails to provide an overall analysis. Therefore, we need to adopt a highly efficient energy efficiency quantitative analysis method to conduct a comprehensive evaluation of the discrete manufacturing system. This quantitative analysis method should be able to convert the multiple quantitative analysis indexes in the quantitative analysis system into one index quantitative analysis value, i.e. the energy efficiency quantitative analysis result. This would help us conduct an object and scientific quantitative analysis of the discrete manufacturing industry' energy efficiency. Scientists both at home and abroad have applied various comprehensive quantitative analysis methods to the energy efficiency quantitative analysis of the discrete manufacturing system, which include the attribute hierarchical model (AHM) [1], the fuzzy synthetic quantitative analysis method [2] and the entropy method [3].

This chapter proposes a discrete manufacturing energy consumption quantitative analysis method based on the reformed principal component analysis method. The goal of the classical principal component analysis method is to conduct a dimension reduction of high dimensional variables, while ensuring the minimum loss of data information. This method simplifies the complex high dimensional index matrix [4], solving the difficult problem of evaluating the discrete manufacturing system's energy efficiency. Furthermore, this chapter reforms the principal component analysis method by introducing the weight function of an index's importance. This revision optimizes the drawback of the traditional principal component analysis, namely its sole emphasis on the weight function of information. Our reformed method achieves an organic combination of the index's objectivity and subjectivity. Besides, it adopts a reformed data standardization method in the dimensionless data processing, which avoids loss of the original data and takes into consideration every factor of the discrete

Y. Wang et al., *Quantitative Analysis and Optimal Control of Energy Efficiency in Discrete Manufacturing System*, https://doi.org/10.1007/978-981-15-4462-0_7

energy consumption analysis, thereby fully and effectively reflecting the energy efficiency condition of the discrete manufacturing system. Finally, we demonstrate the reasonability and stability of the reformed method by case analysis and simulation results.

7.2 Problem Description

The discrete system's energy consumption includes that of the processing equipment, of the manufacturing technique, and of the production resources allocation. Factors such as the equipment performance, technological level and level of resources scheduling directly affect the energy efficiency level of the discrete system, which is the key to evaluating the discrete manufacturing energy efficiency. Based on the discrete manufacturing system's quantitative evaluation index system proposed in the last chapter, this chapter solves the problem of effectively and scientifically evaluating the job shop's energy consumption level. The energy consumption quantitative evaluation index data include energy consumption data of multiple layers, such as that of the equipment, the technique and the product. Such data mainly come from the smart metre and the operating equipment system. Based on the features of the discrete system and the factors of the discrete manufacturing energy consumption evaluation, we propose a reformed principal component analysis method, which analyzes and arranges these complex raw data and achieves a quantitative analysis of the discrete manufacturing system's energy consumption through algorithms.

7.3 Principle of the Principal Component Analysis and Analysis of the Discrete Energy Consumption

The discrete manufacturing energy consumption evaluation index system involves so many indexes, and the analysis is so complex given the mutual influence among the indexes, that the evaluation results are often inaccurate. The goal of the principal component analysis method is to reduce the dimension of high-dimensional variables while ensuring the minimum loss of the data information. Through linear conversion and disposing of a small amount of variable information, this method replaces the original multi-dimensional variables with a few complex variables, thereby simplifying the evaluation model [5] and reducing the mutual influence between the primitive variables, to avoid information repetition.

The comprehensive evaluation is to convert various evaluation index values into one total evaluation value by using the mathematical model. And the principal component analysis method provides such a mathematical model. It converts the primitive component-relevant random variable into a new component-irrelevant variable using one orthogonal transformation. Geometrically speaking, it is to convert the original variable system into a new orthogonal system, so that it points at the orthogonal direction where sample points are most widely scattered, and thus conducts dimensional reduction of the multi-dimensional variable structure. In the energy consumption evaluation index system of the discrete manufacturing progress, the principal component analysis method replaces the primitive indexes with a few comprehensive energy consumption evaluation index, before constructing the comprehensive evaluation function and conducting evaluation and analysis of the discrete manufacturing energy consumption.

7.4 Improving the Principal Component Analysis Method

7.4.1 Improving the Weight Function of the Principal Component Analysis Method

The weight function of the traditional principal component analysis method's comprehensive evaluation originates from the raw data. It considers only the relatively objective information amount weight function, losing sight of the subjective value judgement of the importance weight function upon evaluation index. Consequently, the evaluation results will deviate from the realistic condition in the comprehensive evaluation. To make up for this disadvantage, the discrete manufacturing energy consumption analysis based on the principal component analysis method takes full consideration of the importance weight function of the primitive indexes as well as the objective weight function of the comprehensive indexes.

It endows the primitive index with subjective weight function and endows various evaluation indexes with respective weigh function according to the experience and opinions of experts—i.e. the expert method [6]. It multiplies the primitive variable standardized matrix $\mathbf{Y} = \left(Y_{ij}\right)_{nxp}$ (*n* as the sample volume, *p* as the index volume of the index system) with the corresponding weight function *W*, thereby obtaining the new standardized variable matrix *Z*:

$$
\begin{aligned}
&Z = XW \\
&W = \left\{W_1, W_2, \ldots, W_p\right\}^{\mathrm{T}} \\
&\sum_{j=1}^{p} W_j = p
\end{aligned}
\tag{7.1}
$$

where W_j is the importance weight function value endowed to each index in the index system through the expert method's subjective empowerment.

7.4.2 Improving the Non-dimensionalization Method of the Principal Component Analysis

Due to the difference in the dimension and magnitude of each energy consumption relevant index in the analysis method, it is hardly comparable. Therefore, we need to run standardization of the primitive variable matrix. The traditional principal component analysis is non-dimensionalized, with each index variance be the same value 1. Disregarding the difference in the evaluation index variation; it includes only partial information of the primitive data and is incapable of accurately reflecting all the information included in the primitive data. Therefore, to use the Z-Score conversion to run standardization processing of the primitive variable matrix $X = \left(X_{ij}\right)_{nxp}$ (n as the sample volume, p as the index system's index volume) will lead to better comparability. Z-Score conversion runs data standardisation based on the mean value and standard deviation of the primitive data. It applies to the cases when the maximum and the minimum values of the index attributes are unknown, or cases of discrete data which exceeds the value range. Meanwhile, it also preserves all the information of the primitive data. Through converting the primitive variable matrix X, we get the standardized matrix Y. The conversion formula is as follows:

$$
\begin{aligned}
Y_{ij} &= \frac{X_{ij} - \bar{X}_j}{S_j} \\
\bar{X}_j &= \frac{1}{n}\sum_{i=1}^{n} X_{ij} \\
S_j^2 &= \frac{1}{n-1}\sum_{i=1}^{n} \left(X_{ij} - \bar{X}_j\right)^2 \\
&(i = 1, 2, \ldots, n;\ j = 1, 2, \ldots, p)
\end{aligned}
\tag{7.2}
$$

7.5 Improving the Energy Consumption Analysis Process of the Principal Component Analysis Method

7.5.1 Ascertaining the Energy Consumption Evaluation Index of the Principal Component Analysis

The evaluation index set includes eight indexes at the bottom level of discrete manufacturing progress energy consumption evaluation index. In other words, the evaluation index set is $C = \{C_1, C_2, \ldots, C_8\}\sqrt{a^2 + b^2}$.

7.5.2 Quantification of the Qualitative Indexes

In the energy efficiency quantitative analysis index system, some indexes are qualitative ones. Qualitative index refers to that kind of indexes which cannot be directly quantitatively analyzed, and which requires objective description and analysis to reflect the results of quantitative analysis. When converting the qualitative index into a quantitative one, the commonly adopted methods include classification rating, prognostic description and key events. The classification rating method can clearly define the objects to be quantitatively analyzed via data. It classifies the object into several grades. This method restricts the subjectivity of quantitative analysis, providing an objective basis to the quantitative analysis so that its results would be more reliable. Here, we adopt the classification rating method. We assign a score to each grade. The scores corresponding to the grades of "excellent, good, fair, poor" are "4, 3, 2, 1" respectively.

7.5.3 Uniformization of the Energy Consumption Evaluation Indexes

As for the evaluation index of the energy consumption evaluation system, those of positive effect on the evaluation results are called positive indexes, such as energy savings per unit product, equipment energy efficiency, and the manufacturing technique energy efficiency. Those of negative impact on the evaluation results are called negative indexes, and the bigger their figure, the poorer their influence on the evaluation results. Therefore, we need to run positive processing of the negative index per unit product's comprehensive energy consumption C_1, so that all energy consumption indexes are consistent. C_1^* Is the result after processing:

$$C_1^* = \frac{1}{C_1} \tag{7.3}$$

7.5.4 Constructing the Primitive Variable Matrix

We select n samples of the discrete manufacturing process's energy consumption. Suppose that each sample contains $p = 8$ evaluation index variables, then we can construct the energy consumption evaluation index variable matrix X:

$$X = \left(X_{ij}\right)_{n \times p} = \begin{bmatrix} X_{11} & X_{12} & \ldots & X_{1p} \\ X_{21} & X_{22} & \ldots & X_{2p} \\ \ldots & \ldots & \ldots & \ldots \\ X_{n1} & X_{n2} & \ldots & X_{np} \end{bmatrix}$$

where X_{ij} represents the energy consumption evaluation index variable value of sample no. i, section j.

7.5.5 Assigning Weight Function to the Variable Data

The primitive variable matrix X is converted into the standardized matrix Y. We use the expert method to assign weight functions to the evaluation index $C = \{C_1, C_2, \ldots, C_8\}$ in the evaluation index system. The weight function value is $W = \{1, 1, 0.8, 1.2, 1.2, 0.4, 1.2, 1.2\}$, respectively. Then we get the new variable matrix $u_i = z = [Y_1, Y_2, \ldots, 1.2 \times Y_8]$.

7.5.6 Solving the Principal Component

We calculate the standardized data's relative coefficient matrix R:

$$R = \left[r_{ij}\right]_{8 \times 8} = \frac{Z'Z}{n-1} \tag{7.4}$$

where Z' is the transposition matrix of the standardized matrix Z.

Its feature root $\lambda_1 \geq \lambda_2 \geq \lambda_3 \ldots \geq \lambda_8$ represents the extent of each principal component's impact on the evaluated object. We then solve the corresponding feature vector as $L_k = l_{k1}, l_{k2}, \ldots, l_{k8}$, and solve the principal component according to the standardized index variable:

$$F_k = l_{k1}Z_1 + l_{k2}Z_2 + \cdots + l_{k8}Z_8 \tag{7.5}$$

where $Z = [Z_1, Z_2, \ldots, Z_8]$, F_k is the no. k principal component, set as the new comprehensive energy consumption index.

7.5.7 Ascertaining the Number of Principal Components

When ascertaining the number of principal components, we should make sure that it fulfils the following principle [7]: minimizing the principal component's analysis information, while maintaining the maximum similarity with the primitive variable. Firstly, we give two important definitions:

1. The contribution rate of principal component no. k : $\frac{\lambda_k}{\sum_{i=1}^{p} \lambda_i}$
2. The first P_u principle component's cumulative variance contribution rate: $\frac{\sum_{i=1}^{k} \lambda_i}{\sum_{i=1}^{p} \lambda_i}$.

Based on the variance contribution rate of the first principal component, we select the evaluation model. In real-life situations, when we set the threshold value at 85%, we can ensure the minimum number of principal components when there is the least information loss, and thereby obtain stable evaluation results [8]. If the first principal component's variance contribution rate is larger than 85%, then the similarity adopts only the first principal component to conduct a comprehensive evaluation of the object. If the first principal component's variance contribution rate is no bigger than 85%, then we need to run a linear weighted comprehensive evaluation of the top k principal components according to the size order of the principal components. Such a comprehensive evaluation may include information of all the energy consumption indexes.

According to the size of the cumulative variances contribution rate, we ascertain the number of principal components. This chapter selects the number of principal components according to the contribution rate $\alpha \geq 85\%$.

$$\alpha = \frac{\sum_{i=1}^{k_{\text{principal}}} \lambda_i}{\sum_{i=1}^{p} \lambda_i} \tag{7.6}$$

where $k_{\text{principal}}$ is the number of principal components.

7.5.8 Ascertaining the Comprehensive Evaluation's Functional Analysis Energy Consumption

We calculate the weighted sum of η principal components and get the final evaluation function:

$$\begin{aligned} F = & \frac{\lambda_1}{\lambda_1 + \lambda_2 + \ldots + \lambda_k} F_1 + \frac{\lambda_2}{\lambda_1 + \lambda_2 + \ldots + \lambda_k} F_2 \\ & + \ldots + \frac{\lambda_k}{\lambda_1 + \lambda_2 + \ldots + \lambda_k} F_k \end{aligned} \tag{7.7}$$

where F is the comprehensive result of the discrete manufacturing system's energy consumption sample. Its size represents the energy consumption level. Through the formula of the principal components, we obtain the most closely relevant indexes.

7.6 Case Analysis

The machine tool manufacturing system is a major component of the discrete manufacturing industry. As the basis and the core of the manufacturing industry, this system is a leader during industrialisation and holds an important position in the national economy. Statistics reveal that, in the global machine tool industry, Japan is the number one producer which occupies 22% of the total output value of the global machine tool production. Germany is at the second place, with an occupancy rate of 21%, followed by the US (13%), Italy (11%), Switzerland (6%) and China (3%) [9]. China plays a decisive role in the global machine tool manufacturing industry. It creates the enormous value of output, though with so low an energy utilization ratio that there is much pressure on energy conservation. Effective energy efficiency quantitative analysis and energy conservation analysis are two vital tasks for the discrete manufacturing industry. This chapter chooses the workshop of one machine tool manufacturing industry as the object of analysis, using the principal component analysis method to conduct energy consumption analysis. For the energy consumption evaluation analysis to be effective, first, we need to obtain all the index values in the discrete manufacturing system's energy consumption quantitative evaluation index system. We can acquire the primitive processing energy consumption through the smart metre and the equipment system. The index data thus acquired are shown in Table 7.1. There the unit of energy consumption is KW H.

Table 7.1 The acquired data values

Data variables	Sample 1	Sample 2	Sample 2	Sample 4	Sample 5
$\sum_{i=1}^{n} E_i$	23.3	27.4	20.2	24.7	23.2
E_1	25.0	29.0	21.5	26.1	25.8
M	5	2	10	5	5
C_3	Fair	Fair	Good	Fair	Good
E_2	8.9	10.2	9.6	9.3	8.5
E_3	11.8	12.7	11.5	11.0	10.6
$\sum_{i=1}^{m} E_{2i}$	85.6	93.0	91.5	89.2	88.6
E_4	126	130	120	127	115
$\sum_{i=1}^{u} (C_{1i} M_i)$	110	102	122	131	125
E_5	170	156	185	179	196
E_6	40.1	55.3	45.2	48.9	50.1
E_7	26.0	24.5	28.0	25.0	22.5
E_8	22.5	21.0	22.0	21.0	23.0

Table 7.2 Each index's variable value

Index	Sample 1	Sample 2	Sample 3	Sample 4	Sample 5
C_1	23.3	27.4	20.2	24.7	23.2
C_2	8.5	3.2	13.0	7.0	13.0
C_3	Fair	Fair	Good	Fair	Good
C_4	0.75	0.80	0.83	0.85	0.80
C_5	0.70	0.72	0.76	0.70	0.77
C_6	0.65	0.65	0.66	0.73	0.63
C_7	0.58	0.50	0.45	0.51	0.46
C_8	0.21	0.17	0.27	0.19	0.20

7.6.1 Solving the Evaluation Index Value

Using the index calculation rules, we obtain each evaluation index figure, shown in Table 7.2:

7.6.2 Quantitative Processing of the Qualitative Indexes

The product energy utilization level is a qualitative index in the energy consumption evaluation system, which requires quantitative processing. The product energy utilization level corresponds to the values of 3, 3, 4, 3, and 4, respectively.

7.6.3 Uniformization of the Energy Consumption Evaluation Indexes

The negative index in the energy consumption evaluation system is the comprehensive energy consumption per unit product. We process the variable data using Formula (3.3) and get the following conversion result:

$$C_1^* = \{0.043, 0.036, 0.050, 0.040, 0.043\}$$

7.6.4 Constructing the Primitive Variable Matrix of the Evaluation Index

Based on the five samples and the eight evaluation index values of each sample, we construct the primitive variable matrix X:

$$X = \begin{bmatrix} 0.043 & 8.5 & 3 & 0.75 & 0.70 & 0.65 & 0.58 & 0.21 \\ 0.036 & 3.2 & 3 & 0.80 & 0.72 & 0.65 & 0.50 & 0.17 \\ 0.050 & 13.0 & 4 & 0.83 & 0.76 & 0.66 & 0.45 & 0.27 \\ 0.040 & 7.0 & 3 & 0.85 & 0.70 & 0.73 & 0.51 & 0.19 \\ 0.043 & 13.0 & 4 & 0.80 & 0.77 & 0.63 & 0.46 & 0.20 \end{bmatrix}$$

7.6.5 Standardization Processing of the Primitive Data

Solve the sample mean value of index no. j:

$$\bar{X}_j = \begin{bmatrix} 0.42 & 8.94 & 3.40 & 0.81 & 0.73 & 0.66 & 0.50 & 0.21 \end{bmatrix}$$

Solve the standard deviation of index no. j:

$$S_j = \begin{bmatrix} 0.005 & 4.18 & 0.55 & 0.38 & 0.033 & 0.04 & 0.05 & 0.04 \end{bmatrix}$$

Using Formula (7.2) to run standardization processing of the primitive variables. The result is:

$$Y = \begin{bmatrix} -0.117 & -0.11 & -0.73 & -1.48 & -0.90 & -0.36 & 1.55 & 0.053 \\ -1.25 & -1.37 & -0.73 & -0.16 & -0.30 & -0.36 & 0 & -1.00 \\ 1.48 & 0.97 & 1.10 & 0.63 & 0.90 & -0.10 & -0.97 & 1.65 \\ -0.47 & -0.46 & -0.73 & 1.16 & -0.90 & 1.72 & 0.19 & -0.48 \\ 0.117 & 0.97 & 1.10 & -0.16 & 1.206 & -0.88 & -0.78 & -0.21 \end{bmatrix}$$

7.6.6 Assigning Weight Function to the Variable Data

For the evaluation index $C = \{C_1, C_2, \ldots, C_8\}$, its corresponding weight function value is:

$$\{1, 1, 0.8, 1.2, 1.2, 0.4, 1.2, 1.2\},$$

After processing the variable matrix Y, we get the new variable matrix Z:

$$Z = \begin{bmatrix} 0.117 & -0.11 & -0.58 & -1.78 & -1.08 & -0.14 & 1.86 & 0.063 \\ -1.25 & -1.37 & -0.58 & -0.19 & -0.36 & -0.14 & 0 & -1.20 \\ 1.48 & 0.97 & 0.88 & 0.756 & 1.08 & -0.04 & -1.16 & 1.98 \\ -0.47 & -0.46 & -0.58 & 1.392 & -1.08 & 0.688 & 0.228 & -0.58 \\ 0.117 & 0.97 & 0.88 & -0.19 & 1.447 & -0.352 & -0.94 & -0.25 \end{bmatrix}$$

7.6.7 Solving the Principal Component and the Comprehensive Evaluation Function

1. We calculate the coefficient matrix relevant to the standardized matrix Z: R
   According to Formula (7.4), we solve the relevant coefficient matrix of the sample. The calculation result is as follows:

$$R = \begin{bmatrix} 1 & 0.868 & 0.730 & 0.101 & 0.544 & -0.16 & -0.36 & 0.975 \\ 0.868 & 1 & 0.887 & 0.074 & 0.745 & -0.30 & -0.49 & 0.748 \\ 0.730 & 0.887 & 1 & 0.217 & 0.963 & -0.45 & -0.80 & 0.654 \\ 0.101 & 0.074 & 0.217 & 1 & 0.179 & 0.667 & -0.64 & 0.151 \\ 0.544 & 0.745 & 0.963 & 0.179 & 1 & -0.57 & -0.135 & 0.480 \\ -0.16 & -0.30 & -0.45 & 0.667 & -0.57 & 1 & 0.139 & -0.11 \\ -0.36 & -0.49 & -0.80 & -0.64 & -0.85 & 0.139 & 1 & -0.36 \\ 0.98 & 0.748 & 0.65 & 0.150 & 0.480 & -0.11 & -0.36 & 1 \end{bmatrix}$$

2. Solving the principal component and the evaluation function.

We obtain $\lambda_1 = 4.706, \lambda_2 = 1.823, \lambda_3 = 1.240, \lambda_4 = 0.232$. The remaining feature roots are so small and gravitate towards 0, that they can be ignored here. The first principal component's contribution rate is

$$\frac{\lambda_1}{\sum_{i=1}^{8} \lambda_1} \times 100\% = 58.8\%.$$

According to the selection rules of the evaluation model, and given the fact that the first principal component's contribution rate is smaller than 85%, we need to conduct linear weighting comprehensive processing of the top principal components, based on the size of the principal components' contribution rates. We solve the first two principle components' cumulative variance contribution rate as:

$$\frac{\sum_{i=1}^{2} \lambda_i}{\sum_{i=1}^{8} \lambda_i} \times 100\% = 81.3\%$$

We solve the first three principle components' cumulative variance contribution rate as:

$$\frac{\sum_{i=1}^{3}\lambda_i}{\sum_{i=1}^{8}\lambda_i}\times 100\% = 96.8\% \geq 85\%$$

Since the first three principal components' cumulative variance contribution rate is larger than 85%, we set the number of principal components in the evaluation method as three, namely $k = 3$. The first principal component contains more information than the other two and plays the biggest role in the progress energy consumption analysis. Their corresponding feature vectors, L_1, L_2 and L_3 are shown below:

$$L_1 = \begin{bmatrix} 0.26 \\ -0.02 \\ 0.27 \\ 0.35 \\ 0.48 \\ 0.44 \\ 0.44 \\ 0.27 \end{bmatrix},\ L_2 = \begin{bmatrix} -0.34 \\ 0.30 \\ -0.25 \\ 0.31 \\ 0.13 \\ 0.17 \\ -0.39 \\ 0.54 \end{bmatrix},\ L_3 = \begin{bmatrix} 0.19 \\ 0.37 \\ 0.07 \\ -0.40 \\ 0.20 \\ -0.39 \\ 0.29 \\ 0.39 \end{bmatrix}$$

Based on Formula (7.5), we get the principal component's expression:

$$\begin{aligned} F_1 &= 0.26Z_1 - 0.02Z_2 + 0.27Z_3 + 0.35Z_4 + 0.48Z_5 \\ &\quad - 0.14Z_6 + 0.44Z_7 + 0.27Z_8 \\ F_2 &= -0.34Z_1 + 0.30Z_2 - 0.25Z_3 + 0.31Z_4 + 0.13Z_5 \\ &\quad + 0.17Z_6 - 0.39Z_7 + 0.54Z_8 \\ F_3 &= 0.19Z_1 + 0.37Z_2 + 0.07Z_3 - 0.40Z_4 + 0.20Z_5 \\ &\quad - 0.39Z_6 + 0.29Z_7 + 0.39Z_8 \end{aligned}$$

Based on Formula (7.7), we get the energy consumption comprehensive evaluation function:

$$F = 0.588F_1 + 0.228F_2 + 0.155F_3$$

7.6.8 Results of the Energy Consumption Analysis

We put the sample data into the principal components' expressions, solve the final evaluation of each sample, and rank the results of sample evaluation. The results are listed in Table 7.3:

Table 7.3 Results of the principal components' evaluation

Sample	The first principal component	The second principal component	The third principal component	Comprehensive results	Ranking
1	0.12	1.10	0.01	0.40	1
2	−0.46	−1.09	1.94	−0.24	4
3	1.31	−1.10	−2.60	−0.13	3
4	0.13	0.58	0.63	0.35	2
5	−1.10	0.51	0.02	−0.38	5

Through improving the principal component analysis, we obtain the analysis results:

1. From the ranking of Table 7.3 analysis results, we can see that, during the periods of sample 2, sample 3 and sample 5, the machine tool manufacturing energy consumption levels are so low that effective optimization measures become necessary, to improve the comprehensive utilization ratio of the energy efficiency.
2. From the principal component's expression F_1 we know that the energy consumption evaluation indexes of positive correlation with the first principal component are the collective equipment energy efficiency and the manufacturing technique energy efficiency, which play vital roles in the energy consumption level. Therefore, we should emphasize these two energy efficiency indexes in our optimized method. From the principal component's expression F_2 we know that the energy consumption evaluation index of positive correlation with the second principal component is the manufacturing resources scheduling energy efficiency, which is also an index that demands our attention.

Normally, energy efficiencies of the equipment, the manufacturing technique energy efficiency and the production resources scheduling have a huge impact over the manufacturing industry's energy consumption system [10]. The fact that the evaluation results match the reality testifies the reasonableness of the improved method.

7.7 Simulation Analysis

The purpose of the simulation analysis is to validate the stability of the discrete energy consumption analysis method, which is based on the improved principal component analysis method. We compare the sensitivity of index values, which deviate a lot from the average sample values to the traditional principal component analysis method. We adjust C_4 in Sample 3, from 0.8 to the two abnormal values, 0.4 and 0.1, respectively. Then we apply these values into the MATLAB tool for simulation experiment. The analysis results are shown in Figs. 7.1 and 7.2:

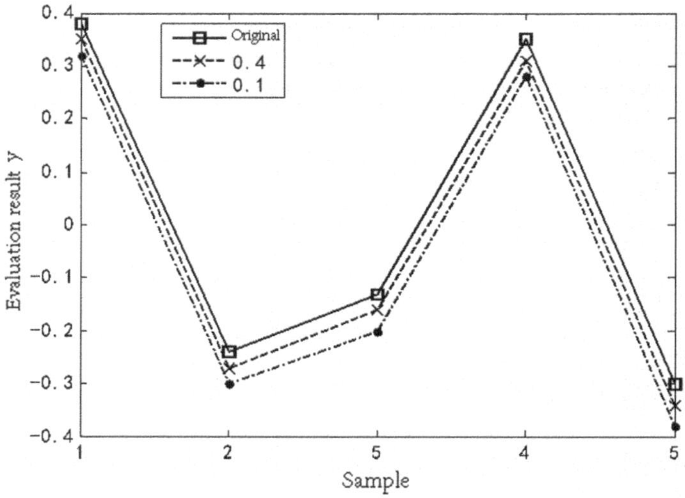

Fig. 7.1 Energy consumption analysis results of the improved principal component method

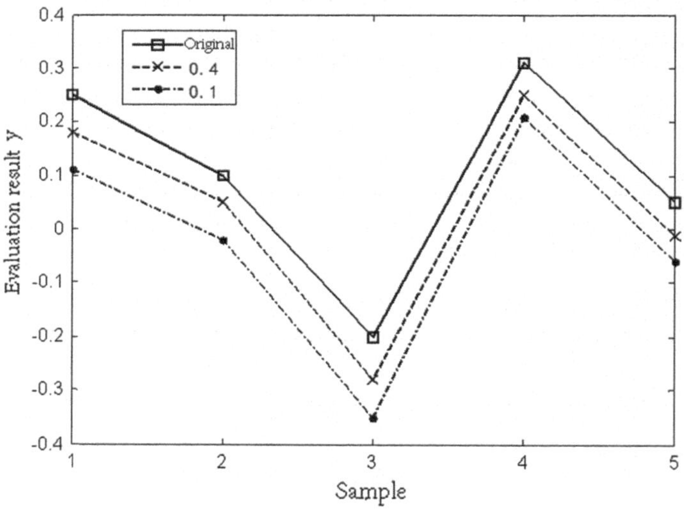

Fig. 7.2 Energy consumption analysis results of the traditional principal component analysis

From Fig. 7.1, we can see how the analysis results of the five samples change. The principal component analysis of energy consumption has stable results, with sensitivity of around 5–13%. From Fig. 7.2, we can see that the results of the traditional principal component analysis fluctuate a lot, with the sensitivity ranging between 15 and 21%. Therefore, one certain abnormal index value has a slight impact on the improved principal component analysis of energy consumption, which proves that this method is more stable. Moreover, since it is the data of sample 3 that we alter, and we do observe the most obvious fluctuation of sample 3 in Fig. 7.1, this proves the reasonableness of this method.

7.8 Conclusion

Energy consumption evaluation of the discrete manufacturing system provides a basis for the job shop to optimize its energy conservation. Taking into consideration the complex features of the discrete manufacturing system's energy consumption, this chapter proposes a discrete energy consumption analysis method based on the improved principal component analysis. The minimum loss of information guaranteed, we conduct dimension reduction of the high-dimensional discrete manufacturing system, while improving the weight function determination method and the non-dimensionalization method in the traditional principal component analysis. Through such improved methods, we can directly reflect the energy consumption condition at each stage of the discrete manufacturing system, while generating closely relevant evaluation indexes which could help enterprises optimize their energy consumption with better focus.

References

1. Han Y, Geng Z, Liu Q (2014) Energy efficiency evaluation based on data envelopment analysis integrated analytic hierarchy process in ethylene production. Chin J Chem Eng 22(11):1279–1284
2. Rad MB, Moghadam MP, Sheikh-El-Eslami MK (2007) Fuzzy evaluation of energy efficiency improvement impact on load shape. In: Power tech (PT). Lausanne: IEEE, pp 1429–1434
3. Cao Z, Ma L, Wang N et al (2011) An entropy-based evaluation method of maintenance support system. In: Reliability, maintainability and safety international conference (ICRMS). Guiyang: IEEE, pp 842–848
4. Li TJ, Du Q (2013) Abstract principal component analysis. Sci China Math 56(12):2783–2798
5. Yan JH, Li L (2013) Multi-objective optimization of milling parameters the trade-offs between energy, production rate and cutting quality. J Clean Prod 52(1):462–471
6. Ruan Y, Chen HW et al (2014) Quantum principal component analysis algorithm. Chin J Comput 37(3):666–676
7. Wang R, Gong X, Xu M et al (2015) Fault detection of flywheel system based on clustering and principal component analysis. Chin J Aeronaut 28(6):1676–1688
8. Min Z, Wang GA, Shuguang HE et al (2014) Modified multivariate process capability index using principal component analysis. Chin J Mech Eng 27(2):249–259
9. Wang Q, Wang X, Yang S (2014) Energy modeling and simulation of flexible manufacturing systems based on colored timed petri nets. J Ind Ecol 18(4):558–566
10. Yang YL, Tai HX, Shi T (2012) Weighting indicators of building energy efficiency assessment taking account of experts' priority. J Central South Univ 19(3):803–808

Chapter 8
Static Optimization and Scheduling of the Discrete Manufacturing System's Energy Efficiency Based on the Integration of Knowledge and MOPSO

8.1 Introduction

Scholars both at home and abroad have undertaken numerous research and proposed many optimization methods to solve the issue of workshop scheduling. Currently, there are such static optimization algorithms for discrete manufacturing system as genetic [1], particle swarm [2], artificial bee colony [3] and ant colony [4] intelligent algorithms, which have accumulated rich and valuable experience for their application in the discrete workshop's scheduling. The core of knowledge reasoning lies in using previous practical experience to solve new problems. Knowledge-based work is to use and reuse knowledge [5]. Through uncovering existing data and knowledge, it reduces the demands for experience in scheme design and field of knowledge, so as to improve design efficiency, shorten development cycle and improve the quality of the scheme. Here, knowledge reasoning has features of similarity and rapidity. The flexibility and practicality of intelligent algorithm can be a good match to make up for the disadvantage of knowledge reasoning.

Targeting at the complexity and restrictiveness of discrete manufacturing workshop, this chapter sets the workshop's minimum total energy consumption as its object of optimization and proposes a discrete knowledge MOPSO algorithm for solving the discrete workshop's energy efficiency optimization. This algorithm introduces parameters during optimization to balance local search and global search, thereby improving algorithm convergence and optimizing capability. By adding a discrete process into the algorithm, we maintain the features of algorithm convergence and optimizing capability while equipping it with the capability to process discrete problems. Through carrying out simulation testing with workshop production data, we compare the optimization results of particle swarm and genetic algorithms with that of this chapter's discrete knowledge MOPSO algorithm, testifying the reasonableness and effectiveness of algorithm proposed in this chapter.

Y. Wang et al., *Quantitative Analysis and Optimal Control of Energy Efficiency in Discrete Manufacturing System*, https://doi.org/10.1007/978-981-15-4462-0_8

8.2 Problem Description

As Fig. 8.1 illustrates, the discrete manufacturing system is made up of the processing task distribution stage prior to the system's operation and the processing stage during the operation. It is complex, has multiple tasks and multiple restraints. Different scheduling task has different scheduling target, producing different scheduling demands and requirements for different resources, hence the difference in scheduling production. Optimization of the discrete manufacturing system's energy efficiency focuses on reducing energy consumption and improving energy consumption efficiency. Through scheduling a reasonable production scheme of manufacturing and processing, it achieves the purpose of optimizing energy efficiency. Therefore, the objectives of optimization are mainly minimization of the facility's total energy consumption, minimization of production and utility rate of the facility, followed by the solution of scheduling problems for purpose of optimizing energy efficiency.

8.2.1 *Assumed Conditions*

We make the following assumed conditions for the workshop's production process:

1. The order of processing workpiece and procedure at a machine is determined exclusively by the process route;
2. Before a workpiece or procedure has finished processing procedure, the processing task cannot be terminated (unless for machine fault);
3. A workpiece can be processed only at one machine at one moment;
4. The processing priority of all workpieces is identical at time zero;
5. For different workpieces, the processing order priory of each of their procedure is identical.

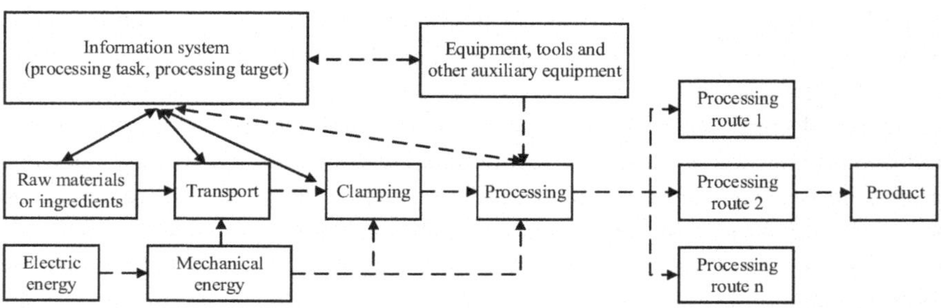

Fig. 8.1 Flow diagram of energy consumption in discrete manufacturing system's course of processing

8.2.2 Definitions of Relevant Parameters

m: represents the number of processing equipment
N: represents the number of workpieces to be processed
c: represents the abbreviation of $ijegk$, i, e represents the workpiece number, j, g represents the procedure number, k presents the device number
h: represents the device number
$M = \{M_k \mid 1 \leq k \leq m\}$: represents the device set
$J = \{J_i \mid 1 \leq i \leq n\}$: represents the workpiece set
s: represents the largest procedure number of all workpieces
O_{ij}: represents procedure j of workpiece J_i
$M_{ij} = \{M_k \mid X_{ijk} = 1\}$: represents the available device set of procedure O_{ij} for workpiece J_i
P_{ijk}: represents a certain procedure of the workpiece can be processed at multiple devices
R_c: two workpieces can be processed at one device
t_{ijk}: represents the processing time of procedure O_{ij} at device M_k
S_{ijk}: represents the start time for processing procedure O_{ij} at device M_k
E_{ijk}: represents the finish time for processing procedure O_{ij} at device M_k
w_k: represents the energy consumption of a certain procedure of a certain workpiece at device M_k
E_i: represents the total energy consumption of all workpieces at device M_k
$E_{\text{total}-ec}$: represents the total energy consumption for processing all workpieces;O_i: represents the procedure set of workpiece J_i
S_{ijk}: represents the start time for processing procedure O_{ij} at device M_k
E_{ijk}: represents the finish time for processing procedure O_{ij} at device M_k

$$x_{ijk} = \begin{cases} 1, & \text{Process } O_{ij} \text{ is processed by equipment } M_k \\ 0, & \text{else} \end{cases}$$

$$R_c = \begin{cases} 1, & O_{ij} \text{ and } O_{eg} \text{ are processed on } M_k,\ O_{ij} \text{ is processed first} \\ 0, & \text{else} \end{cases}$$

8.2.3 Description of Energy Consumption Knowledge

In current scheduling optimization of the manufacturing industry, how to dig, store and utilize data in the practical production through the data integration technology is the trend for the contemporary development of intelligent manufacturing. These

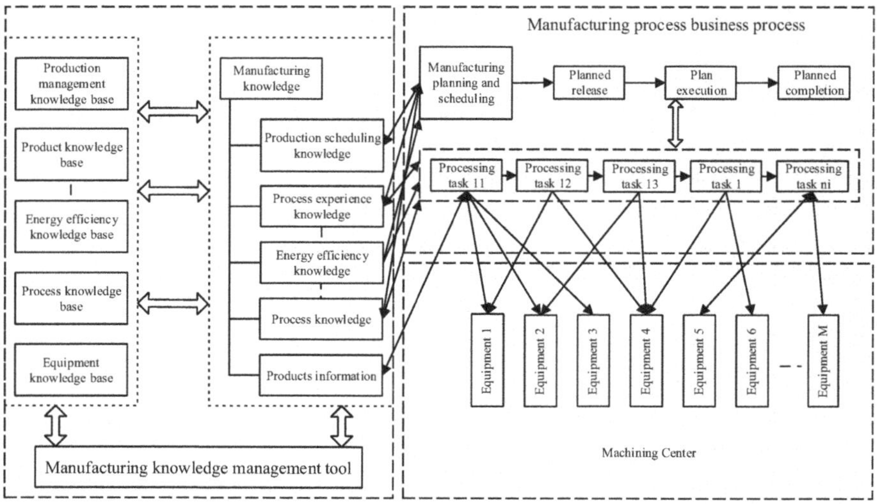

Fig. 8.2 Integration diagram based on production knowledge

production data contain values and rules related to the optimization process of production energy efficiency, which can help enterprises to solve optimization problems in actual production. Figure 8.2 illustrates the integration of production knowledge. When solving complex optimization problems, this chapter undermines relevant knowledge of to-be-settled problems from historical experience, before arriving the best solution via the knowledge integration mechanism.

The basic principle of knowledge reasoning is preload large amounts of existing and effective data set into knowledge base, then search knowledge data group according to the similarity with the requirements and features of a certain scheme, conduct necessary revision, combination and processing, and thus forming a new scheme while expanding the new knowledge data group to the knowledge base.

1. **Methods of Representing of Knowledge Related with Discrete Manufacturing Production Energy Consumption**

The discrete manufacturing system is a set of combinatorial optimization of complex problems. This chapter uses parameters in the Triple Representation of Object-Attribute-Value ($a|b|c$ presentation method) to represent the problem of discrete manufacturing energy consumption. Here, $a = (a_1, a_2, a_3, a_4, a_5, \ldots, a_n)$ represents workpiece feature, a_1 represents workpiece name, a_2 represents the number of workpieces, a_3 represents the workpiece process matrix, a_4 represents the processing time matrix of workpiece procedure, a_5 represents the processing energy consumption of workpiece procedure; b represents the serial number of available processing machine in the workshop; $c = (c_1, c_2, c_3, \ldots, c_n)$ represents the production processing matrix; c_1 represents the minimalized maximum completion time, c_2 represents the total volume of minimalized production energy consumption, c_3 represents the minimalized machine's total load.

Since in the discrete manufacturing workshop, there are m sets of available machines for n workpieces waiting for processing, each workpiece has k procedure, with different operation paths for different workpiece, and different processing machines equipped for different operational paths, even if the identical processing procedure can choose difference machines to process. Due to the difference among machines, the different processing routes and scheduling plans for to-be-processed workpiece would result in different processing time, processing energy consumption, standby time and standby energy consumption, thereby creating space for energy conservation for optimization of scheduling energy efficiency in discrete workshop. Therefore, how to reasonably and effectively arrange the machine and processing order corresponding to the workpiece's processing technique directly influences the optimizing efficiency of energy efficiency in the discrete workshop. Based on the features of corresponding arrangement between the discrete manufacturing workshop's procedure and machine, we construct a knowledge data system in a method of hierarchical design. Then, based on the features of the discrete manufacturing processing, we set the processing route matrix made up of each and every workpiece's processing route A (i as workpiece serial number, j as each procedure number, k as processing machine, without position zero-padding) as

$$\begin{bmatrix} a_{11k} & a_{12k} & \cdots & a_{1jk} & \cdots \\ \vdots & \vdots & \ddots & \vdots & \vdots \\ a_{i1k} & a_{i2k} & \cdots & a_{ijk} & \cdots \\ \vdots & \vdots & \ddots & \vdots & \vdots \\ a_{n1k} & a_{n2k} & \cdots & a_{njk} & \cdots \end{bmatrix} \tag{8.1}$$

The processing time matrix of each operation T (i as workpiece serial number, j as procedure number, k as processing machine, without position zero-padding) as

$$\begin{bmatrix} T_{11k} & T_{12k} & \cdots & T_{1jk} & \cdots \\ \vdots & \vdots & \ddots & \vdots & \vdots \\ T_{i1k} & T_{i2k} & \cdots & T_{ijk} & \cdots \\ \vdots & \vdots & \ddots & \vdots & \vdots \\ T_{n1k} & T_{n2k} & \cdots & T_{njk} & \cdots \end{bmatrix} \tag{8.2}$$

The energy consumption matrix made up of each workpiece's processing route W (i as workpiece serial number, j as procedure number, k as processing machine, without position zero-padding) as

$$\begin{bmatrix} W_{11k} & W_{12k} & \cdots & W_{1jk} & \cdots \\ \vdots & \vdots & \ddots & \vdots & \vdots \\ W_{i1k} & W_{i2k} & \cdots & W_{ijk} & \cdots \\ \vdots & \vdots & \ddots & \vdots & \vdots \\ W_{n1k} & W_{n2k} & \cdots & W_{njk} & \cdots \end{bmatrix} \tag{8.3}$$

2. **Discrete Manufacturing Production Knowledge Retrieval Framework**

Knowledge retrieval requires a method of hierarchical retrieval into set threshold to discover the information group of maximum matching degree with the sought target. The inquiry method first conducts component sets' matching, then components' procedure set matching, next device set matching and finally object set matching, so as to arrive at the target with matching information of maximum similarity.

3. **Evaluation of Discrete Manufacturing Production Knowledge**

If G is target component set, H is scheduling component set in history, d is the distance between object component set and scheduling component set in history, then we adopt the following formula to calculate the similarity of knowledge.

$$d = \sum_{i=1}^{N} \omega_i \left| p_{i\text{target}} - \bar{p}_i \right| \tag{8.4}$$

where N is the size of the component set, i.e. the total number of components, ω_i is weight value of no. i component, $p_{i\text{target}}$ is no. i component's target value at the best level, $\bar{p}_i$ is the historical value of no. i component's priority level. The smaller d is the higher matching degree between the scheduling component set in history and the current target component set.

4. **Revision of Discrete Manufacturing Production Knowledge**

Knowledge revision is to search the mismatch between the matched historical scheduling information with the current scheduling information. We need to find a data set or multiple data set of the highest similarity in the sought knowledge database. And through revising and adjusting the knowledge data group scheme, for the purpose of being applicable to solve the current problem, we obtain a solution agenda. Here, we adopt the heuristic rule (HR algorithm) to revise and adjust the inquired knowledge information, so as to make it fulfil the current production scheduling demands at the fastest speed and with the highest efficiency. Integrate the heuristic rule according to the workshop's scheduling tasks, we revise and adjust the knowledge data set so that meets the current production scheduling demands at the fastest speed and with the highest efficiency. The heuristic rule knowledge revision is as illustrated in Fig. 8.3.

From historical scheduling information, we know that in $\begin{pmatrix} 1\ 2\ 3 \\ 2\ 2\ 3 \end{pmatrix} \frac{\text{Workpiece number}}{\text{Process quantity}}$, workpiece number ID is $\begin{pmatrix} 1\ 2\ 3 \end{pmatrix}$, while the current scheduling workpiece ID is

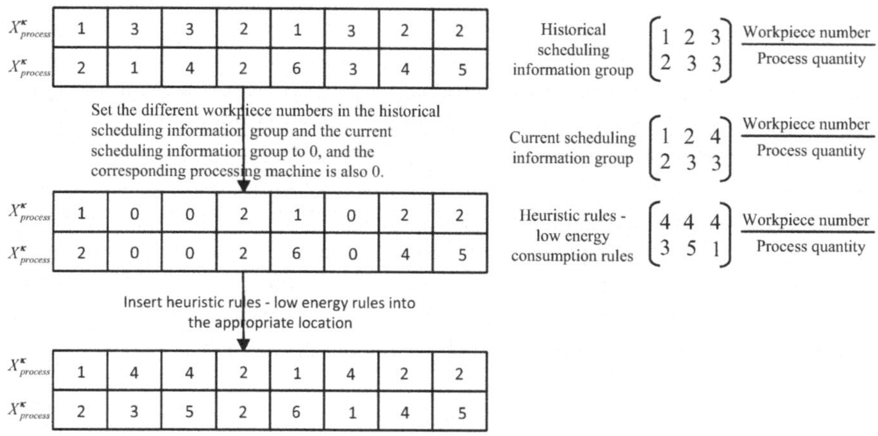

Fig. 8.3 Heuristic rule knowledge revision

$\left(1\ 2\ 4\right)$, hence the need to revise historical scheduling information to satisfy the current scheduling demands. Here, we use the low energy consumption rule in the heuristic rule to retrieve machine sequence $\left(3\ 5\ 1\right)$ which is of the lowest energy consumption consumed by three procedures of processing workpiece 4. Then, we insert it into workpiece $ID = 3$ which is incompatible with the current scheduling information and arrive at the best initial population of the current scheduling scheme in the historical scheduling information via knowledge retrieval.

8.2.4 Energy Efficiency Optimization Objective Function

Based on the production features of discrete manufacturing workshop, we construct objective functions from the two respective perspectives of order completion time and total production energy consumption. The order completion time uses the device's minimalized maximum completion time C_M, minimalized total production energy consumption $E_{\text{min-total}}$ and minimalized machine's total load W_T. The mathematical description is as follows:

$$C_M = \min\left\{\max\left\{t_{ijk} | M_k \in \{M_1, M_2, L, M_m\}\right\}\right\} \tag{8.5}$$

$$E_{\text{min-total}} = \min\left(\sum_{i=1}^{m} E_i\right) \tag{8.6}$$

$$W_T = \min\left\{\sum_{k=1}^{m}\sum_{i=1}^{n}\sum_{j=1}^{n_i} t_{ijk} x_{ijk}\right\} \tag{8.7}$$

Based on the energy consumption model formula from (3.18) to (3.20) in Chap. 3, we know that energy consumption during manufacturing has a positive correlation with the critical machine load. As for production process comprising multiple procedures and working steps, during the stage of successive workpiece processing, the load of processing machine plays a decisive role in the process's energy consumption. Notably, the heavier the load, the larger the energy consumption. Considering that currently little research has been devoted to solve the FJSP issue of energy efficiency index optimization, this chapter converts the minimalized total production energy consumption E into the optimization of the critical machine load index W_M, for the convenience of comparability between multiple-target optimization algorithm with other algorithms. Critical machine load (W_M) can be represented as

$$W_M = \max_{1 \le k \le m} \sum_{i=1}^{n} \sum_{j=1}^{n_i} t_{ijk} x_{ijk} \tag{8.8}$$

8.2.5 *Constraints of Energy Consumption Optimization*

1. Machine Constraint

$$E_{ijk} - E_{ejk} \ge t_{egk}, \; R_v = 1, \; P_{ijk} = P_{egk} = 1 \tag{8.9}$$

Formula (8.9) indicates that each device can process only one workpiece at the same moment.

2. Processing Constraint

$$E_{ijk} - S_{ijk} = t_{ijk}, \; P_{ijk} = 1 \tag{8.10}$$

Formula (8.10) indicates that when a certain procedure begins processing, it cannot be terminated.

3. Operational Path Constraint

$$S_{ijk} - E_{i(j-1)h} \ge 0, \; P_{ijk} = P_{i(j-1)h} = 1 \tag{8.11}$$

Formula (8.11) indicates that the same workpiece procedure is processed according to the operational path.

8.3 The Process of Integrating Energy Consumption Knowledge Retrieval with Intelligent Optimization Algorithm

For the sole target of energy efficiency optimization in the discrete workshop, we design the optimization mechanism combining knowledge with MOPSO algorithm. First, we use hierarchical knowledge retrieval to conduct workpiece set matching, procedure set matching, machine set matching and priority set matching, retrieving the information group of the highest historical similarity. Then, based on the implementation result or satisfaction level, we evaluate the solved knowledge scheme, so as to verify whether the historical scheduling information data set can fulfil the demands of knowledge retrieval. If the answer is yes, then push scheduling knowledge; if no, then conduct knowledge revision according to the heuristic rule until it meets the demands, then export knowledge results and convert them into optimized initial population. Finally, for the energy consumption target function, we use MOPSO algorithm to conduct algorithm optimization of the initial population so that it reaches the optimal status. The process of integrating knowledge retrieval with optimization algorithm is shown in Fig. 8.4.

The mechanism of integrating energy consumption knowledge retrieval with intelligent optimization algorithm illustrated in Fig. 8.4 first generates a group of initial optimized information befitting this discrete manufacturing system in the historical scheduling information, so as to make energy efficiency practical and effective in the initial optimization. As a result, this method is better optimized and has faster convergence than the primary colony algorithm at the beginning stage of the optimized algorithm. But experiments find out that the initial optimized population revised by heuristic rule has not reached the best solution, therefore after obtaining the initial optimized population, it can be mixed with the initial optimized pigeon population. On the one hand, this mixture endows this optimization with overall population information, which ensures the comprehensiveness of optimization; on the other hand, further optimization of the MOPSO algorithm enhances the potential of energy efficiency optimization. As a result, the optimization design mechanism which integrates knowledge with MOPSO algorithm can combine their strengths of both knowledge retrieval and algorithm optimization, enabling energy efficiency optimization to reach its best.

8.3.1 Scheduling Knowledge Retrieval

Knowledge retrieval theory uses knowledge retrieval method to achieve storage, retrieval and application of knowledge. Knowledge retrieval can find the knowledge required among massive fragments of data and information. In this chapter, the discrete manufacturing workshop's knowledge retrieval system adopts the idea of

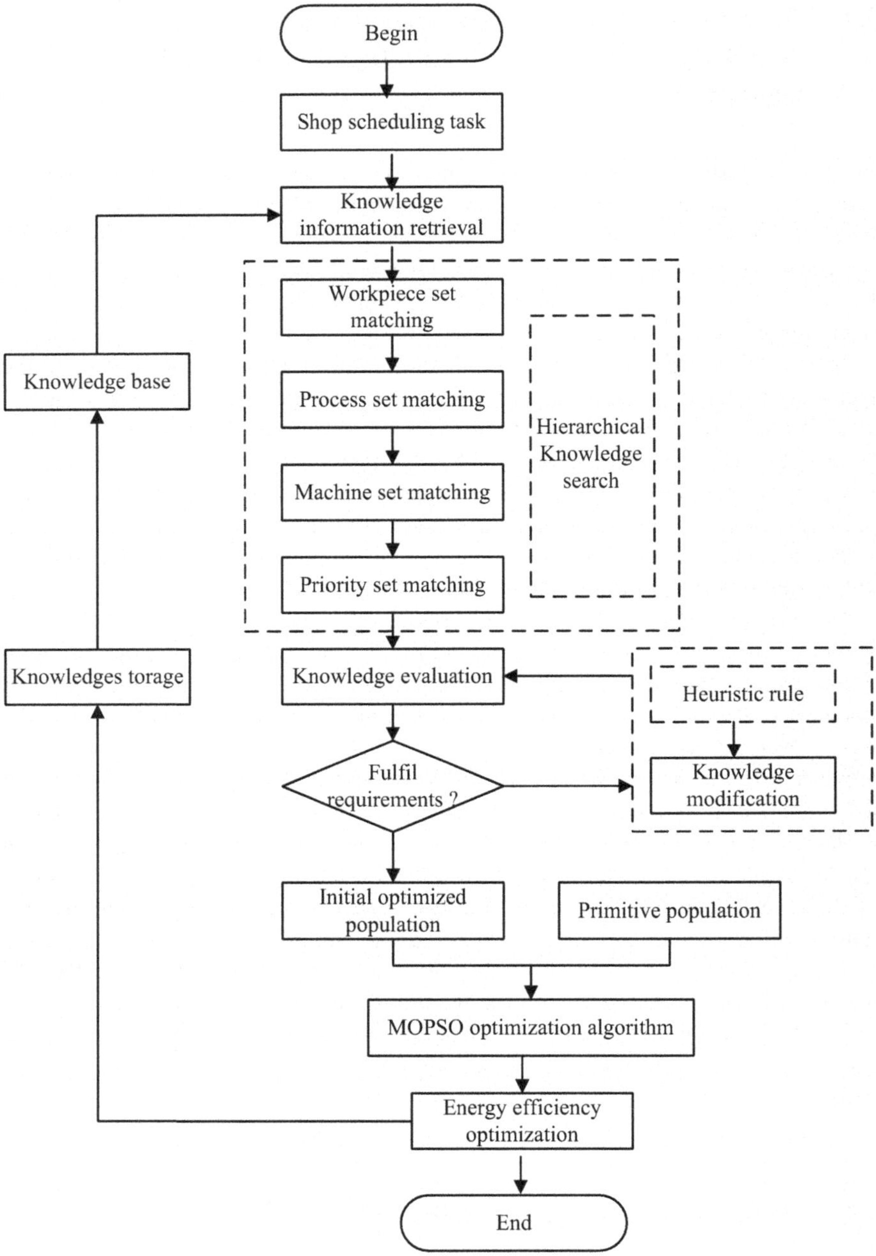

Fig. 8.4 Flow chart of integrating knowledge retrieval with MOPSO algorithm energy efficiency optimization

hierarchical knowledge retrieval, retrieving historical scheduling information of high similarity with the current scheduling from historical scheduling database.

1. **Workpiece set matching**

We define the target component set as G, component set of historical scheduling as H, then we define the similarity of these two sets as

$$\mathrm{Sim}(G, H) = \frac{G \cap H}{|G|} \tag{8.12}$$

Formula (8.12) reveals that the similarity between set G and set H is the ratio of the number of components shared by both sets to the target component sets G. This formula is asymmetric, i.e. $\mathrm{Sim}_{ID}(G, H) = \mathrm{Sim}_{ID}(H, G)$. When setting the similarity threshold, we should set relatively high threshold value (the closer to 1, the high degree of similarity). The reference value of a too low threshold historical scheduling knowledge is quite trivial.

2. **Workpiece set matching**

After ascertaining the degree of similarity between the target component set and the historical scheduling component set, we should ascertain whether the to-be-processed procedures are similar. Since for even one single component, the to-be-finished procedure at each time of scheduling is different. Even if the two components to be processed are the same, their procedure set may not identical. Then, we define the degree of similarity between these two sets as

$$\mathrm{Sim}(G, H) = \frac{G \cap H}{|G|} \tag{8.13}$$

Formula (8.13) reveals that the similarity between set G and set H is the ratio of the number of procedures shared by both sets to the target procedure sets G. The threshold value of the procedure set can be relatively low.

3. **Machine set matching**

Taking into consideration reasons like failure or maintenance of discrete manufacturing workshop devices, the machine tools available are not identical at each time of scheduling, hence the need to match machine set. Then, we define the degree of similarity between these two sets as

$$\mathrm{Sim}(G, H) = \frac{G \cap H}{|G|} \tag{8.14}$$

Formula (8.14) indicates that the degree of similarity between set G and set H is the ratio of the number of machines shared by both sets to the target machine sets G. The threshold value of the procedure set adopts only the historical scheduling of the highest degree of similarity, without setting the threshold value.

4. **Priority set matching**

Since different components require different processing time and delivery time when scheduling the production arrangement, we need to conduct priority matching. The priority matching formula is

$$P_i = \frac{\sum t_{ij}}{t_{\text{day}} \times t_{\text{work}}} \in (0, 1) \tag{8.15}$$

where $\sum t_{ij}$ represents the length of time to complete producing this type of component, t_{day} represents the number of days prior to the completion date, t_{work} represents the daily working hours. From Formula (8.15), we get $P_i \in (0, 1)$. The closer this parameter approximates 1, the more urgent is the processing for corresponding components, the processing of which should be prioritized; the closer this parameter approximates 0, the less urgent is the processing for the corresponding components, which can be postponed for consideration.

8.3.2 *Scheduling Knowledge Evaluation*

After knowledge retrieval, we need to examine the historical scheduling information data set based on the solved knowledge scheme's effect of implementation or level of satisfaction, thereby making the judgement whether it fits the requirements of knowledge retrieval.

Assume the target component set is $G = \{(A, 0.9), (B, 0.5), (C, 0.1)\}$

Component set of historical scheduling: $H = \{(A, 0.6), (C, 0.1), (D, 0.3)\}$,

Then, the component set after their respective unification is

$$G = \{(A, 0.9), (B, 0.5), (C, 0.1), (D, 0)\}$$

$$H = \{(A, 0.6), (B, 0), (C, 0.1), (D, 0.3)\}$$

For G, the component's weight evaluation formula is

$$w_i = \frac{p_i}{\sum p_i} \tag{8.16}$$

where p_i represents the priority level of component no. i.

Then, the priority level corresponding to component no. G is $w = (0.6, 0.3333, 0.6666, 0)$.

The distance between G and H is

$$\begin{aligned} d &= \sum \omega_i \left| P_i^{\text{Objective}} - P_i^{\text{History}} \right| \\ &= 0.6|0.9 - 0.6| + 0.33|0.5 - 0| \end{aligned}$$

$$+ 0.66|0.1 - 0.1| + 0|0 - 0.3|$$
$$= 0.3466 \quad (8.17)$$

8.3.3 Scheduling Knowledge Revision

Knowledge revision is to search the mismatch between the matched historical scheduling information with the current scheduling information. We need to find a data set or multiple data set of the highest similarity in the sought knowledge database. And through revising and adjusting the knowledge data group scheme, for the purpose of being applicable to solve the current problem, we obtain a solution agenda. Here, we adopt the heuristic rule (HR algorithm) to revise and adjust the inquired knowledge information, so as to make it fulfil the current production scheduling demands at the fastest speed and with the highest efficiency.

1. **Rule of low energy consumption**

For the purpose of optimizing energy efficiency, we prefer to choose a machine with the lowest processing energy consumption for processing when rescheduling, so as to ensure low energy consumption during the entire course of processing the components. The priority level of the procedure is expressed as follows:

$$Z_i = M_{w_\tau} \quad (8.18)$$

where $w_\tau = \min(w_m), m \in$ {Machinable machine set}.

2. **Rule of short processing time**

The course of processing can effectively shorten processing time, achieving the purpose of reducing the energy consumption of processing. When scheduling arrangements, we adopt the principle of maximum processing time t, choosing the machine with minimum processing hours for processing. The mathematical expression for the priority level of procedure is

$$Z_i = t_\tau \quad (8.19)$$

where $t_\tau = \min(t_m)$, $m \in$ {Machinable machine set};

3. **Rule of short relaxation time**

Compared with delivery time d_i, the shorter the relaxation time applied between the different component processing procedures after disturbance time t, the higher the priority level. The mathematical expression for the procedure's priority level is

$$Z_i = d_i - t - \sum_{q=j+1}^{n_i} \max(t_{iq}) \quad (8.20)$$

where n_i represents the procedure number of the workpiece i, $\max(t_{iq})$ represents the maximum processing time of procedure q;

8.3.4 Update of Scheduling Knowledge

When the system solves new scheduling information in a new issue, it needs to load high-quality information set betting the new issue into the database. During this process, redundant storage of knowledge may easily occur and lead to decreased retrieval efficiency. At this time, we need to conduct filtering for information data group.

Assume the knowledge information base as $U = (X_1 \ldots X_i \ldots X_m), i \in [1, m]$.

where X_i is the source information data set, the degree of similarity between target information data set Y and X_i is $\varepsilon_i \in [0, 1]$, η is the defined threshold.

There are following cases of knowledge storage strategies:

1. $\forall \varepsilon_i = 0$, incompatibility between the newly found knowledge and the knowledge base, update knowledge base;
2. $\forall \varepsilon_i > \eta$, wide discrepancy between the newly found knowledge and the knowledge base of the same knowledge type, update knowledge base;
3. $\forall \varepsilon_i < \eta$, basic similarity between the newly found knowledge information and the knowledge information base, rewrite $\max(\varepsilon_i)$ knowledge as newly found knowledge;
4. $\forall \varepsilon_i = 1$, incompatibility between the newly found knowledge and the knowledge base, update knowledge base.

8.4 Discrete MOPSO Intelligent Optimization Algorithm

The issue of flexible workshop scheduling belongs to the issue of discrete optimization, and a discretized particle swarm optimization algorithm is needed to solve this kind of problems. Meanwhile, since this scheduling problem involves the optimization of multiple indexes, among which there exist mutual conflicts, we need to design a method capable of dealing with the evaluation and update of multiple targets. Therefore, in this chapter, we propose a multi-objective particle swarm optimization (MOPSO) algorithm based on variable neighbourhood search, in order to solve three issues of target flexible workshop scheduling—the longest machine completion time f_1, the production process energy consumption f_2 and the machine's total load f_3.

8.4.1 Implementation of the MOPSO Algorithm

The basic particle swarm optimization algorithm is only fit for solving continuous optimization problems. Therefore, we need to design a particle swarm optimization algorithm with discrete practice to solve the flexible workshop's scheduling problem. Drawing on the ideas of document [6], this chapter designs a discrete operator based on flexible workshop scheduling to process discrete variables, based on which we propose a multi-objective particle swarm optimization algorithm. This algorithm adopts, rather than particle speed update, the current location of the particle. And it updates the particle's location through information exchange between the individual best location and overall best location with the discrete operator. The location update formula of the particle i is as follows:

$$x_i^{t+1} = \omega \otimes f_1\left(x_i^t\right) + c_1 \otimes f_2\left(x_i^t, p_{best,i}^t\right) + c_2 \otimes f_2\left(x_i^t, g_{\text{best}}^t\right) \tag{8.21}$$

where $\otimes$ represents probabilistic operations. $+$ represents finishing the current operation and begins successive operations. ω represents selection probability. c_1 and c_2 represent individual cognitive probability and social cognitive probability. In this algorithm, c_2 equals $\bar{c}_1$, namely, it cannot satisfy the probability of rand $\leq c_1$. f_1 represents selection operator. f_2 is the discrete operator based on flexible workshop scheduling, including such two processes as procedure order discrete operation and machine assignment discrete operation. As for particle location, its update is mainly through information exchange between f_2 operator and the individual best location plus the overall best location. The improved precedence operation crossover (IPOX) operator and multipoint preservative crossover (MPX) operator raised by Zhang et al. [7, 8] are, respectively, very effective procedure order discrete operator and machine assignment discrete operator. Therefore, the two processes of f_2 operator shall be, respectively, operated by IPOX operator and MPX operator. Meanwhile, to guarantee the diversity of particle, we introduce perturbed operator f_3. Specifically, the update process of particle location is as follows:

The update process of particle location
01: Input: x^t: location of generation t of all particles N: population size
02: Output: x^{t+1}: location of generation $t+1$ of all particles
03: For $t = 1\ to\ G$
04: For $i = 1\ to\ N$
05: If rand $< \omega$.
06: $x_i^{t+1} = f_1\left(x_i^t\right)$;
07: If rand $< \omega$
08: $x_i^{t+1} = f_2\left(x_i^t, p_{best,i}^t\right)$;

(continued)

(continued)

The update process of particle location
09: Else
10: $x_i^{t+1} = f_2\left(x_i^t, g_{\text{best},i}^t\right)$;
11: End If
12: Else
13: $x_i^{t+1} = f_3(x_i^t)$;
14: End If
15: End For
16: End For

Below are the specific steps of MOPSO algorithm: Step ① introduces the coding method; step ② introduces the method of initialization; step ③ introduces the method of updating particle location, i.e. specific discrete operation of f_1 operator, f_2 operator and f_3 operator; step ④ introduces variable neighbourhood search algorithm; step ⑤ introduces the definition and update of individual best solution set and overall best solution set; step ⑥ introduces update of individual best location and overall best location; step ⑦ introduces the standards for terminating the algorithm.

① Coding methods

This algorithm adopts two-vector coding methods, i.e. the particle location code includes both procedure order vector and the machine assignment vector. The procedure order vector is used for determining the processing order of all procedures, whereas the machine assignment vector is used for assigning processing machines to all procedures. Since the procedure-based coding method can always obtain a feasible solution, this algorithm adopts the procedure-based coding method to conduct coding for the procedure order vector. In this coding method, the length of the procedure order vector equals the total number of all workpiece procedures, with each workpiece represented by a workpiece number. The no. k time's appearance of the same workpiece number represents the no. k procedure of this workpiece. Machine assignment vector assigns machine to each workpiece procedure according to the ascending order of workpiece number.

For instance, in Fig. 8.5, procedure processing order vector $\left[2\ 1\ 1\ 3\ 2\ 1\ 2\ 3\right]$ represents that the workpiece procedure processing order is $[O_{21}, O_{11}, O_{12}, O_{31}, O_{22}, O_{13}, O_{23}, O_{32}]$, machine assignment vector $\left[\ 1\ 3\ 2\ 1\ 3\ 1\ 3\ 2\ \right]$ represents that the machine assignment is as follows: (O_{11}, M_1), (O_{12}, M_3), (O_{13}, M_2), (O_{21}, M_1), (O_{22}, M_3), (O_{23}, M_1), (O_{31}, M_3), (O_{32}, M_2). Table 8.1 provides the processing timetable of four workpieces and five machines. Then, the processing time for Table 8.1, arranged according to the ascending order of workpiece number, is $\left[5\ 2\ 1\ 1\ 4\ 5\ 3\ 4\right]$.

② Population Initialization

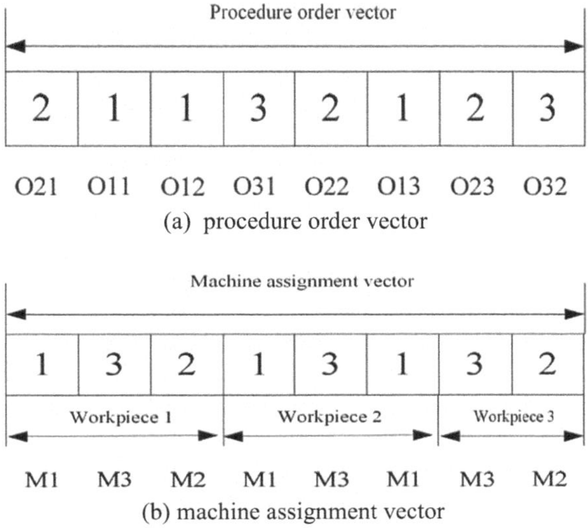

Fig. 8.5 Codes of the procedure order vector and the machine assignment vector

Table 8.1 Processing timetable of four workpieces and five machines

Workpiece	Procedure	M_1	M_2	M_3
J_1	O_{11}	5	3	–
	O_{12}	–	1	2
	O_{13}	3	1	–
J_2	O_{21}	1	–	4
	O_{22}	–	5	4
	O_{23}	5	–	6
J_3	O_{31}	–	6	3
	O_{32}	5	4	5

In the application scene of flexible workshop scheduling, the quality of population initialization will affect the algorithm's best performance. Therefore, we need to make use of field knowledge in designing an effective initialization method. Among particle swarm optimization algorithms, a good initialization method can produce initial population of rich diversity and many solutions of prospect so that the algorithm can have better retrieval orientation. Population initialization of flexible workshop's scheduling problem can be divided into initialization of the procedure order vector and that of the machine assignment vector.

This algorithm adopts the method of knowledge scheduling proposed in Sect. 8.3 to optimize the population initialization stage, using hierarchical knowledge retrieval to conduct workpiece set matching, procedure set matching, machine set matching and priority set matching in order to retrieve the information group of the highest historical similarity. Then, based on the implementation result or satisfaction level, we evaluate the solved knowledge scheme, so as to verify whether the historical scheduling information data set can fulfil the demands of knowledge retrieval. If the answer is yes, then push scheduling knowledge; if no, then conduct knowledge revision according to the heuristic rule listed in Sect. 8.3.3 until it meets the demands, then export knowledge results and convert them into the optimized initial population.

③ **Update particle location, i.e. conduct operation of f_1 operator, f_2 operator and f_3 operator**

f_1 means choosing operator. It decides whether the particle conducts f_2 operation or f_3 operation. In Formula (8.21), particle conducts f_2 operator action at probability ω; otherwise, particle conducts f_3 operator action. Then, f_2 operator obtains information at probability c_1 from individual best locations or obtains information at probability c_2 from overall best location. The adaptability of other individuals—made up of partial genetic fragments from individuals of good adaptability—would also of high probability be good as well [9]. Therefore, f_2 operator should be able to learn the excellent features of either individual best location or overall best location, thereby obtaining a solution of better adaptability. f_2 operator's implementation involves two stages: discrete operator for procedure order vector and discrete operator for machine assignment vector. In other words, we adopt a workpiece IPOX operator for procedure order vector [8] and MPX operator for machine assignment vector [7]. f_2 operator has the following detailed progress:

In IPOX operator, Fig. 8.6a, F_1 represents the current individual, F_2 represents the individual best location or overall best location, S_1 represents an individual of the next generation. Keep the machine assignment vector unchanged, and the operation progress is as follows:

1. For the two parent individuals F_1 and F_2, divide all workpieces into two subsets J_1 and J_2 at random.
2. For procedure order vector in F_1, copy workpiece procedure including J_1 to the same location in the filial individual S_1. For procedure order vector of F_2, copy workpiece procedure including J_2 to the same location in filial individual S_2.
3. For procedure order vector in F_2, copy workpiece procedure including J_2 in the same order to the remaining locations in filial individual S_1. For procedure order

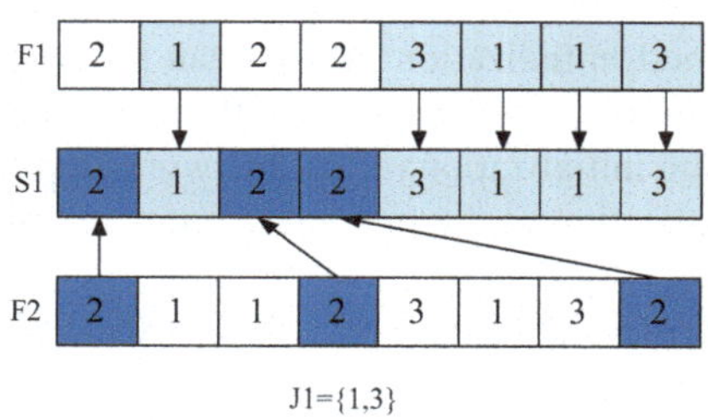

(a) IPOX operator

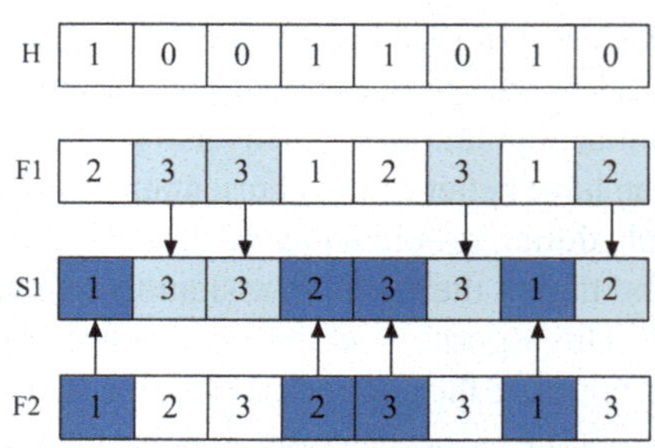

(b) MPX operator

Fig. 8.6 Operational process of IPOX operator and MPX operator

vector in F_1, copy workpiece procedure including J_1 in the same order to the remaining location in the filial generation individual S_2.

In IPOX operator, Fig. 8.6b, F_1 represents the current individual, F_2 represents the individual best location or overall best location, S_1 represents the filial individual. Keep the procedure order vector unchanged, and the operation progress is as follows:

1. For the two parent individuals F_1 and F_2, produce a decision vector H which shares identical length with the parent individual, and the decision vector includes only 0 and 1 elements at random.
2. Record locations including 0 in decision vector H, copy machines of same location in F_1 and F_2, respectively, to S_1 and S_2, record locations including 1 in decision vector H, then copy machines of the same location in F_2 and F_1, respectively, to S_2 and S_1.

F_1 and F_2 are parents, while S_1 is of the filial generation; then, the operation progress of IPOX operator and MPX operator is illustrated in Fig. 8.6a, b, respectively.

MOPSO algorithm overcomes the problem of late-stage particles' reduced diversity with perturbed operator f_3. The rule of Earliest Completion Machine (ECM) can select a machine for processing procedure according to the earliest completion time, and it has satisfactory results for reducing the maximum machine completion time. But the ECM rule has so large calculated amount that it consumes lots of CPU time. Under smaller probability c_3, the MOPSO algorithm treats the ECM rule as perturbed operator f_3 for conduction machine re-assignment, which improves optimization effect without consuming too much CPU time. f_3 operator, i.e. ECM rule, has the following explicit process: assign processing machine for procedures according to the order in the procedure order vector; each time a processing machine is to be arranged, calculate the completion time of the procedure at all machines, then select the machine with minimum completion time as the processing machine for the current procedure.

④ **Variable neighbourhood search**

Flexible workshop scheduling can be represented by using disjunctive graph $G = (N, A, E)$. In disjunctive graph G, the longest route from the start point to the end point is called the critical route. The length of the critical route equals the maximum machine completion time of the scheduling scheme [10]. The procedure at the critical route is called the critical procedure. The common procedure at all critical routes is called the common critical procedure. The largest connecting procedure of neighbouring common critical procedures at the same machine is called the common critical block. Without changing the common critical procedure, the maximum machine completion time would not be reduced. Neighbourhood search of the common critical block can effectively reduce the search space. This algorithm defines three neighbourhood structures based on common critical block, then adopts neighbourhood search algorithm to enhance the local development capability of the algorithm. Here, among the three neighbourhood structures, two are machine assignment neighbourhood based on the common critical block (NH_1 and NH_2), and one is procedure order neighbourhood based on the common critical block (NH_3).

Machine assignment neighbourhood (NH_1): find out the machine set with maximum machine completion time, and from machine set M_s, randomly select a machine M_k, at machine M_k randomly select a common critical procedure O_{ij}, from candidate machine sets O_{ij}, randomly select a machine M'_k different from M_k, then at machine M'_k, randomly select an insertion point which satisfies the procedure constraints, insert procedure O_{ij} to this point.

Machine assignment neighbourhood (NH_2): randomly select a common critical procedure O_{ij} with at least two candidate machines, then arrange all candidates machine in the ascending order of machine processing time, from the first half of candidate machines, randomly select a machine M'_k different from M_k and assign procedure O_{ij} to machine M'_k.

Procedure order neighbourhood (NH_3): randomly select a common critical block π with at least three procedures, from π, randomly select a procedure O_i^{π} (not including procedures of the block's head and end in π). If the procedures contained in π equals 3, then when procedure O_i^{π} does not belong to the same workpiece as that of the block's head or end, switch the order between procedure O_i^{π} and the block's head or end procedure. If the procedures contained in π exceed 3, then from common critical block π, randomly select a position different from the block's head or end procedure in common critical block and insert the block's head or end procedure prior to this position. The operation process of neighbourhood NH_3 is as illustrated in Fig. 8.7.

Through defining three neighbourhood search structures, the neighbourhood search algorithm is as follows:

Variable neighbourhood search algorithm
01: Input: Ω overall best solution set K: neighbourhood varieties
02: Output: Ω
03: For all $s \in \Omega$
04: $k = 1$;
05: While ($k \leq K$)
06: $s' \leftarrow$ produces s, a NH_k neighbourhood.
07: $s'' \leftarrow$ conduct local search for s'.
08: If $s'' \succ s$
09: $s \leftarrow s''$; $k = 1$;
10: Else
11: $k = k + 1$.
12: End If
13: End While

(continued)

(continued)

Variable neighbourhood search algorithm
End For

In above, K is the number of neighbourhood varieties. $S'' \succ S$ represents S'' dominates S. In MOPSO algorithm, K chooses the value of 3.

In the neighbourhood search algorithm, we need to define a local search. Its application progress is as follows:

Local search
Input: S'; individual NH_k: neighbourhood k N_s : size of local search

Output: S'
For $k = 1$ to N_s
$S'' \leftarrow$ produces S', a NH_k neighbourhood
If $S'' \succ S'$
$S' \leftarrow S''$
End If
End For

In above, N_s is the size of local search. Based on experience, N_s is normally set at 20.

⑤ Update of the individual best solution set and overall best solution set

The individual best solution set and overall best solution set are used to store the non-dominated solutions obtained by each particle and those obtained by the entire population. In order to guarantee the convergence and uniformity of the non-dominated solutions obtained, we need to design a selection strategy for the non-dominated solution, in order to update the individual best solution set and the overall best solution set. At present, many selection strategies of multi-objective problem non-dominated solution have been raised, such as SPEAII [11], NSGA-II [12] and MOEA/D [13]. The evolution algorithm based on decomposition adopts the method of weight vector and weighted sum to decompose MOP into a series of scalar problems and has achieved satisfactory results in solving multi-objective problems [14]. In the MOPSO algorithm, we propose a non-dominated strategy which is based on Pareto dominance relationship and which treats scalar problem as non-dominated solution, and we update individual best solution set and overall best solution set through this strategy.

The update strategy of the individual best solution set is as follows: suppose the maximum storage number of the individual best solution of particle i is N_a, we add particle i to the new solution set Ω formed by current individual best solution sets. Then, through non-dominated set's update strategy, we select and preserve N_a non-dominated solutions from Ω. The selection strategy of overall best solution set is as follows: suppose the maximum storage number of overall best solution set is N_g, we add all the non-dominated solutions of the current population to the new solution

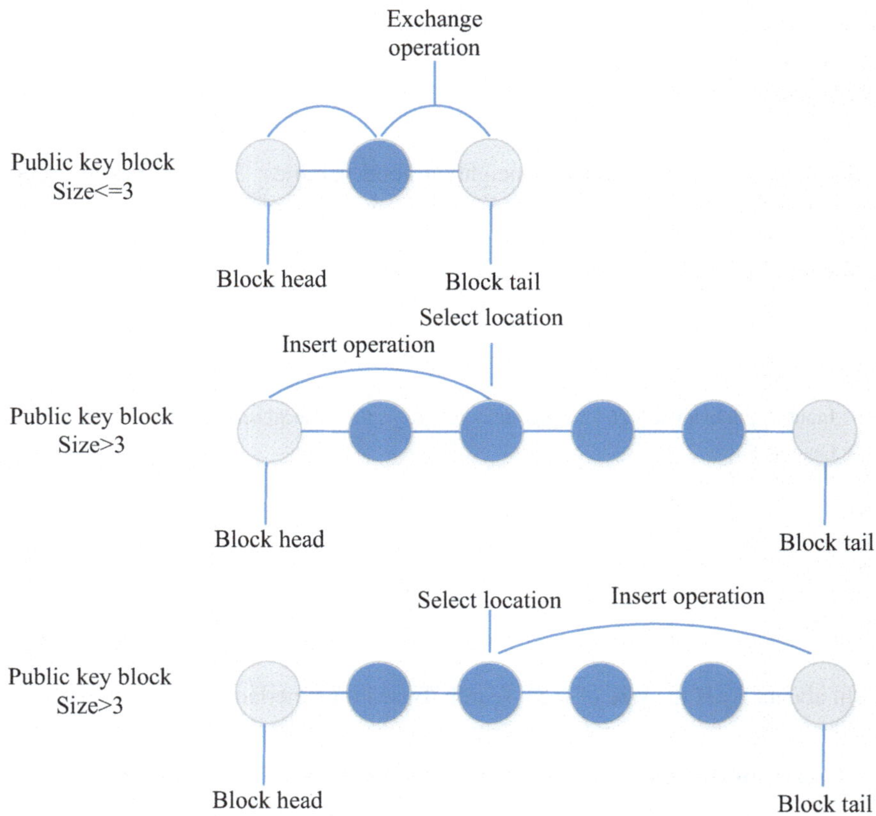

Fig. 8.7 Operation process of NH_3

set Ω' formed by overall best solution set; then, through non-dominated set's update strategy, we select and preserve N_g non-dominated solutions from Ω'.

Non-dominated set's update strategy: randomly generate three numbers within [0, 1], λ_1, λ_2 and λ_3. These weights can increase the randomness of target weights. Suppose the non-dominated solution set is Ω, its storage size is N_a, then the steps of the non-dominated set's update strategy are as follows:

1. Obtain the to-be-updated solution set Ω, the maximum storage number is N_a.
2. Delete identical individual in the solution set Ω, calculate the number of individuals Count in the updated solution set Ω.
3. Decide whether Count is larger than N_a. If yes, carry out steps 4 and 5; if no, carry out step 6.
4. Generate three random weights and arrange them in descending order $[\lambda_1, \lambda_2, \lambda_3]$, normalize all weights into per unit value $\lambda_i / \sum \lambda$, calculate all individuals' scalar problems SC in the solution set Ω according to their weights.

5. Arrange in ascending order according to SC, select the first N_a individuals as a new non-dominated solution set Ω.
6. Export Ω.

⑥ **Update of the individual best location and overall best location**

Choices of the individual best location and overall best location, as well as particle location update, are as illustrated in Fig. 8.8. After updating the individual best solution set and the overall best solution set, we need to select the individual best location and overall best location, in order to update particle location. For each particle, the

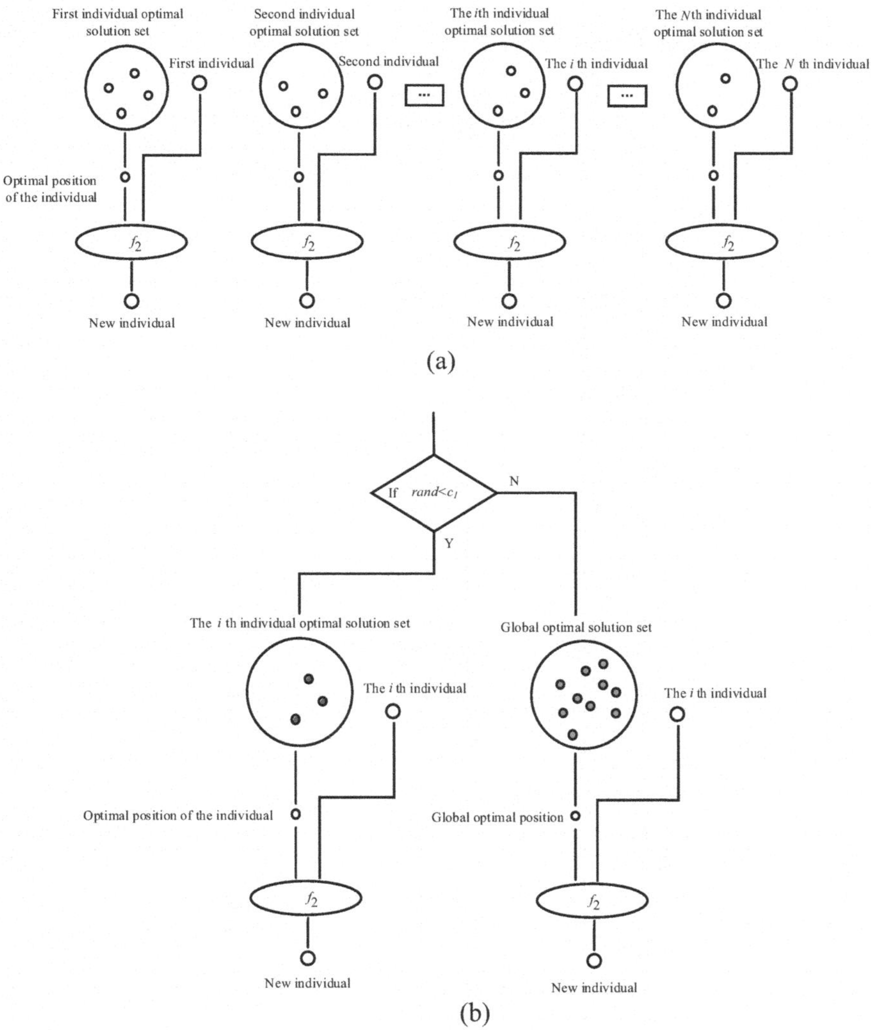

Fig. 8.8 Choices of the individual best location and overall best location as well as function of f_2

method to select the individual best location is as follows: from the individual best solution set, randomly select a non-dominated solution as the individual best location of this particle. The method to select the overall best location is as follows: from the overall best solution set, randomly select a non-dominated solution as the overall best location.

⑦ **Termination standard**

MOPSO algorithm terminates operation if it reaches the maximum number of iterations.

8.4.2 *Algorithm Flow Chart*

To sum up, the flow chart of MOPSO algorithm is as illustrated in Fig. 8.9.

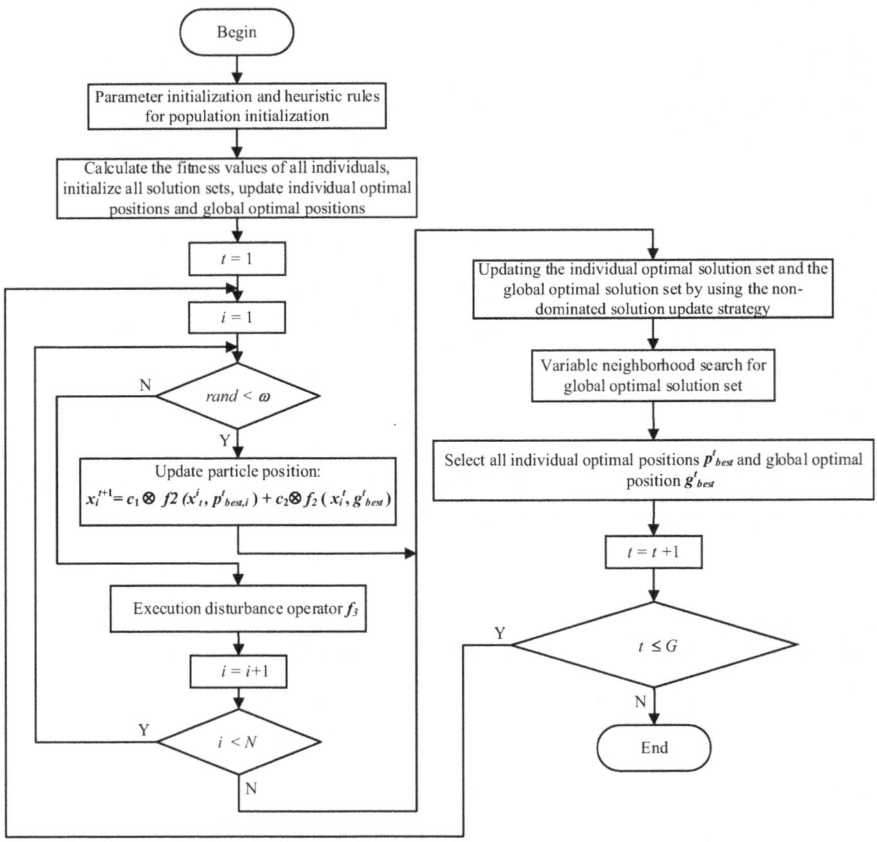

Fig. 8.9 MOPSO Algorithm flow chart

8.5 Experimental Verification and Results Analysis

This section adopts Kacem test function [15] and Brdata test function [16] to analyse the performance of MOPSO algorithm. Here, the software platform for experimental simulation is MATLAB2011a on Windows 7 system. The hardware configuration for the experiment PC is Intel(R) core(TM) i3-3240 processor, of 3.4GHZ frequency, with 4 GB memory. The parameter setting of MOPSO is as shown in Table 8.2, with each test function optimized ten times.

1. The experiment of Kacem test function

This section selects four Kacem test functions to verify the optimal performance of MOPSO algorithm. The algorithms for comparison include HTSA raised by Li et al. [17], ATA raised by Bagheri et al. [18], Xing algorithm raised by Xing et al. [19], MOGA raised by Wang et al. [20], P-DABC raised by Li et al. [21] and SEA raised by Chiang et al. [22]. Table 8.3 lists the results of non-dominated solution obtained by seven algorithms' optimization Kacem test function. For the examples of 4 workpieces × 5 machines, 8 workpieces × 8 machines and 10 workpieces × 10 machines, all the non-dominated solutions obtained by the seven algorithms are of non-dominated relationship. For the example of 15 workpieces × 10 machines, HTSA algorithm, Xing algorithm and MOPSO algorithm have all arrived at the same non-dominated solutions. This solution dominates the non-dominated solutions obtained by AIA algorithm, MOGA algorithm and P-DABC algorithm.

Table 8.4 lists the number of non-dominated solutions obtained by seven algorithms' optimization of Kacem test function. The bar chart in Fig. 8.10 describes the comparison data of Table 8.4. From Table 8.4 and from the bar chart in Fig. 8.10, we can clearly observe that, in the optimization of the four Kacem test functions, MOPSO algorithm has achieved more non-dominated solutions than HTSA algorithm, AIA algorithm, Xing algorithm and P-DABC algorithm. As for the example 8 workpieces × 8 machines, only MOPSO algorithm and SEA algorithm have found four non-dominated solutions. As for the example of 15 workpieces × 10 machines, MOGA algorithm obtains three non-dominated solutions, but two of them are both dominated by the non-dominated solution obtained by MOPSO algorithm. Only MOPSO algorithm, HTSA algorithm and Xing algorithm have found two non-dominated solutions (11, 91, 11) and (11, 93, 10) for the example of 15 workpieces × 10 machines. As for Kacem test function, MOPSO algorithm has obtained all the non-dominated solutions, none of which are dominated by solutions obtained by other algorithms. Therefore, MOPSO algorithm has comparatively excellent general performance.

Figure 8.11a, b are the Gantt charts of partial non-dominated solutions of the four

Table 8.2 Parameter setting

Test function	N	G	ω	(c_1, c_2)	N_a	N_g
Kacem	100	300	0.98	(0.6, 0.4)	5	15
Brdata	100	300	0.98	(0.6, 0.4)	5	15

Table 8.3 Optimization results of Kacem examples

Kacem ($m \times n$)	4×5			8×8			10×10			15×10		
Optimization objective	C_M	W_T	W_M	C_M	W_T	W_M	C_M	W_T	W_M	C_M	W_T	W_M
HTSA	11	32	10	14	77	12	7	43	5	11	91	11
	12	32	8	15	75	12	7	42	6	11	93	10
AIA		N/A		14	77	12	7	43	5	11	93	11
Xing	12	32	8	14	77	12	7	42	6	11	91	11
		N/A		15	76	12	8	42	5	11	93	10
MOGA	11	32	10	15	81	11	8	42	5	11	91	11
	12	32	8	15	75	12	7	42	6	12	95	10
	11	34	9	16	73	13	8	41	7	11	98	10
P-DABC	11	32	10	14	77	12	8	41	7	12	91	11
	12	32	8	15	75	12	7	43	5	11	93	11
	13	33	7	16	73	13	8	42	5		N/A	
		N/A			N/A			N/A			N/A	
SEA		N/A		14	77	12		N/A			N/A	
		N/A		15	75	12		N/A			N/A	
		N/A		16	73	13		N/A			N/A	
		N/A		16	77	11		N/A			N/A	
MOPSO	11	32	10	16	73	13	8	41	7	11	91	11
	13	33	7	14	77	12	8	42	5	11	93	10
	12	32	8	16	77	11	7	43	5		N/A	
		N/A		15	75	12	7	42	6		N/A	

Table 8.4 Number of non-dominated solutions obtained by seven algorithms in the Kacem example

Kcem ($m \times n$)	HTSA	AIA	Xing	MOGA	P-ABC	SEA	MOPSO
4 × 5	2	N/A	1	3	3	N/A	3
8 × 8	2	1	2	3	3	4	4
10 × 10	2	1	2	3	3	N/A	4
15 × 10	2	1	2	3	2	N/A	2

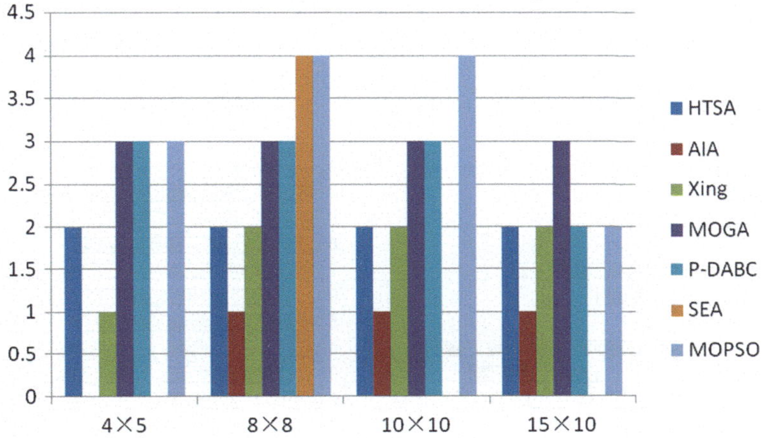

Fig. 8.10 Number of non-dominated solutions obtained by seven algorithms in the Kacem example

Kacem test functions obtained by the MOPSO algorithm

2. The experiment of Brdata test function

This section uses the ten Brdata test function from 10 workpieces × 6 machines to 20 workpieces × 15 machines proposed by Brandimarte for simulation experiment. The algorithms selected for comparison include Xing's algorithm [23], MOGA algorithm [24], HTSA algorithm [25], HSFLA algorithm [26] and AIA algorithm [18]. The optimization results of algorithm comparison come from documents [17, 27].

Table 8.5 lists the experiment results of the minimum non-dominated solution of maximum machine completion time obtained by these algorithms. The data in bold show that this solution is dominated by other partial solutions. From Table 8.5, we can see that, for example, $MK02$, $MK05$, $MK06$ and $MK07$, the non-dominated solutions obtained by MOPSO algorithm dominate those obtained by Xing's algorithm. For example of $MK01$ and $MK03$, the non-dominated solutions obtained by MOPSO algorithm dominate those obtained by MOGA algorithm. For $MK07$ example, the non-dominated solutions obtained by MOPSO algorithm dominate those obtained by HTSA algorithm. For example, $MK02$, $MK05$ and $MK07$, the non-dominated solutions obtained by MOPSO algorithm dominate those obtained by HSFLA algorithm. For example, $MK01$, $MK02$, $MK03$, $MK05$, $MK06$, $MK07$,

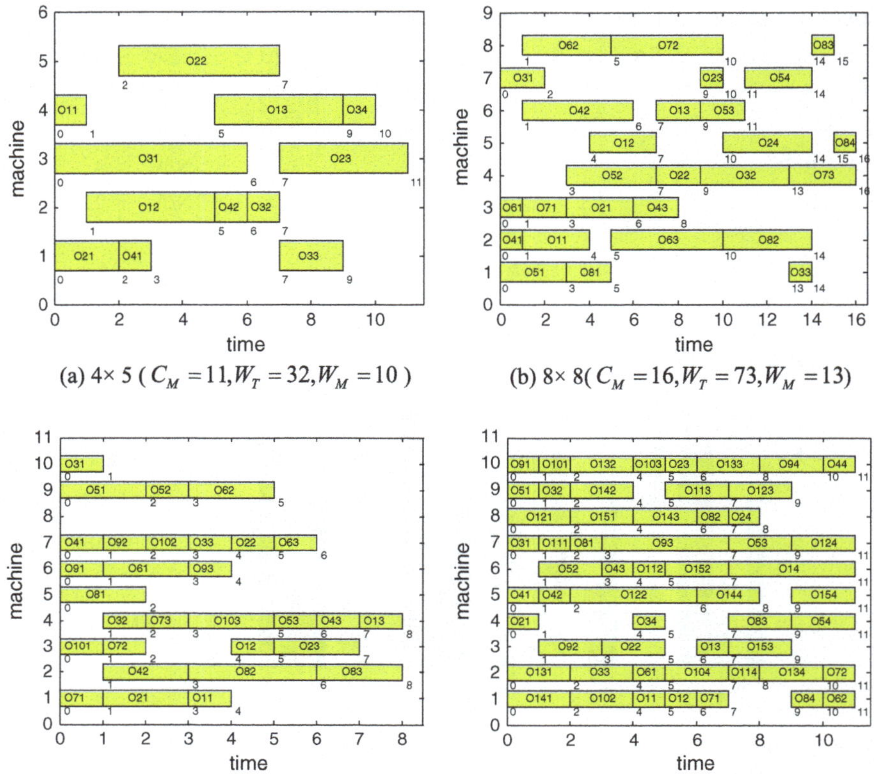

(a) 4× 5 ($C_M = 11, W_T = 32, W_M = 10$)

(b) 8× 8($C_M = 16, W_T = 73, W_M = 13$)

(d) 10× 10($C_M = 8, W_T = 41, W_M = 7$)

(e) 15× 10 ($C_M = 11, W_T = 91, W_M = 11$)

Fig. 8.11 Gantt chart of the results and partial solutions of the four Kacem test examples obtained by MOPSO

$MK09$ and $MK10$, the non-dominated solutions obtained by MOPSO algorithm dominate those obtained by MOGA algorithm. Only for $MK08$ example, the non-dominated solutions obtained by MOGA algorithm dominate those obtained by MOPSO algorithm. From Table 5.5, we can see that, for the ten Brdata examples, MOPSO algorithm still achieves high-quality non-dominated solutions.

Table 8.6 lists experiment results of the best maximum machine completion time (C_M), average CPU time (Av(CPU)), the mean value (Av(C_M)) and standard deviation (Std(C_M)) of the best maximum machine completion time obtained by MOPSO algorithm, SEA algorithm [28] and TSPCB algorithm [29]. The computing formula for the MOPSO algorithm's performance upgrade rate ($imp\%$) is as follows:

$$imp\% = \frac{C_M^{\text{com}} - C_M^{\text{pro}}}{C_M^{\text{pro}}} \times 100\% \quad (22)$$

Table 8.5 Optimization results of Ten Brdata examples

Test function		Xing	MOGA	HTSA	HSFLA	AIA	MOPSO
$MK01$	C_M	42	40	40	40	40	40
	W_T	162	169	167	165	171	167
	W_M	42	36	36	37	36	36
$MK02$	C_M	28	26	26	26	26	26
	W_T	155	151	151	152	154	151
	W_M	28	26	26	26	26	26
$MK03$	C_M	204	204	204	204	204	204
	W_T	852	855	852	852	1207	852
	W_M	204	199	204	204	204	204
$MK04$	C_M	68	66	61	62	60	61
	W_T	352	345	366	364	403	382
	W_M	67	63	61	61	60	60
$MK05$	C_M	177	173	172	173	173	173
	W_T	702	683	687	685	686	683
	W_M	177	173	172	173	173	173
$MK06$	C_M	75	62	65	64	63	62
	W_T	431	424	398	403	470	424
	W_M	67	55	62	55	56	55
$MK07$	C_M	150	139	140	141	140	139
	W_T	717	693	695	696	695	693
	W_M	150	139	140	141	140	139
$MK08$	C_M	523	523	523	523	523	523
	W_T	2524	2524	2524	2524	2524	2524
	W_M	523	515	523	523	523	523
$MK09$	C_M	311	311	310	311	312	310
	W_T	2374	2290	2294	2275	2591	2514
	W_M	299	299	301	299	306	299
$MK10$	C_M	227	214	214	215	214	214
	W_T	1989	2082	2053	1957	2121	2082
	W_M	221	204	210	198	206	204

where C_M^{com} and C_M^{pro} represent the best maximum machine completion time obtained, respectively, by comparison algorithm and MOPSO algorithm. $imp\%$ is the performance upgrade percentage of MOPSO algorithm compared with other algorithms. Table 8.6 lists simultaneously the performance upgrade rate of MOPSO algorithm compared with SEA algorithm and TSPCB algorithm ($imp_1\%$ and $imp_2\%$).

In Table 8.6, from $imp_1\%$ data, we can see that MOPSO algorithm has achieved better C_M in $MK06$, $MK09$ and $MK10$ than SEA algorithm. As for the ten Brdata

Table 8.6 Comparison between the Ten Brdata examples' best C_M and CPU time

Test function	EA	TSPCB		MOPSO					
	C_M	C_M	Av(CPU)	C_M	$imp_1\%$	$imp_2\%$	Av(C_M)	Std(C_M)	Av(CPU)
$MK01$	40	40	2.80	40	0.00	0.00	40.00	0.00	10.22
$MK02$	26		19.31	26	0.00	0.00	26.40	0.52	30.35
$MK03$	204	204	0.89	204	0.00	0.00	204.00	0.00	20.07
$MK04$	61	62	40.82	61	0.00	+1.61	62.35	0.47	58.22
$MK05$	173	172	20.23	173	0.00	−0.58	173.75	0.81	27.04
$MK06$	65	65	27.18	62	+4.62	+4.62	62.34	0.36	75.22
$MK07$	140	140	35.29	139	0.00	+0.71	139.30	0.46	52.51
$MK08$	523	523	4.65	523	0.00	0.00	523.75	0.68	45.20
$MK09$	311	310	70.38	310	+0.32	0.00	312.60	1.53	104.42
$MK10$	255	214	89.83	214	+16.08	0.00	214.55	0.65	156.73

examples, MOPSO algorithm has achieved C_M value no worse than SEA algorithm. From $imp_2\%$ data, we can see that, compared with TSPCB algorithm, MOPSO algorithm has achieved fairly good C_M value in $MK04$, $MK06$, and $MK07$; only in the example of $MK07$, the C_M value obtained by MOPSO algorithm is worse than that by TSPCB algorithm. In Table 8.6, the Av(C_M) values of the ten Brdata examples obtained by MOPSO algorithm are close to the C_M values obtained, and all Std(C_M) values are smaller than 1. Therefore, data demonstrate that MOPSO algorithm has stable optimization performance. But judging from Av(CPU) data, MOPSO algorithm consumes more CPU tie than TSPCB algorithm. This may be caused by the different software platform and design language since TSPCB algorithm is achieved by using C++ language and 1.6 Hz Pentium processor. The Gantt chart of the non-dominated solution of $MK01$ example's best maximum machine completion time achieved by MOPSO algorithm is as illustrated in Fig. 8.12. The approximate Pareto front ends of examples $MK03$ and $MK04$ obtained by MOPSO algorithm are as shown in Fig. 8.13.

3. Energy Consumption Analysis

For the optimization solution set sought by MOPSO algorithm, we adopt the method of hierarchical analysis to select the best processing parameters from the solutions obtained. In decision-making, the importance of each target is arranged in the order of maximum completion time, product energy consumption amount (critical machine load) and machine total load. We use Figs. 8.1, 8.2, 8.3, 8.4, 8.5, 8.6, 8.7, 8.8 and 8.9 to represent the degree of mutual significance between each target function, thereby getting the judgement matrix:

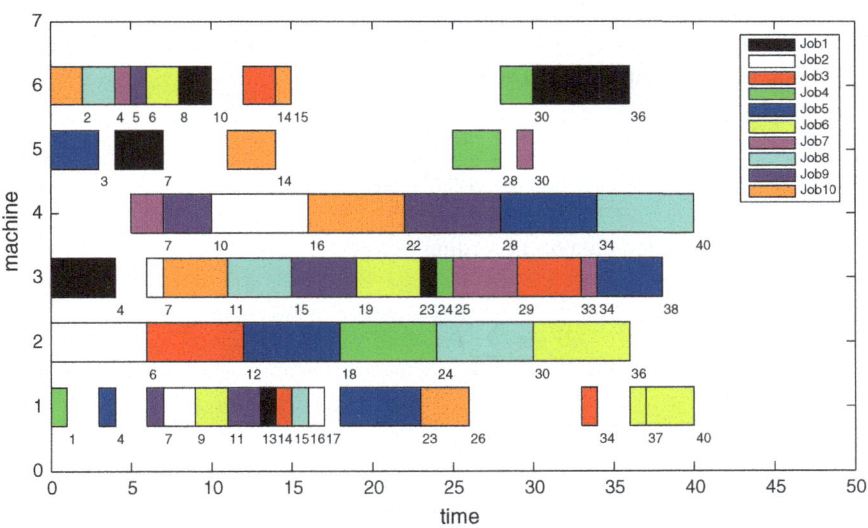

Fig. 8.12 Gantt chart of $MK01$ obtained by MOPSO

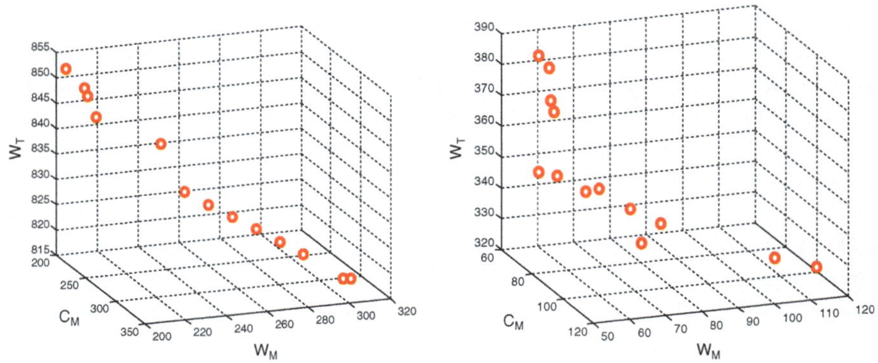

(a) MK03 (b) MK04

Fig. 8.13 Approximate Pareto front ends of examples $MK03$ and $MK04$ obtained by MOPSO

$$A = \begin{bmatrix} & E & T_p \\ E & 1 & 1/5 \\ T_p & 5 & 1 \end{bmatrix} \tag{23}$$

Taking into consideration the different units (dimension) of each index—for example, the unit for the maximum completion time is minute, whereas the unit for energy consumption is watt—there is no uniform reference in the choice of best processing parameters, and hence, it is impossible to compare between the two. Therefore, before selecting the best processing parameter set, we should first conduct normalization, to ensure that all indexes are at the same order of magnitude and convenient for comparison. The actual method is deviation standardization and to conduct linear conversion of the A matrix data. Then, we have

$$b_{ij} = \frac{a_j^{\max} - a_{ij}}{a_j^{\max} - a_j^{\min}} \tag{8.24}$$

To sum the converted data b_{ij} according to rows:

$$\widetilde{W}_j = \sum_{i=1}^{p} b_{ij},\ q \ge j \ge 1 \tag{8.25}$$

Normalize the summed-up vectors:

$$W_j = \frac{\widetilde{W}_j}{\sum_{j=1}^{q} \tilde{W}_j} \tag{8.26}$$

Table 8.7 Energy conservation percentage of this chapter's algorithm compared with those of other algorithms

Algorithm	MTLBO	Method of experience
Energy conservation ratio (%)	15.8	34.9

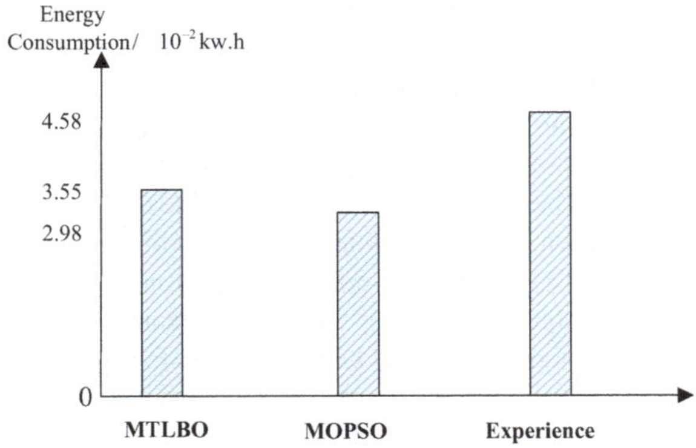

Fig. 8.14 Energy consumption amount of each Algorithm

where $a_j^{\max}$ and $a_j^{\min}$ are, respectively, the maximum and minimum values to the corresponding column j in the judgement matrix A; p is the number of evaluation schemes; q is the number of indexes, $i = 1, 2, \ldots, p$, $j = 1, 2, \ldots, q$; a_{ij} is the value of index in the Pareto scheme. From the above formula, we can obtain the scheme's decision matrix $B = \left(b_{ij}\right)_{p\times q}$. Then, multiply the decision matrix B with weight vector $W^T = [W_1, W_2, \ldots, W_4]$ to get the best satisfaction matrix, i.e. $D = (D_i) = BW^T = \sum_{j=1}^{q} b_{ij} W_j$, before finally getting the evaluation scheme's satisfaction matrix $D = (D_i)_p = \left[\, D_1 \; D_2 \; \ldots \; D_p \,\right]^{\mathrm{T}}$.

Run a random test of the algorithm proposed in this chapter for 20 times, and compare the energy consumption after optimization scheduling with the energy consumption values under personal experience scheduling and multi-objective teaching-and-learning optimization algorithm MTLBO scheduling, as shown in Table 8.7. The method of experience means that processing parameters are selected by operators according to their years of processing experience and reference manuals. In comparison with the other two methods, the energy consumption amount and energy conservation amount of the algorithm proposed in this chapter are as illustrated in Fig. 8.14 and Table 8.7.

8.6 Conclusion

This chapter first establishes a flexible job shop optimization scheduling model with minimalized maximum machine completion time, task-level product energy consumption and machine total load as optimization objectives. Then, considering the comparability between optimization algorithm and results of existing materials and the positive correlation between the product's energy consumption and critical machine load, we convert the product energy consumption target into critical machine load target. Later, we propose a neighbourhood search discrete particle swarm optimization algorithm based on knowledge integration and common critical block to solve this kind of problems. In this discrete particle swarm optimization algorithm, we introduce the population initial value method based on knowledge scheduling and design discrete operators. We define individual best solution set and overall best solution set, as well as the update strategy of the corresponding multi-objective dominated set. We also design a method of selecting the overall best location and the individual best location. Finally, we define three neighbourhood operations based on the common critical block and adopt neighbourhood search based on common critical block to search the overall best solution set. Comparison verification with other classical algorithms using the Kacem example and Brdata example demonstrates that this chapter's MOPSO algorithm, which is based on knowledge integration, has better optimization results in solving multi-objective flexible workshop's scheduling problems and can achieve better energy conservation effect.

References

1. Roberge V, Tarbouchi M, Labonte G (2012) Comparison of parallel genetic algorithm and particle swarm optimization for real-time UAV path planning. IEEE Trans Indust Inf 9(1):132–141
2. Zhao SZ, Suganthan PN, Pan QK et al (2011) Dynamic multi-swarm particle swarm optimizer with harmony search. Expert Syst Appl 38(4):3735–3742
3. Liu QW (2014) Research on artificial bee colony and its application in scheduling problems. Beijing Jiaotong University
4. Silva RR, Brandã£O CRF (2010) Morphological patterns and community organization in leaf-litter ant assemblage. Ecol Monog 80(1):107–124
5. Gui WH, Chen XF, Yang CH et al (2016) Knowledge automation and its industrial application. Sci Sini (Inform) 46(8): 1016
6. Sha DY, Lin HH (2010) A multi-objective PSO for job-shop scheduling problems. Expert Syst Appl 37(2):1065–1070
7. Zhang CY, Rao YQ, Li PG et al (2007) Bilevel genetic algorithm for the flexible job-shop scheduling problem. Chin J Mech Eng 43(4):119–124
8. Zhang C, Li P, Rao Y, Li S (2005) A new hybrid GA/SA algorithm for the job shop scheduling problem. In: Evolutionary computation in combinatorial optimization. Springer, Berlin Heidelberg, pp 246–259
9. Zhang CY, Guan ZL, Liu Q et al (2008) New scheduling type applied to solving job-shop scheduling problem. J Mech Eng 44(10):24–31
10. Balas E (1969) Machine sequencing via disjunctive graphs: an implicit enumeration algorithm. Oper Res 17(6):941–957
11. Zitzler E, Laumanns M, Thiele L (2001) SPEA2: improving the strength Pare-to evolutionary algorithm for multiobjective optimization. TIK-report, 103, Swiss Federal Institute of Technology, pp 1–21
12. Deb K, Pratap A, Agarwal S et al (2002) A fast and elitist multiobjective genetic algorithm: NSGA-II. IEEE Trans Evol Comput 6(2):182–197

13. Li H, Zhang Q (2009) Multiobjective optimization problems with complicated Pareto sets, MOEA/D and NSGA-II. IEEE Trans Evol Comput 13(2):284–302
14. Zhang Q, Li H (2007) MOEA/D: a multiobjective evolutionary algorithm based on decomposition. IEEE Trans Evol Comput 11(6):712–731
15. Kacem I, Hammadi S, Borne P (2002) Pareto-optimality approach for flexible job-shop scheduling problems: hybridization of evolutionary algorithms and fuzzy logic. Math Comput Simul 60(3):245–276
16. Brandimarte P (1993) Routing and scheduling in a flexible job shop by tabu search. Ann Oper Res 41(3):157–183
17. Li J, Pan Q, Liang YC (2010) An effective hybrid tabu search algorithm for multi-objective flexible job-shop scheduling problems. Comput Ind Eng 59(4):647–662
18. Bagheri A, Zandieh M, Mahdavi I et al (2010) An artificial immune algorithm for the flexible job-shop scheduling problem. Future Gener Comput Syst 26(4):533–541
19. Xing LN, Chen YW, Wang P et al (2010) A knowledge-based ant colony optimization for flexible job shop scheduling problems. Appl Soft Comput 10(3):888–896
20. Wang X, Gao L, Zhang C, Shao X (2010) A multi-objective genetic algorithm based on immune and entropy principle for flexible job-shop scheduling problem. Int J Adv Manufac Technol 51(5–8):757–767
21. Li JQ, Pan QK, Gao KZ (2011) Pareto-based discrete artificial bee colony algorithm for multi-objective flexible job shop scheduling problems. Int J Adv Manuf Technol 55(9–12):1159–1169
22. Chiang TC, Lin HJ (2013) A simple and effective evolutionary algorithm for multiobjective flexible job shop scheduling. Int J Prod Econ 141(1):87–98
23. Xing LN, Chen YW, Yang KW (2009) An efficient search method for multi-objective flexible job shop scheduling problems. J Intell Manuf 20(3):283–293
24. Li J, Pan Q, Xie S (2012) An effective shuffled frog-leaping algorithm for multi-objective flexible job shop scheduling problems. Appl Math Comput 218(18):9353–9371
25. Li JQ, Pan QK, Suganthan PN et al (2011) A hybrid tabu search algorithm with an efficient neighborhood structure for the flexible job shop scheduling problem. Int J Adv Manuf Technol 52(5–8):683–697
26. Ishibuchi H, Murata T (1998) A multi-objective genetic local search algorithm and its application to flowshop scheduling. IEEE Trans Syst Man Cybernet. Part C: Appl Rev 28(3):392–403
27. Adibi MA, Zandieh M, Amiri M (2010) Multi-objective scheduling of dynamic job shop using variable neighborhood search. Expert Syst Appl 37(1):282–287
28. Wang XJ, Guo YJ (2004) Aggregate analysis of group decision making based on G1-method. Chin J Manage Sci, 14–16
29. Huang ZT, Yang J, Zhang CY et al (2016) Energy-oriented CNC milling process modelling and parameter optimization. China Mech Eng 27(18):2524–2532

Chapter 9
Discrete Manufacturing System's Dynamic Intelligent Optimization Scheduling Method for High-Dimensional and Multi-objective Optimization of Production/Energy Consumption

9.1 Introduction

Current documents are mostly research targeting at low-dimensional multi-objective flexible workshop's scheduling problem. The number of optimization targets does not exceed three, and these targets are normally related to completion time, machine load and delivery period. But actual discrete manufacturing system also involves some realistic objectives, including energy consumption, cost and processing quality, which represent the decision expectation and evaluation standards of the different departments in an enterprise towards scheduling scheme. Besides, dynamism is an ubiquitous feature in discrete manufacturing systems, mainly reflected in these cases: (1) in the actual course of processing, generation of new tasks and completion of on-going tasks; (2) in the course of production scheduling, due to impacts of such unforeseen factors as urgent order insertion, machine breakdown and delivery period modification, which make the original production plan destroyed and unenforceable. Therefore, it is necessary to take the dynamic event factor and production/energy consumption high-dimensional complex targets into consideration of the flexible job shop's scheduling problems. We need to design reasonable scheduling mechanism to reduce the impact of dynamic emergencies and to ensure that production steadily proceeds.

Although some achievements have been made with regard to solving the dynamic job shop's scheduling problem with pre-actor scheduling methods, there are still insufficient aspects: (1) the research work mostly concentrates on the two targets of optimizing scheduling efficiency and stability, with little consideration for other optimization targets such as energy consumption; (2) the methods of solution are mostly to add weights of multiple optimization targets, yet unable to obtain a group of Pareto compromise solution for the scheduling administrator's decision-making; (3) normally there is only one dynamic event taken into consideration, with little research considering multiple dynamic interfering incidents; (4) existing research normally take account of single rescheduling moment, with little attention given to the course of continuous production where workpieces arrive in successive manner;

Y. Wang et al., *Quantitative Analysis and Optimal Control of Energy Efficiency in Discrete Manufacturing System*, https://doi.org/10.1007/978-981-15-4462-0_9

(5) previous research mostly target at the aspect of optimization, scarcely considering the decision-making progress.

For the above reasons, this chapter takes into consideration three dynamic events—urgent order insertion, machine breakdown and machine repair. Here, we construct a high-dimensional multi-objective flexible job shop scheduling model which simultaneously contains four aspects of performance indexes, i.e. scheduling efficiency, machine load, rescheduling stability and energy consumption, and study the high-dimensional multi-objective discrete manufacturing system's dynamic pre-response scheduling method based on NSGA-III.

9.2 Discrete Manufacturing System's Dynamic Intelligent Optimization Scheduling Model for High-Dimensional and Multi-objective Optimization of Production/Energy Consumption

The dynamic scheduling problem which this chapter attempts to solve targets at the continuous progress of workpiece processing, including the continuous production of new workpiece processing tasks and the successive completion of workpieces in-process. In the workpiece processing production, we need to consider such dynamic events as machine breakdown and urgent order insertion. The task of scheduling is to arrange a suitable machine for processing for each available procedure in accordance with information of the system's current status at each moment when rescheduling is triggered, and to reasonably arrange the processing order of the procedures assigned to each machine available, so as to ensure that the set optimization targets reach their best. At each moment of rescheduling, take into consideration relevant processing information, including: (1) the currently released machines available be for processing; (2) procedures which have been processed in previous scheduling moments and are now available for processing; (3) procedures available for processing contained in the newly arrived workpieces.

It needs to be taken into consideration that each machine has the same processing speed set $\mathcal{V} = \{v_1, v_2, \ldots, v_d\}$ containing d different values, and that machine speed should remain unchanged during each procedure's course of processing. The basic processing time for the jth procedure of workpiece J_i at machine M_k is η_{ijk}. When procedure O_{ij} is processed at speed v_l at the machine M_k, its actual processing time is $p_{ijkl} = \eta_{ijk} / v_l$, and the constraints fulfilled by this procedure during the course of processing are identical with Sect. 8.2, all of which are listed on the first row in Table 9.1.

Through optimizing the four objectives, we generate the new scheduling scheme, the details of which are as follows:

$$\min F = [f_1(t_l), f_2(t_l), f_3(t_l), f_3(t_l)]^{\mathrm{T}} \tag{9.1}$$

Table 9.1 Notations used for describing the mathematical model

t_0	Initial moment
t_l	Rescheduling time point $l = 1, 2, \ldots$
$n(t_l)$	Rescheduling time point t_l number of workpieces with machinable operations
$J_i(t_l)$	The i of the t_l moment contains the workpieces of the machinable process
$m(t_l)$	Number of machines that can be released at t_l time
$M_k(t_l)$	k machine released at t_l time, $k = 1, 2, \cdots, m(t_l)$
$C_i(t_l)$	t_l time, the completion time of all processable parts of the workpiece $J_i(t_l)$
$e_i^l(t_l)$	Number of machinable processes included in workpiece $J_i(t_l)$ at t_l time
$DD_i(t_l)$	t_l time, the delivery date of all processable parts of $J_i(t_l)$
$S_i(t_l)$	t_l time, workpiece $J_i(t_l)$ the start time of the first process in the process
$n^*(t_l)$	Rescheduling time point t_l has the number of workpieces that can be processed and processed
p^{k_r}	t_l time, the remaining processing time of the processing process on $M_k(t_l)$
$I_i(t_l)$	t_l time, workpiece $J_i(t_l)$ subscript of the first process in all processable processes
$c_{i(I_i(t_l)-1)}$	Process $J_i(t_l)$, the completion time of the last process in the process that was completed before the t_l time
$O_{ij}(t_l)$	Step j in the workpiece $J_i(t_l)$,
$M_{ij}(t_l)$	Process $O_{ij}(t_l)$ is a collection of machinable machines at t_l
$p_{ijk}(t_l)$	Process $O_{ij}(t_l)$ processing time on the machine $M_k(t_l) \in M_{ij}(t_l)$, $k = 1, 2, \cdots, \lvert M_{ij}(t_l) \rvert$, ($\lvert \bullet \rvert$ indicates the size of the collection)
$s_{ij}(t_l)$	Start processing time of process $O_{ij}(t_l)$
$c_{ij}(t_l)$	Finishing time of process $O_{ij}(t_l)$
$c^{k\_/textlast}(t_{l-1})$	The completion time of the last process of the machine $M_k(t_l)$ completed before t_l
$n_{M_k}(t_l)$	Number of operations assigned to the machine $M_k(t_l)$ at t_l, $k = 1, 2, \ldots, m(t_l)$
$O^{r,M_k(t_l)}$	The r process assigned on the machine $M_k(t_l)$, $r = 1, 2, \ldots, n_{M_k(t_l)}$
$p^{r,M_k(t_l)}$	Process $O^{r,M_k(t_l)}$ processing time
$c^{r,M_k(t_l)}$	Finishing time of process $O^{r,M_k(t_l)}$
K_i^*	The delivery period elastic coefficient of the workpiece $J_i(t_l)$, which satisfies the positive distribution of 1.5 with a variance of 0.5
E_k	Machine M_k processing energy per unit time
$A_k(t_l)$	Machine $M_k(t_l)$ in rescheduling period t_l initial release time
$X_{ijk}(t)$	Binary variable

$$f_1(t_l) = 5\left(\max_{i=1,2,\ldots,n(t_l)}(C_i(t_l)) - \min_{i=1,2,\ldots,n(t_l)}(S_i(t_l))\right)$$
$$+2\left(\sum_{i=1,2,\ldots,n(t_l)} \max(0, C_i(t_l) - DD_i(t_l))\right) \quad (9.2)$$

$$f_2(t_l) = \max_{k=1,2,\ldots,m(t_l)}\left(\sum_{r=1}^{nM_k(t_l)} p^{r,M_k(t_l)} + \chi_k p^{k\_r}\right) \quad (9.3)$$

$$f_3(t_l) = \sum_{O_{ij} \in \text{rush starting}} \gamma\left|s_{ij}(t_l) - s_{ij}(t_{l-1})\right| + \sum_{O_{ij} \in \text{delay starting}} \gamma\left|s_{ij}(t_l) - s_{ij}(t_{l-1})\right|$$
$$+\sum\left|c_{ij}(t_l) - c_{ij}(t_{l-1})\right| \quad (9.4)$$

$$f_4(t_l) = \int_{t_l}^{C_i(t_l)}\left(\sum_{i=1}^{n^*(t_l)}\sum_{j=1}^{e_i^*}\sum_{k=1}^{m(t_l)} K_k X_{ijk}(t) + \sum_{k=1}^{m(t_l)} SK_k z_k(t)\right) \quad (9.5)$$

$f_1(t_l)$—Scheduling efficiency objective; this chapter's scheduling efficiency adopts the two indexes of makespan and delay period for measurement. In Formula (9.1), makespan means the deviation between the latest processing completion time and the earliest processing start time of all workpieces containing procedures available for processing at moment t_l. Delay period refers to the deviation sum between the completion time and delivery period of all delayed workpieces. Here, $DD_i(t_l)$ is the delivery period for the ith workpiece containing procedures available for processing. Since the procedure in $J_i(t_l)$ may not necessarily be all the procedures of its yet unprocessed procedures, here, we revise the TWK rule raised in document [1] which is based on solving delivery period, in order to get the delivery period which we mark as

$$DD_i(t_l) = R_i + K_i^* \sum_{j=1,2,\ldots,I_i(t_l)+e_i'(t_l)-1} A\_p_{ij} \quad (9.6)$$

From Formula (9.6) we can see that $DD_i(t_l)$ is the time when the ith workpiece $J_i(t_l)$ arrives at the workshop plus the average processing time from its first procedure to the last procedure currently being processed then multiplied by factor K_i^*. The attribute of K_i^* is the tightness factor of $J_i(t_l)$ and is related to $J_i(t_l)$. This chapter adopts the method proposed by Zhang et al. [2] to conduct weighting of the two optimized objectives, and assign weight values 5 and 2 respective to the two objectives according to the suggestions of Ishibuchi and Murata [3].

$f_2(t_l)$—Machine load objective. It represents the total amount of time for all procedures to be processed by machines which can be released for processing. χ_k is binary variant. If at the moment t_l, machine $M_k(t_l)$ has procedure currently being processed, then its value is 1, otherwise 0.

$f_3(t_l)$—Rescheduling stability objective [4]. From Formula (9.4), we can see that stability comprises three parts. Its major concern is procedures which have been scheduled during the last rescheduling period yet still unprocessed and can be processed at this rescheduling moment. First, for the first part of the Formula, if these procedures start processing prior to this rescheduling moment, then multiply their processing start time deviation with factor γ. In this chapter, the value of γ is 1.5. For those procedures whose processing start time or completion time are advanced or delayed, solve the sum of their deviation time in the latter two parts of the formula, respectively. Here, different from literature [4], we take into consideration not only procedures with delayed completion time but also those with advanced completion time.

$f_4(t_l)$—Machine energy consumption objective. By introducing the energy consumption objective into traditional work shop scheduling, we design a reasonable scheduling scheme which, when responding to dynamic events, can effectively reduce materials' consumption and environmental impact during the course of workshop production. The energy consumption in this chapter deals mainly with the energy consumption of a machine in the processing and standby modes. In Formula (9.5), e_i^* represents, at the moment t_l, all the processing procedures of each workpiece $J_i(t_l), i = 1, 2, \ldots n^*(t_l)$ in the workpiece set which contains procedures available to be processed and procedures currently being processed. For $J_i(t_l)$, procedures which have completed processing before t_1 moment and which cannot be processed in the current scheduling cycle can be set aside. $X_{ijk}(t)$ is the binary decision variant. If at the moment t, machine $M_k(t_l)$ in the available machine set which each procedure in $J_i(t_l)$ corresponds to is in the processing status, the $X_{ijk}(t) = 1$; otherwise, $X_{ijk}(t) = 0$. If machine $M_k(t_l)$ is idle at moment t, then $z_k(t) = 1$; otherwise, $z_k(t) = 0$.

9.3 Pre-response Dynamic Scheduling Method

9.3.1 *Rescheduling Strategy Driven by Cycle and Event*

In the pre-response scheduling method, there are three commonly adopted rescheduling methods: cyclical rescheduling, rescheduling driven by events and rescheduling strategy combining the previous two. Cyclical rescheduling means rescheduling at intervals or every certain number of times, for instance, to conduct rescheduling when production and processing reaches a certain period, when the number of new workpieces reaches a certain number, or when the processed workpieces reach a certain number. The conditions for triggering cyclical rescheduling, such as temporal intervals, need to be determined by a combination of the enterprises' production scheme, rules and number of tasks. Rescheduling based on events means that rescheduling is triggered in case of emergencies, which normally include urgent order insertion and machine breakdown (repair).

Rescheduling based on events can actively response to dynamic events, yet lacks the pre-judgement capability of events. While cyclical rescheduling can improve the production system's stability, it fails to timely handle and lacks responsiveness to emergencies. Therefore, this chapter combines the strengths of these two scheduling strategies, to maintain the scheduling system's stability while increasing its responsiveness and capability to handle emergencies. For cyclical rescheduling, set the rescheduling cycle as ΔT, namely start rescheduling at every interval of ΔT. For dynamic event rescheduling, this chapter considers three emergencies: machine breakdown, machine repair and arrival of urgent workpieces. In production scheduling, when any machine breaks down or finishes repair, start rescheduling immediately. When urgent workpieces arrive, allocate weights 1, 2 or 4 to all the newly arrived workpiece according to their importance. When weight 4 workpiece reaches the workshop, trigger urgent workpiece arrival rescheduling. Here, the weight ratios of these three kinds of workpieces are 60%, 30% and 10%, respectively.

9.3.2 Construction of Static Scheduling Window

For each trigger moment of rescheduling, regard dynamic scheduling as a static scheduling window containing multiple workpieces and machines. Therefore, the entire scheduling progress of successive arrival of various workpieces can be regarded as made up of multiple static scheduling. At each trigger moment of rescheduling, it is especially necessary to update the current status of the workshop, considering the following three points:

(1) Information of machines available for processing: both machines which are out of order at present and machines which have not been repaired belong to machines unavailable for processing, marked as M^*. Machines under the category of M^* all belong to inoperable machines. All inoperable machines need to be removed from the set of machines available for processing; otherwise, the repaired machines need to be added to the set of machines available for processing.
(2) Information of procedures available for processing: procedures which need to be taken into consideration at the rescheduling moment include procedures previously unable to be processed, procedures yet unprocessed in the last rescheduling cycle, and procedures contained in the newly arrived workpieces. Merge these three types of procedure and get the procedure set ϑ^*. For any procedure in ϑ^*, so long as its machines available for processing belong to M^* or its previous procedure fails to be processed, then this procedure cannot be processed at this moment. Notice that, if the procedure O_{ij} cannot be processed at the current moment, then none of the succeeding procedures of this same workpiece can be processed.

(3) The processing start time available for each procedure. Suppose after (1) and (2) operations, we know that currently there are $n(t_l)$ workpieces containing procedures available for processing, $m(t_l)$ machines available to process procedures, then the processing start time available for each workpiece and each machine are [5]:

$$R_i(t_l) = \max\left(t_l, c_{i(I_i(t_l)-1)}\right), \quad i = 1, 2, \ldots, n(t_l) \tag{9.7}$$

$$A_k(t_l) = \max\left(t_l, c^{k\_\text{last}}(t_{l-1})\right), k = 1, 2, \ldots, m(t_l) \tag{9.8}$$

Based on the above two formula, taking into consideration the processing restraint for the same workpiece's procedure, we can get each time's rescheduling strategy. Suppose $O_{ij}(t_l)$ is the no. r procedure of scheduling at the machine M_k, then its processing start time is calculated as follows:

$$s_{ij}(t_l) = \begin{cases} \max\left(t_l, c_{i(I_i(t_l)-1)}, c^{k\_\text{last}}(t_{l-1})\right), & j = I_i(t_l), \quad r = 1 \\ \max\left(t_l, c_{i(I_i(t_l)-1)}, c^{r-1, M_k(t_l)}\right), & j = I_i(t_l), \quad r = 2, \ldots, nM_k(t_l) \\ \max\left(t_l, c_{i(j-1)}(t_l), c^{k\_\text{last}}(t_{l-1})\right), & j = I_i(t_l) + 1, \ldots, I_i(t_l) + e_i'(t_l) - 1, \quad r = 1 \\ \max\left(t_l, c_{i(j-1)}(t_l), c^{r-1, M_k(t_l)}\right), & j = I_i(t_l) + 1, \ldots, I_i(t_l) + e_i'(t_l) - 1, \quad r = 2, \ldots, nM_k(t_l) \end{cases} \tag{9.9}$$

Note 1: if the machine breaks down at t_l, and if there is any procedure undergoing processing at this machine, then immediately stops processing this procedure and wait for the malfunctioning machine to recover. When the malfunctioning machine which this procedure corresponds to is repaired, resume processing.

9.3.3 Process of Pre-response Scheduling

Based on NSGA-III algorithm, this chapter solves the problem of high-dimensional multi-objective flexible job shop scheduling by designing chromosome encoding and decoding methods based on the features of scheduling problems as well as effective evolutionary operators. The process of using the pre-response scheduling method to solve high-dimensional multi-objective optimization scheduling problem is shown in Fig. 9.1. At the initial scheduling moment, we use the NSGA-III algorithm to optimize scheduling efficiency, machine load and energy consumption, thereby obtaining a pre-response scheduling scheme. Then, in the course of implementing the entire scheduling system, when cyclical scheduling moment or three dynamic events (machine breakdown, urgent order insertion and machine repair) occur, trigger rescheduling. Through optimizing the four objectives of scheduling efficiency, maximum machine load, stability and energy consumption, we generate the scheduling scheme of the next rescheduling cycle. This scheduling scheme shall operate in the

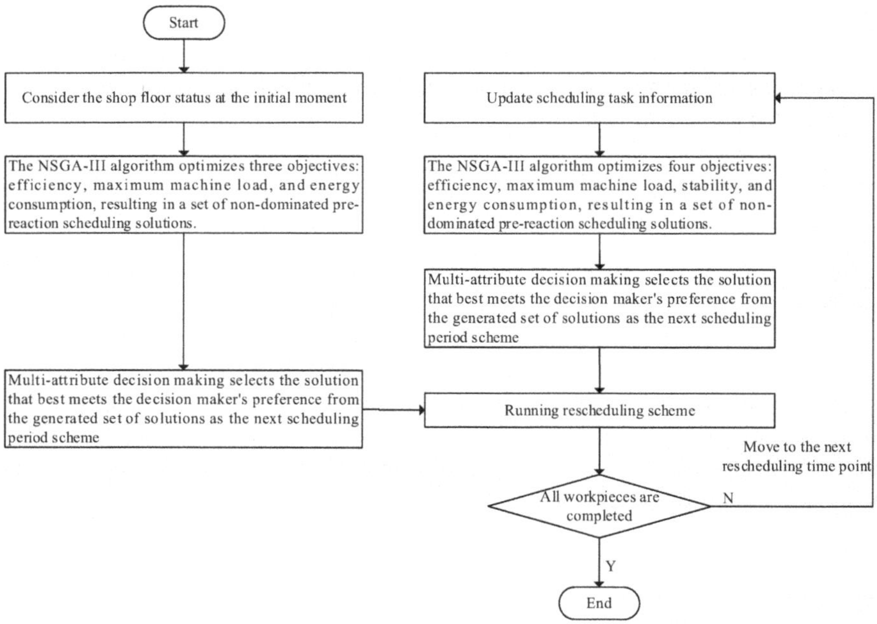

Fig. 9.1 Flowchart of the proposed predictive-reactive scheduling

next rescheduling cycle until the new rescheduling moment is generated. In short, the high-dimensional multi-objective flexible job shop scheduling problem is dynamic scheduling progress, which is made up of a series of static scheduling windows, each containing different machines and workpieces. At each rescheduling moment, when NSGA-III solves the high-dimensional multi-objective scheduling problem in correspondence with static windows, it generates a group of non-dominated solutions. Therefore, this chapter shall also use decision methods to select a solution to the best satisfaction of the decision-maker's preference.

Note 2: When solving high-dimensional multi-objective optimization problems, NSGA-II encounters great difficulties due to two reasons. First, when the number of optimization objectives exceeds three, most—or even all—of the solutions in the evolutionary population are non-dominated solutions. Yet the individual comparison mechanism of NSGA-II under the dominance of Pareto cannot provide an evolutionary population with enough pressure of selection, which results in the retarded evolutionary progress of the population. Second, in high-dimensional situations, the crowding distance operators of NSGA-II cannot truthfully reflect the distribution of Pareto front end in the space of high-dimensional objective. However, NSGA-III algorithm revises the selection operator in the NSGA-II framework and gives priority to those individuals of strong dominance and short distance to the reference line when selecting environments. It ensures convergence while paying particular attention to diversity. Besides, NSGA-III contains adaptive normalization which can effectively solve such actual optimization problem as the different dimensions of various target function values, and is therefore very suitable for solving high-dimensional multi-objective optimization problems.

9.4 Solving Static Scheduling Windows Based on NSGA-III

9.4.1 Algorithm Framework

The progress of NSGA-III algorithm is as illustrated in Algorithm 9.1. First, generate a group of reference points $\Lambda = \{\lambda_1, \lambda_2, \ldots, \lambda_N\}$. Here, $\lambda_j, j \in \{1, 2, \ldots, N\}$ is the vector of a dimension M. After generating the reference point, random initialization generates an evolutionary population $P(0)$ which contains N solutions $x_1, x_2, \ldots, x_N$. Then, initialize the ideal point z^* through solving the minimum value of all solutions in the initial population at every dimension, and this ideal point shall be updated in step 15 by $S(t)$. Steps 6–26 reiterates until the termination conditions are fulfilled. In the evolution of each generation of the population, first, generate filial population $Q(t)$ from the current parent population $P(t)$ by evolutionary operators (crossover and mutation), then merge populations $P(t)$ and $Q(t)$ to get a temporary population $R(t)$ of the size 2N. Steps 8–13 are the same non-dominated sorting process as that of the NSGA-II algorithm. In population $S(t) = \bigcup_{i=1}^{\tau} F_i$, F_i is the Pareto non-dominated individual at level no. i, τ satisfies $\sum_{i=1}^{\tau} |F_i| < N$ and $\sum_{i=1}^{\tau} |F_i| \geq N$. Steps 19–24 are selection process, in which all individuals of the first $\tau - 1$ levels enter the next generation of evolutionary population $P(t+1)$. The feature of NSGA-III lies in its selection of some individuals form F_τ and places them in $P(t+1)$. First, in step 21, individuals in population $S(t)$ need to undergo objective normalization, at which moment the reference point becomes the initial point. Then, use the initial point and reference point to construct the reference line. In steps 22–23, NSGA-III's cluster operators will attach each individual to specific reference line by calculating the distance between each individual and all other reference lines. Finally, select some individuals through implementing niches selection operator in step 24 so as to make the number of $P(t+1)$ reaches N.

In order to effectively solve the dynamic optimization problems while maintaining the system's continuity and stability, and in order to effectively catch original scheduling information, NSGA-III algorithm makes three aspects of design: (1) adopt mixed strategy to generate initial population; (2) design evolution operators based on specific questions; (3) add stability index into optimization objectives to keep the original scheduling scheme from major revision.

Algorithm 9.1 Proposed NSGA-III algorithm framework

1: Generate a set of weight vectors Λ
2: Initialize the population $P(0)$
3: Initialize Ideal point z^*

$t \leftarrow 0$
while the termination criterion is not satisfied **do**
Performing an evolutionary operation on the parent population $P(t)$ produces a child population $Q(t)$
Obtain mixed population $R(t) \leftarrow P(t) \cup Q(t)$
Non-dominated sorting hierarchy $\{F_1, F_2, \ldots\}$ of mixed population $R(t)$ $S(t) \leftarrow 0$, $i = 1$
while $|S(t)| + |F_i| \leq N|$ **do**
$S(t) \leftarrow S(t) \cup F_i$
$i \leftarrow i + 1$
end while
The last layer of individuals to be considered $F_\tau = F_i$ $F_{-} = F_i$
Update the ideal point with $S(t)$
if $|S(t)|{=}N$ | **then**
$P(t+1) = S(t)$, break
else
$P(t+1) = \bigcup_{J=1}^{\tau-1} F_j$
$K \leftarrow N - |P_{t+1}|$
/* K: The number of individuals to be selected in F_τ */
$[S(t), z^r] \leftarrow$ Normalize $(S(t), z*, \Lambda)$
/* z^r: Normalized reference point on the hyperplane*/
$[\pi(s), d(s)] \leftarrow$ Associate $(S(t), z^r)$
/*s: The distance between $d(s)$: s and $\pi(s)$, the nearest reference point of a solution $\pi(s)$: in $S(t)$
$\rho_j = \sum_{s \in S(t)/F_l} ((\pi(s) = j)?1 : 0)$
/* ρ_j: Number of niches at reference point $j \in z^r$ */
$P(t+1) \leftarrow= \left(K, S(t), P(t+1), F_\tau, z^r, \rho_j, \pi(s), |d(s)\right)$
end if
$t \leftarrow t + 1$
end while

9.4.2 Generation of Reference Point

NSGA-III adopts the system method of Das and Dennis [6] to generate a group of reference points evenly distributed in the objective space. This method has also been used by MOEA/D to generate a group of evenly distributed weight vectors. The number of reference points generated with this method is determined by both the number of objective dimensions M and the positive integer H.

First, consider the following equation

$$\sum_{i=1}^{M} x_i = H, \quad x_i \in \mathbb{N}, \quad i = 1, 2, \ldots, M \tag{9.10}$$

The number of solutions for this equation is

$$N = \binom{H + M - 1}{M - 1} \tag{9.11}$$

Suppose $\left(x_j^1, x_j^2, \ldots, x_j^M\right)^{\mathrm{T}}$ is the no. j solution of this equation, then the no. j reference point $\lambda_j = \left(\lambda_j^1, \lambda_j^2, \ldots, \lambda_j^M\right)^{\mathrm{T}}$ can be obtained from the following formula

$$\lambda_j^k = \frac{x_j^k}{H}, \quad k = 1, 2, \ldots, M \tag{9.12}$$

where H is the target partition value of each dimension. From the geometrical perspective, $\lambda_1, \lambda_2, \ldots, \lambda_N$ are all positioned above the hyperplane $\sum_{i=1}^{M} f_i = 1$. What needs special attention is that, when $H < M$, this method cannot generate the midpoint. When M value is large, $P(t + 1)$ will generate a large number of reference points which results in a massive number of evolution populations. To avoid such a phenomenon, when solving the scheduling problems of this chapter, NSGA-III also adopts the two-layered reference points method [7], which adopts relatively small H value. Let H_1 and H_2 be respectively the number of partition on the margin and inner layers, then the population size N can be calculated with the following formula

$$N = \binom{H_1 + M - 1}{M - 1} + \binom{H_2 + M - 1}{M - 1} \tag{9.13}$$

Figure 9.2 is the distribution of two-layered reference points in the 3D objective space, six of which are on the margin layer ($H_1 = 2$), three in the inner layer ($H_2 = 1$).

9.4.3 Chromosome Encoding and Decoding

The chromosome encoding method still adopts the MSOS encoding method. Suppose machine set: $\{M_1, M_2, \ldots, M_{10}\}$. Suppose machine set available to be released at the rescheduling moment: $\{M_1, M_3, M_4, M_7, M_9\}$. The newly arrived workpieces include J_6 and J_7, containing procedures $\{O_{61}, O_{62}, O_{63}, O_{71}, O_{72}\}$. The unfinished procedure of the last rescheduling cycle includes $\{O_{25}, O_{33}, O_{43}, O_{45}, O_{52}, O_{53}\}$. The chromosome encoding is as illustrated in Fig. 9.3.

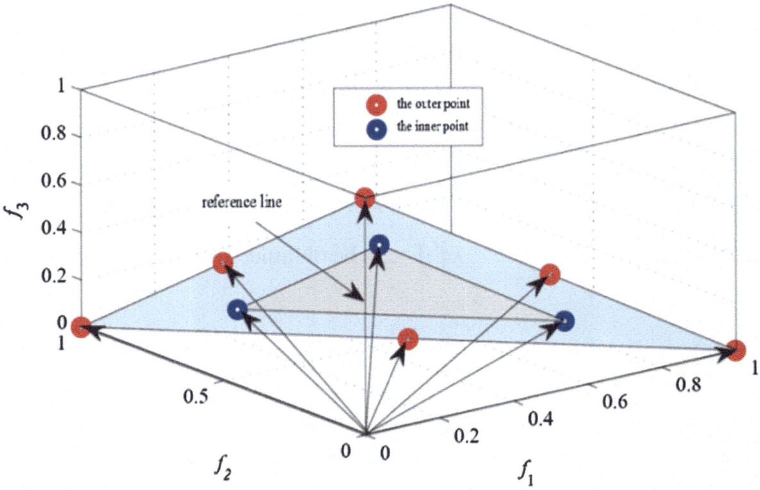

Fig. 9.2 Distribution of two-layered reference points in the 3D objective space

Operation	O_{25}	O_{33}	O_{43}	O_{44}	O_{45}	O_{52}	O_{53}	O_{61}	O_{62}	O_{63}	O_{71}	O_{72}
Distribution machine	M_1	M_3	M_4	M_4	M_7	M_9	M_1	M_7	M_9	M_1	M_9	M_7

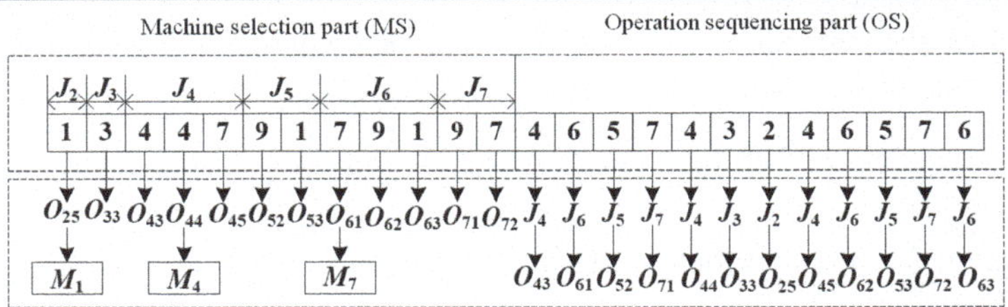

Fig. 9.3 Illustration of chromosome encoding

9.4.4 Population Initialization

At each rescheduling moment, construct a high-dimensional multi-objective flexible job shop scheduling problem which contains workpieces and machine set available to be processed after the update. In order to accelerate the convergence of NSGA-III while ensuring the system's stability, NSGA-III adopts the method of combining partial initialization [5] and random initialization, which avoids the potential destruction which complete random initialization may bring to the system's stability. The usage ratio of partial initialization and random initialization are 20% and 80%, respectively.

(1) Partial initialization

Partial initialization is expansion of the method for dynamic job shop scheduling proposed by Zhang et al. [8] on the issue of flexible job shop scheduling. This method preserves the processing priority order of unprocessed procedures in the original

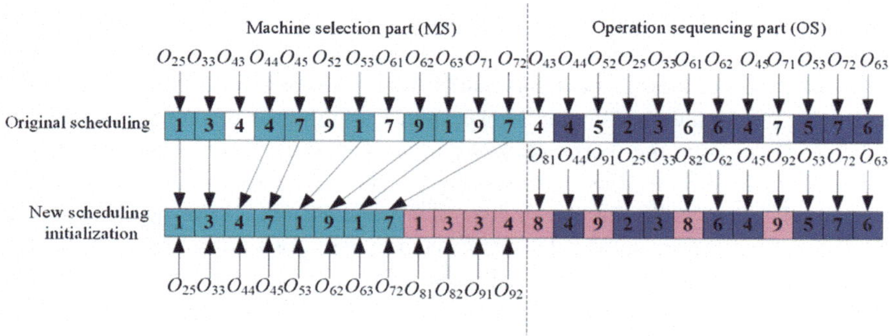

Fig. 9.4 Example of partial initialization

scheduling scheme, while maintaining the machine assignment scheme of unprocessed procedures at the same time. For the newly arrived workpieces' procedures available to be processed, randomly insert them to OS and randomly assign processing machines for them. As shown in Fig. 9.4, procedures O_{43}, O_{52}, O_{61} and O_{91} have all finished processing the last rescheduling cycle. The machines assigned to procedures O_{25}, O_{33}, O_{44}, O_{45}, O_{52}, O_{62}, O_{63} and O_{72} remain the same, and so does these procedures' order of processing $O_{44} \rightarrow O_{25} \rightarrow O_{33} \rightarrow O_{62} \rightarrow O_{45} \rightarrow O_{53} \rightarrow O_{72} \rightarrow O_{63}$. While procedures O_{81}, O_{82}, O_{91}, and O_{92} in respective correspondence to new workpieces J_8 and J_9 are randomly inserted into the OS part encoding and are randomly assigned to machines in the MS part.

(2) Random initialization

For the randomly initialized OS, the processing order of all procedures schedulable for this rescheduling cycle is randomly assigned. But MS adopts the GLR method proposed by Zhang et al. [9].

9.4.5 Evolutionary Operator

The evolutionary operator adopted in this chapter is different from solving continuous optimization problems. When solving the discrete manufacturing system's multi-objective optimization scheduling problem, since chromosome encoding includes the MS and OS parts, we need to design effective crossover and mutation operation for both parts in order to achieve individual evolution. The MS part adopts uniform crossover and two-point mutation. The OS part adopts POX crossover and exchange mutation.

For MS, we adopt the uniform crossover operation proposed in document [10]. First, generate a random number within range [1, *D*]. Then, generate r random numbers which are different from each other within the range [1, *D*]. For the two parental individuals P1 and P2, exchange their genes at corresponding positions of the r random numbers, while genes at other positions remain unchanged. For the crossover in procedure order arrangement part, adopt the POX crossover proposed by Zhang et al. [10]. First, randomly divide the workpiece set into two workpiece sub-set JS1 and JS2. Then, copy the workpieces of parent generation chromosome P1 and P2 which are in JS1/JS2 to filial chromosome C1 and C2, and keeping their positions unchanged. Later, copy the workpieces of P1 and P2 which are not included in JS1/JS2 to filial chromosome C2 and Ca, keeping their positions unchanged as well. Figures 9.5 and 9.6 are respective illustrations of MS part's uniform crossover and OS part's POX crossover. Adopt two-point mutation [11] for the MS part. First randomly select two genetic positions, then from the containable machine set of chromosome's corresponding procedures, the chosen genetic positions select a different machine to replace the current one. The OS part uses exchange mutation, namely to randomly select two genetic positions and exchange the procedures in correspondence to the genetic positions. The mutation operation of single chromosome's MS and OS parts is as illustrated in Fig. 9.7.

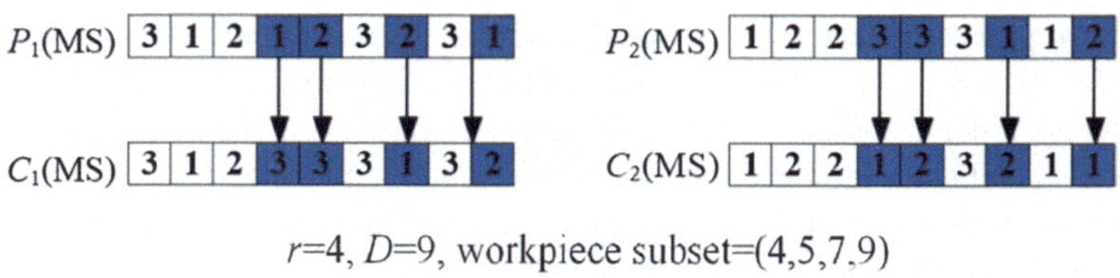

Fig. 9.5 Uniform crossover for the MS

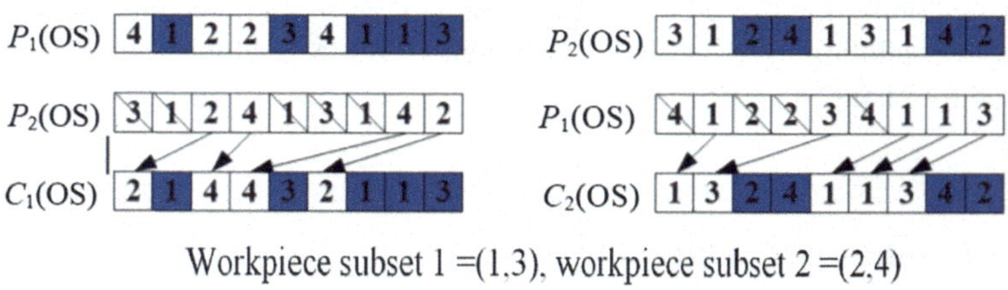

Fig. 9.6 POX crossover for the OS

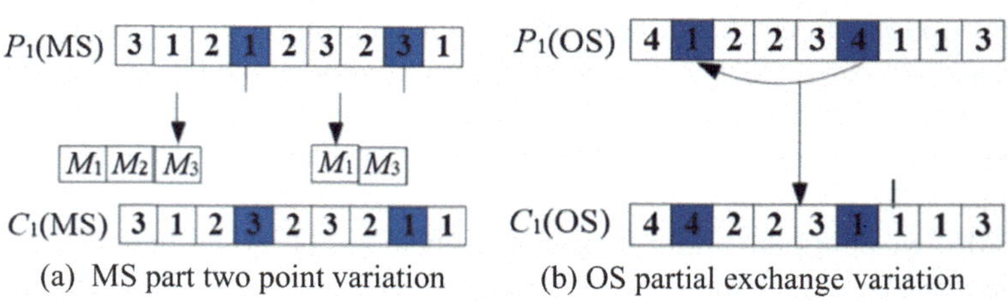

Fig. 9.7 Illustration of mutation

9.4.6 Normalization of Adaptive Objectives

One feature of NSGA-III is that the insertion of adaptive objectives normalization for the effective solution of non-normalization problems. And this chapter's targets of optimization also differ in terms of dimension. Therefore, before selecting an environment, conduct target normalization of all individuals in $S(t)$. We take into consideration the fact that every dimension z^* in the ideal point $z^* = \left(z_1^*, z_2^*, \ldots, z_M^*\right)^{\mathrm{T}}$ has been estimated by the minimum f_i, $i = 1, 2, \cdots, M$ sought by the target prior to target normalization. Thus, the major task of target normalization is to estimate the worst point $z_{\mathrm{nad}} = \left(z_1^{\mathrm{nad}}, z_2^{\mathrm{nad}}, \ldots, z_M^{\mathrm{nad}}\right)^{\mathrm{T}}$.

First, convert the target value $f_i(x)$ which each individual in $S(t)$ corresponds to into $f_i'(x) = f_i(x) - z_i^*$. Then, make $w_j = \left(w_{j1}, w_{j2}, \ldots, w_{jM}\right)^{\mathrm{T}}$ the axis direction of the target axis f_j, then w_{ij} fulfils

$$w_{ji} = \begin{cases} 10^{-6}, & i \neq j \\ 1, & i = j \end{cases} \tag{9.14}$$

Later, search for a solution $x \in S(t)$ in population such that

$$\mathrm{ASF}\left(x, w_j\right) = \max_{i=1}\left\{\frac{f_i'(x)}{w_{ji}}\right\} \tag{9.15}$$

Then, the extreme point e_j in the direction of the target axis f_j is endowed as the target vector corresponding to x, i.e. $e_j = f(x)$.

After reviewing all the targets, M target axes' corresponding extreme points $e_1, e_2, \ldots, e_M$ then make up a hyperplane. Suppose matrix $E = (e_1 - z^*, e_2 - z^*, \ldots, e_M - z^*)^{\mathrm{T}}$ and $u = (1, 1, \ldots, 1)^{\mathrm{T}}$, then the hyperplane's intercept $a_{1,} a_2, \ldots, a_M$ in the direction of $\left(1, z_2^*, \ldots, z_M^*\right)^{\mathrm{T}}, \left(z_1^*, 1, \ldots, z_M^*\right)^{\mathrm{T}}, \ldots, \left(z_1^*, \ldots, z_{M-1}^*, 1\right)^{\mathrm{T}}$ can be calculated through the following formula

$$\begin{pmatrix} (a_1 - z_1^*)^{-1} \\ (a_2 - z_2^*)^{-1} \\ \cdots \\ (a_M - z_M^*)^{-1} \end{pmatrix} = E^{-1}u \tag{9.16}$$

At this time, a given dimension z_i^{nad} of the worst point $z_{\mathrm{nad}} = \left(z_1^{\mathrm{nad}}, z_2^{\mathrm{nad}}, \ldots, z_M^{\mathrm{nad}}\right)^{\mathrm{T}}$ can be endowed with the value a_i. Finally, for every individual x_i in the population, the no. j dimensional representation of its normalized target function value is

$$f_j(x) = \frac{f_j'(x)}{z_j^{\mathrm{nad}} - z_j^*} \tag{9.17}$$

Here, what needs special attention is the fact that, during the process of constructing a hyperplane, it is possible that M extreme points cannot form a hyperplane because the order of matrix $\boldsymbol{E}$ is smaller than M. Besides, even if a hyperplane can be established, in some directions the intercept would be unobtainable or the obtained intercept could not satisfy $a_i > z_i^*$. The original version of NSGA-III does not give the course of processing for the above problems. Here, we use the processing method proposed in document [12], namely to establish f_i as the maximum value of all non-dominated individuals in $S(t)$ towards target f_i.

9.4.7 *Niches Preserving Operator*

After target normalization of individuals in the population $S(t)$, the ideal point now becomes the original point. Then, link every reference point with the ideal point to form reference lines and calculate the vertical distance between each solution and all the reference lines. Suppose λ_j is the no. j reference line, then the vertical distance between solution x and reference line λ_j is as illustrated in Fig. 9.8 and can be obtained through the following formula

$$d(x, \lambda_j) = \left\| \left(x - \lambda_j^T x \lambda_j / \left\| \lambda_j \right\|^2 \right) \right\| \tag{9.18}$$

Each individual shall be attached to the reference line of shortest vertical distance to it. When every individual has its own attaching reference line, calculate the niches number of each reference line. The niches number ρ_j of no. j reference point's

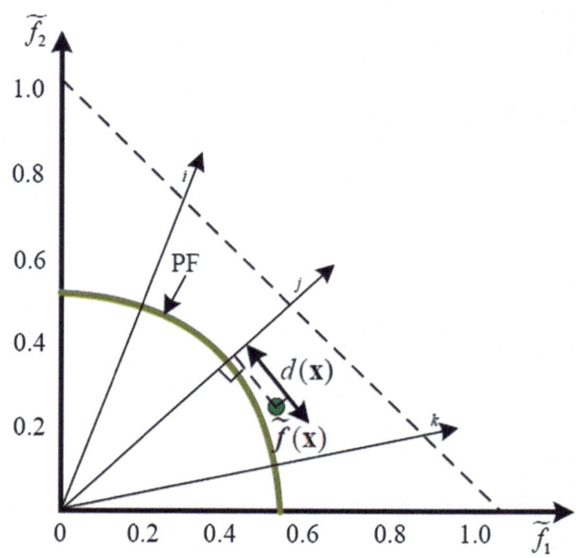

Fig. 9.8 Illustration of the perpendicular distance from population member to reference line

corresponding reference line represents the number of individuals attached to it in population $S(t)\backslash F_\tau$. From now on, implement the niches selection operator, the actual course of which is as follows. First, select the reference point set $J_{\text{min}} = \{j : \arg\min_j \rho_j\}$, the reference line corresponding to each reference point in which has a minimum ρ_j. When $\|J\| > 1$, randomly select a reference point $\bar{j} \in J_{\text{min}}$ from J_{min}. For the chosen reference point $\bar{j}$, if there is no individual in F_τ related with it, then discard this reference point and neglect it in the following selection. Then, recalculate J_{min} and select $\bar{j}$ again. If there is any individual F_- which is related with it, then take $S(t)$ into consideration, which also has two alternative situations: (1) if $S(t)$, then select an individual from F_τ which is attached to this reference point and has the smallest vertical distance to the reference line, and add it to $P(t+1)$; (2) if $\rho_j \geq 1$, then randomly select an individual from individuals attached to this reference point and put it into $P(t+1)$. After implementing the two situations ρ_j and $\rho_j + 1$. The above niches selection operator will repeat implementation until the size of the population $P(t+1)$ reaches N.

9.4.8 Hierarchical Analysis Method's Decision-Making

Since this chapter's optimization problem takes the objective of stability into consideration, at a given moment, the Pareto non-dominated solution obtained by using NSGA-III may have a value of 0 in correspondence to this objective. This section adopts the AHP hierarchical analysis method [13] to select a solution to the best satisfaction of the decision-maker's preference. First, $W = (w_1, w_2, \ldots, w_M)^{\mathrm{T}}$ is the weight which each target corresponds to; then, for every solution x_i^*, $i = 1, 2, \ldots, |X^*|$ in the Pareto best solution set X^*, its corresponding target vector $F_i = \left(f_i^1, f_i^2, \ldots, f_i^M\right)^{\mathrm{T}}$ undergoes such target normalization as follows:

$$\tilde{f}_i^j = \frac{f_j^{\max} - f_i^j}{f_j^{\max} - f_j^{\min}}, \quad j = 1, 2, \ldots, M \tag{9.19}$$

where $f_j^{\min}$ and $f_j^{\max}$ are, respectively, the minimum and the maximum values of all solutions' corresponding target vector at each dimension in X^*. Then, through computing the weighted sum of every target value $\tilde{F}_i$ after normalization, we can get each solution's corresponding degree of satisfaction ξ_i. Finally, select the solution with maximum degree of satisfaction as the most satisfactory scheduling solution x_{best}. The above process is summarized in Algorithm 9.2.

Algorithm 9.2 AHP-based multi-attribute decision-making framework

1: Structural judgment matrix ζ^*
2: Find the target weight vector W by judging the matrix ζ^*
3: $F_i = f(x_i), i = 1, 2, \ldots, |X^*|$

4: Targeting $\{F_1, F_2, \ldots, F_{|X^*|}\}$ to normalize, get $\left\{\tilde{F}_1, \tilde{F}_2, \ldots, \tilde{F}_{|X^*|},\right\}$
5: **for** every solution x_i^* in X^* **do**
6: Calculate the satisfaction degree $\zeta^* = W^T \tilde{F}_i$ of the solution x_i^*
7: **end for**
8: Choose the solution with the greatest satisfaction as x_{best}

9.5 Experimental Design and Analysis

9.5.1 Experiment Settings

In order to effectively validate the effectiveness of the proposed method, this section imitates a real flexible job shop environment for simulation. The proposed solution-seeking method based on NSGA-III adopts MATLAB programming language, its operating environment is: Intel Corei5-6500 3.20 GHz processor with 8 GB storage. The initial workshop has 10 workpieces and 10 machines. For workpieces' relevant information, see document [5]. The simulation experiment terminates when the number of workpieces in the workshop scheduling reaches 300, or in other words, when the number of new workpieces reaches 290. To ensure the fairness of comparison, the data used for generating workshop scheduling examples are summarized in Table 9.2. All the practical examples of problems adopted in the following experiments derive from data in this table.

The parameter setting of NSGA-III algorithm is as illustrated in Table 9.3.

Use three performance indexes-HV measurement, IGD measurement and DCI measure—to evaluate the performance of comparison algorithms. When evaluating an algorithm, HV and IGD measurements are prioritized in consideration, while DCI measurement is considered at the last. Due to the prioritized targets' difference in dimension, when calculating the IGD and HV measurements, we normalize all Pareto solutions' corresponding target vectors according to Formula (9.20) before calculating the three measurements.

$$\tilde{f}_i(x) = \frac{f_i(x) - f_i^{\min}}{f_i^{\max} - f_i^{\min}} \tag{9.20}$$

where $f_i^{\max}$ and $f_i^{\min}$ take respectively all comparison algorithms' maximum and minimum values of $f_i(x)$ in all the results obtained from workshop examples at the current rescheduling moment.

Therefore, when calculating the HV measurement, the reference point is set as $(1.1, 1.1, 1.1, 1.1)^{\mathrm{T}}$.

Through four groups of experiments, this chapter shall verify the performance of its proposed algorithm from different perspectives.

Table 9.2 Summary of the flexible job shop data adopted in the experiments

Processing parameter name	Parameter value
Job shop	
Number of machines (m)	10
Machine mean time to failure (MTBF)	U[100, 300][a]
Machine repair time (MTTR)	U[20, 120][a]
TBF distribution function	Exponential distribution a with mean MTBF
TTR distribution function	Exponential distribution a with mean MTTR
Distribution of elastic modulus K	Normal distribution with mean 1.5 and variance 0.5
Processing time per process	Exponential distribution with mean 1
Machine energy consumption per unit time E_k	Round(U[2,4])
Rescheduling period ΔT	12
Unit energy consumption per machine idle time SE	1
Workpiece	
Number of operations per workpiece	U[1, m][b]
Number of machine able machines per process	U[1, m]
Processing time of each process on the corresponding processing machine	Exponential distribution with mean 1
Average time interval for workpiece arrival (MTBJA)	0.6.25
Time distribution of new workpiece arrival	Exponential distribution with mean MTBJA
Delivery tightness coefficient distribution	Normal distribution with mean 1.5 and variance 0.5
Workpiece weight	30% of the workpiece weights are 1, 60% of the workpiece weights are 2, and 10% of the workpiece weights are 4
Number of new workpieces	920
Performance indexes	Efficiency
	Maximum machine load
	Stability

[a]Relevant data from the literature [5]
[b]$U(a; b)$ represents a randomly generated random number that satisfies evenly distributed in the interval $[a, \ b]$

Table 9.3 Parameter settings of NSGA-III

Parameter settings	Parameter value	Parameter settings	Parameter value
Population size (N)	120	Number of divisions (H)	7
Cross probability (P_c)	1.0	Mutation probability (P_m)	0.1
Maximum objective function evaluation times	15,000	Algorithm independent run times	15

(1) Comparison with variants of its own algorithm.
(2) Performance comparison with different MOEAs (including comparison of initial scheduling and dynamic scheduling).
(3) Comparison with the rule-based complete reaction scheduling method.
(4) Impact of different scheduling cycle ΔT on the scheduling performance.

9.5.2 Comparison of Initialization Scheduling

For the initial moment $t_0 = 0$, there are 10 machines available for processing in total, the number of workpieces to be processed is 10, and there are 50 procedures in total. Conduct a performance comparison between NSGA-III algorithm and the three classic MOEAs respectively in initialization scheduling. The three algorithms for comparison are EFR-RR, crEA and ε-MOEA. The previous two MOEAs are especially suitable for calculating MaOPs, and we have applied EFR-RR and crEA to solve the high-dimensional multi-objective flexible job shop's scheduling problem in the last chapter. Here, special emphasis should be given to ε-MOEA, which has been successfully applied to solving multi-objective dynamic flexible job shop's scheduling problem in document [5]. Since it is initialization scheduling, we need not consider the objective of stability, and there are only three optimization objects, namely efficiency, machine load and energy consumption. Then, set the partition value H at the moment of calculating initialization scheduling to be 14, so as to maintain the size of the population at 1230, and randomly generate initialization population. In EFR-RR, we still set a solution's adaptability function number K permissible for collation at 2. Both crEA and EFRRR adopt the same evolutionary operator and the same population initialization strategy as those of NSGA-III. As for ε-MOEA, apart from the randomly generated initialization population, all other evolutionary operators and parameter setting are identical as document [5]. In order to fairly compare the four algorithms and overcome random error, each algorithm is independently operated for 15 times, and terminates each time when it reaches the maximum function evaluation number of times.

Table 9.4 gives the operation results of NSGA-III and other three MOEAs in initialization scheduling, including the best, median and worst values of HV, IGD and DCI measurements, and mark the optimum results in bold. Besides, conduct Wilcoxon rank sum test of significance level 0.05 towards the results obtained by the two algorithms in comparison. From Table 9.4 we can see that, as for HV measurement, while NSGA-III achieves better median and worst values, but in significance test, it does not differ from the HV measurement values obtained by the other three algorithms. Whereas IGD measurement and DCI measurement values are rather similar, and NSGA-III is obviously superior to the other three algorithms. To a certain extent, it indicates that, when solving the initialization scheduling, NSGA-III can obtain Pareto solution set with better convergence and diversity. But in terms of CPU

Table 9.4 Best, median and worst values with respect to three metrics for four MOEAs on initial static scheduling

Algorithm	HV	Wt1	IGD	Wt2	DCI	Wt3	CPU time (s)
NSGA-III	1.1149	–	**0.0869**	–	**0.7541**	–	128.3610
	0.8567		**0.1674**		**0.4952**		
	0.6783		**0.2458**		**0.1458**		
EFR-RR	1.0144	a	0.1088	b	0.3958	b	108.6855
	0.8283		0.2079		0.1852		
	0.5758		0.3177		0.0000		
crEA	1.1102	a	0.0985	b	0.4167	b	105.8385
	0.7711		0.1930		0.2722		
	0.5022		0.3879		0.0000		
ε-MOEA	**1.1539**	a	0.1263	b	0.4583	b	60.9021
	0.8564		0.2196		0.1250		
	0.5323		0.3336		0.0000		

[a]Indicates that the result is significantly inferior to NSGA-III
[b]Indicates that the results are significantly different from NSGA-III

time, NSGA-III has the most calculation cost, whereas ε-MOEA has the least calculation cost and highest efficiency. Figure 9.9 is the Gantt graph of the best scheduling scheme decided by AHP hierarchical analysis at the initial scheduling moment.

9.5.3 Comparison of Dynamic Scheduling Processes

This section shall discuss in details the performance comparison of different MOEAs during the entire rescheduling progress. Rescheduling differs from initialization scheduling mainly in two aspects: (1) apart from objectives of scheduling efficiency, machine load and energy consumption optimization, there is an additional objective of stability; (2) the initializations of NSGA-III, EFR-RR and crEA adopt the mixed strategy proposed in Sect. 9.4.3, whereas ε-MOEA adopts the initialization strategy proposed in document [5]. Furthermore, in order to verify the effectiveness of the proposed population initialization method, we also compare NSGA-III with NSGA-III*. The only difference between NSGA-III and NSGA-III* lies in that the latter's initialization population is randomly generated. In the entire course of dynamic scheduling, there are 77 times of effective rescheduling (including rescheduling containing procedures available to be processed) in total. At each rescheduling moment, we run each of the five MOEAs for 15 times independently and conduct significance test of the HV, IGD and DCI measurements obtained by these five MOEAs. Table 9.5 summarize results of the significance tests of NSGA-III and other algorithms in HV, IGD and DCI indexes at 77 times of rescheduling moments.

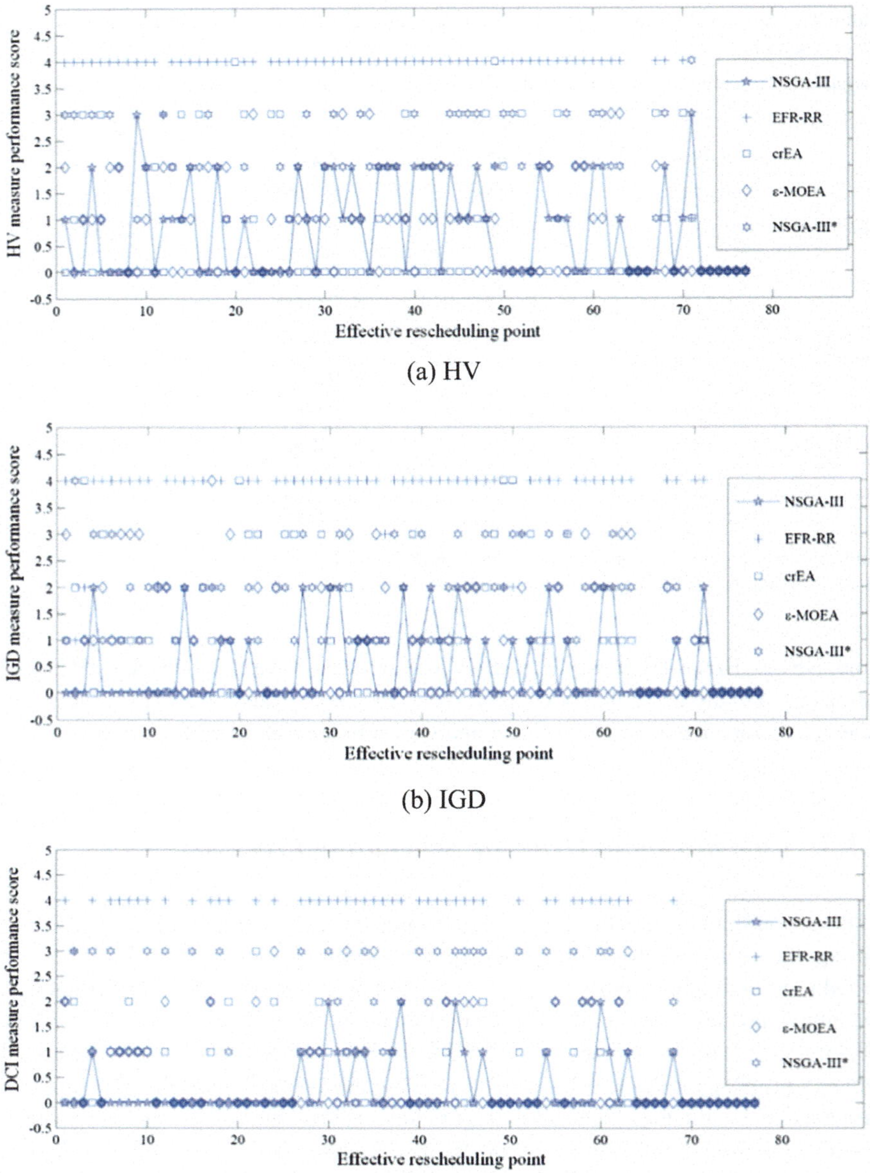

(a) HV

(b) IGD

(c) DCI

Fig. 9.9 Performance score of IGD, HV and DCI metrics at each rescheduling point for the compared five algorithms

Table 9.5 Significance test between NSGA-III and the other algorithm

	Results	NSGA-III*	EFR-RR	crEA	ε-MOEA
NSGA-III versus (HV)	B	43	68	22	21
	W	1	1	34	29
	E	33	8	21	27
NSGA-III versus (IGD)	B	44	68	36	32
	W	0	0	20	18
	E	33	9	21	27
NSGA-III versus (DCI)	B	36	48	16	22
	W	0	0	12	10
	E	41	29	49	45

Figure 9.9 gives the performance marks of NSGA-III and other MOEAs for HV, IGD and DCI measurements at different rescheduling moments. The lower the performance mark, the better the overall performance of HV, IGD and DCI. In Fig. 9.10, we use straight lines to connect all the marks obtained by NSGA-III. Furthermore, Fig. 9.11 is the average performance of all comparison algorithms at all rescheduling moments, with their ranking of performance given in brackets.

Figure 9.11 displays the CPU time of four MOEAs solved at each rescheduling moment. From Fig. 9.11, we can see that ε-MOEA has the highest calculation efficiency. This is mainly because ε-MOEA adopts a stable method to produce filial individuals, whereas the other three MOEAs adopt the producing mode to generate individuals. Besides, since NSGA-III adopts non-dominated ranking hierarchy and niches' choice of individuals mechanism based on reference points, it has the highest computation consumption, although its calculation efficiency is still acceptable.

By analysing all the above experiment results we can conclude that, compared with the other MOEAs' pre-response scheduling methods, the pre-response scheduling method based on NSGA-III is highly competitive in solving high-dimensional multi-objective dynamic flexible job shop scheduling problems, and holds certain advantage in consideration of Pareto solution set's convergence and diversity, achieving rather excellent balance between convergence and diversity to a certain extent.

9.5.4 Comparison with the Rule-Based Complete Reaction Scheduling Method

For further testing of NSGA-III's performance, this section shall compare NSGA-III with the rule-based rescheduling method. Considering sub-problems of machine selection and procedure sorting which are inherent of FJSSP, we set reasonable machine allocation rule and procedure priority scheduling rule to these two sub-problems respectively in rescheduling. Here, the machine allocation rule adopted

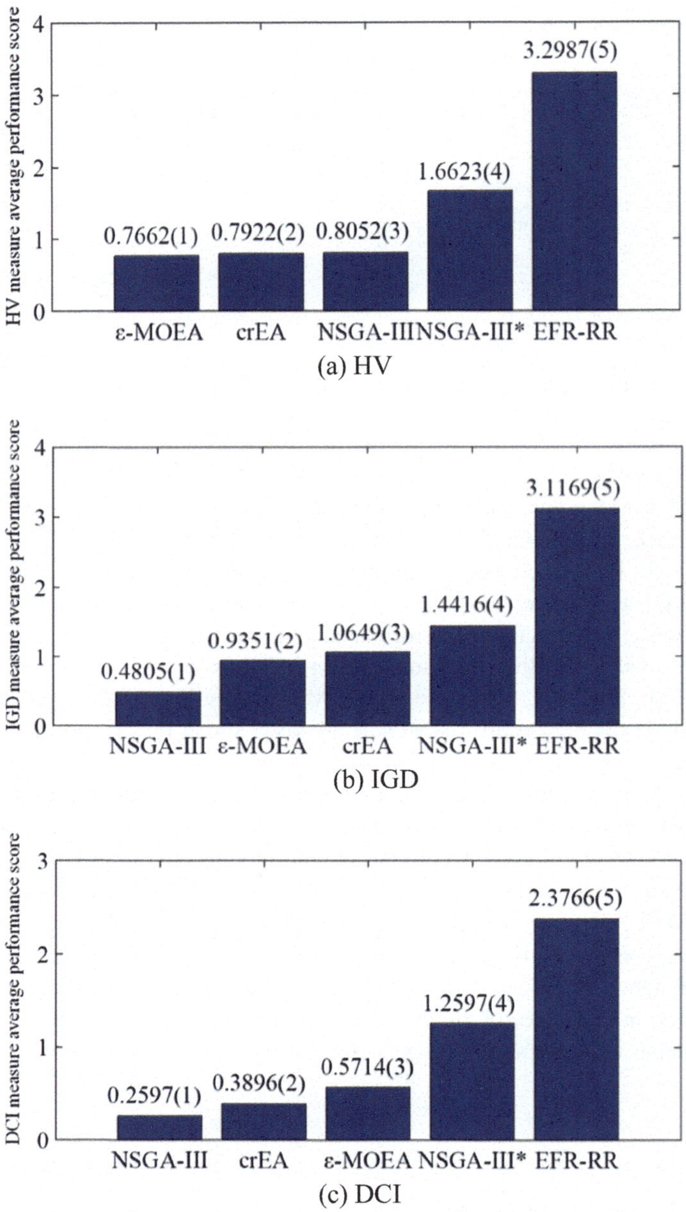

Fig. 9.10 Ranking in the average performance score over all rescheduling points for the compared five algorithms

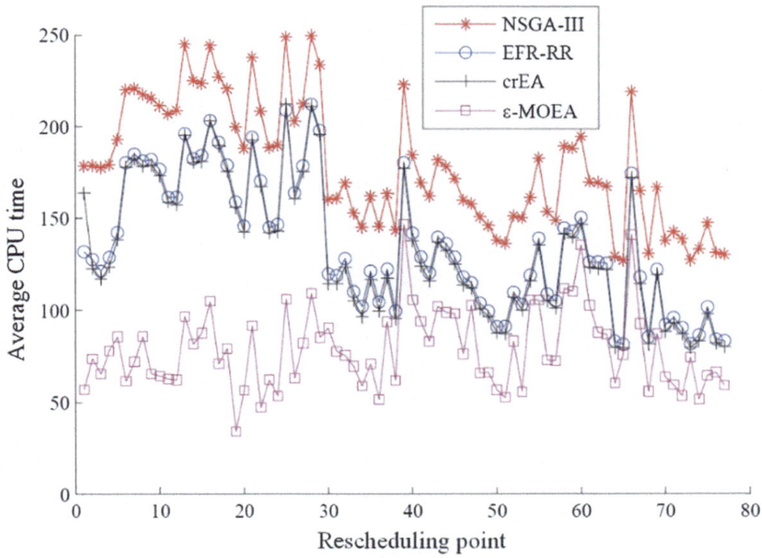

Fig. 9.11 CPU time (in seconds) comparisons of four MOEAs at each rescheduling point

is as follows: (1) from the currently available machine set, select the machine with shortest processing time (SPT) for each procedure; (2) from the currently available machine set, select the machine with least load for each procedure, and update the machine load in time once the procedure has selected a machine; (3) randomly select an available machine for each procedure. Here, we term these three machine allocation rules as MAR1, MAR2 and MAR3, respectively.

When one machine is in idle status, the procedure priority rule decides which procedure in its procedure waiting list is given processing priority. This experiment adopts four classical procedure priority rules [14], namely shortest processing time rule (SPT), first-in-first-out rule (FIFO), last-in-first-out rule (LIFO) and random rule (Random). Combine all machine allocation with procedure priority rules, we get 12 mixed scheduling rules, all of which are listed on the first row in Table 9.6.

Adjustment of partial information can neither give an exact scheduling method nor make an overall decision for scheduling scheme adjustment at each rescheduling moment via optimizing multiple indexes, as the MOEAs-based scheduling methods do. Therefore, efficiency, stability, machine load and energy consumption indexes designed for each rescheduling moment are not applicable. Here, with regard to the entire course of scheduling, namely once the entire scheduling of the 290 workpieces which arrive at the workshop successively has finished, we compare the method based on scheduling rule with the NSGA-III method in terms of the three indexes of efficiency, average machine load and total energy consumption. Use the NSGA-III and the 12 scheduling rules to run independent operation of the entire scheduling course for 15 times, respectively, and the average values and standard deviations of the three objectives obtained are listed in Table 9.6. From Table 9.6, we can clearly

Table 9.6 Comparison of NSGA-III against the existing dynamic scheduling methods

Rescheduling method	Efficiency	Average machine load	Energy consumption
MAR1 + SPT	5.0996e+4 ±1.5063e−11	68.1407 ± 1.4710e−14	7.7447e+3 ± 0
MAR2 + SPT	5.2413e+4 ± 3.8546e+2	83.6820 ± 0.6434e+0	8.0624e+3 ± 1.3387e+1
MAR3 + SPT	5.9375e+4 ± 7.1959e+2	123.4930 ± 1.7962e+0	8.9089e+3 ± 3.7072e+1
MAR1 + FIFO	5.2849e+4 ± 7.5313e−12	67.9350 ± 1.4710e−14	7.7420e+3 ± 2.8242e−12
MAR2 + FIFO	5.3924e+4 ± 7.0292e+2	84.4639 ± 0.8408e+0	8.0734e+3 ± 2.2512e+1
MAR3 + FIFO	7.8649e+4 ± 1.7450e+3	159.7179 ± 2.3204e+0	9.6833e+3 ± 5.6074e+1
MAR1 + LIFO	5.2668e+4 ± 1.5063e−11	68.3825 ± 1.4710e−14	7.7438e+3 ± 1.8828e−12
MAR2 + LIFO	5.4755e+4 ± 6.4467e+2	88.4318 ± 1.4604e+0	8.1558e+3 ± 2.9250e+1
MAR3 + LIFO	8.7948e+4 ± 3.0866e+3	159.7195 ± 4.1469e+0	9.6495e+3 ± 1.0039e+2
MAR1 + Random	5.2547e+4 ± 1.4535e+2	68.3564 ± 0.3164e+0	7.7472e+3 ± 5.5229e+0
MAR2 + Random	5.4050e+4 ± 6.3746e+2	87.1207 ± 1.2124e+0	8.1318e+3 ± 2.4477e+1
MAR3 + Random	7.5357e+4 ± 2.2649e+3	159.4949 ± 2.8777e+0	9.6783e+3 ± 6.1976e+1
NSGA-III	5.0475e+4 ± 1.0207e+2	73.8696 ± 0.3967e+0	7.8263e+3 ± 7.8082e+0

see that, for these three objectives, the objective values obtained by NSGA-III are far superior to most of the scheduling rules, especially in the case of efficiency objective. But we could also see that, when adopting the four mixed rules of MAR1 + SPT, MAR1 + FIFO, MAR1 + LIFO and MAR1 + Random, the average machine load and energy consumption values are smaller than the results obtained by using NSGA-III. This is mainly because MAR1 rule assigns each procedure to machine with shortest processing time in the available machine set. But to a certain extent, it also results in excessive number of procedures waiting for processing at a certain number of machines, and the prolonged waiting hours affect efficiency. Comprehensive analysis of the above results leads to the conclusion that the rescheduling method based on NSGA-III is more effective than methods based on scheduling rules, and it can improve scheduling efficiency to a great extent.

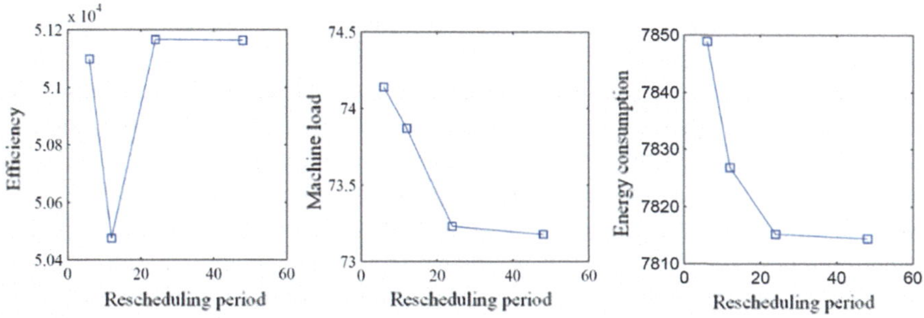

Fig. 9.12 Influence of different rescheduling period

9.5.5 *Impact of Different Scheduling Cycles*

To test the impact of different scheduling cycles ΔT on scheduling performance, we set four different ΔT values at 6, 12, 24 and 48, and we can see that the lengths of these four scheduling cycles grow geometrically. Here, adopting the identical method of comparison as in Sect. 9.5.4, we run independent operation of the entire scheduling course for 15 times under different ΔT, respectively, and record the average values of the three indexes—scheduling efficiency, average machine load and energy consumption. Figure 9.12 displays the trend of the three indexes' average values. From this figure we can see that, as the scheduling cycle increases, the scheduling efficiency value first decreases and then increases, achieving the best scheduling efficiency value at $\Delta T = 12$ moment. As for the other two indexes, they tend to become smaller as the scheduling cycle increases. Therefore, we conclude that the increase of the times of cyclical scheduling can improve scheduling efficiency to a certain extent, yet excessive number of times of scheduling would reduce efficiency. Meanwhile, increasing times of cyclical scheduling would not help reduce the average machine load and energy consumption. This chapter sets ΔT at 12, which is very suitable for cyclical rescheduling.

9.6 Conclusion

This chapter takes into consideration such common interfering emergencies as machine breakdown (repair) and the arrival of urgent workpiece in the discrete flexible manufacturing system. It studies the problem of production/energy consumption high-dimensional multi-objective dynamic optimization and provides the pre-response scheduling method based on NSGA-III. This method adopts the rescheduling strategy which combines cyclical with dynamic events, thereby converting the dynamic scheduling process into a continuous series of static scheduling windows. At each static scheduling window, we use the NSGA-III algorithm in combination with mixes initialization strategy and the evolutionary operators based on scheduling problem to simultaneously optimize the four objectives of scheduling efficiency, maximum machine load, stability and energy consumption, while using the hierarchical analysis method to select a scheduling scheme to the best satisfaction of the decision-maker's preference at the rescheduling moment. For the experiment part, through creating a simulation environment with 10 machines and 300 successively arriving workpieces, we compare NSGA-III with three pre-response scheduling methods based on different MOEAs and with scheduling methods based on mixed rules. The results verify the superiority of the method proposed in this chapter.

References

1. Han Y, Geng Z, Liu Q (2014) Energy efficiency evaluation based on data envelopment analysis integrated analytic hierarchy process in ethylene production. Chin J Chem Eng 22(11):1279–1284
2. Adibi MA, Zandieh M, Amiri M (2010) Multi-objective scheduling of dynamic job shop using variable neighborhood search. Expert Syst Appl 37(1):282–287
3. Ishibuchi H, Murata T (1998) A multi-objective genetic local search algorithm and its application to flowshop scheduling. IEEE Trans Syst Man Cybern Part C: Appl Rev 28(3):392–403
4. Shen XN, Yao X (2015) Mathematical modeling and multi-objective evolutionary algorithms applied to dynamic flexible job shop scheduling problems. Inf Sci 298:198–224
5. Liu F, Wang QL, Liu GJ (2013) Content architecture and future trends of energy efficiency research on machining systems. J Mech Eng 49(19):88–92
6. Das I, Dennis JE (1998) Normal-boundary intersection: a new method for generating the Pareto surface in nonlinear multicriteria optimization problems. SIAM J Optim 8(3):631–657
7. Deb K, Jain H (2014) An evolutionary many-objective optimization algorithm using reference-point-based nondominated sorting approach. IEEE Trans Evol Comput 18(4):577–601
8. Zhang LP, Gao L, Li XY (2013) A hybrid intelligent algorithm and rescheduling technique for job shop scheduling problems with disruptions. Int J Adv Manuf Technol 65(5–8):1141–1156
9. Zhang GH, Gao L, Shi Y (2011) An effective genetic algorithm for the flexible job-shop scheduling problem. Expert Syst Appl 38(4):3563–3573
10. Draganescu F, Gheorghe M, Doicin CV (2003) Models of machine tool efficiency and specific consumed energy. J Mater Process Technol 141(1):9–15
11. Wang X, Gao L, Zhang C et al (2010) A multi-objective genetic algorithm based on immune and entropy principle for flexible job-shop scheduling problem. Int J Adv Manuf Technol 51(5–8):757–767
12. Yuan Y, Xu H, Wang B et al (2016) A new dominance relation-based evolutionary algorithm for many-objective optimization. IEEE Trans Evol Comput 20(1):16–37
13. Wang YM, Chin KS (2011) A linear programming approximation to the eigenvector method in the analytic hierarchy process. Inf Sci 181(23):5240–5248
14. Xiong J, Xing L, Chen Y (2013) Robust scheduling for multi-objective flexible job-shop problems with random machine breakdowns. Int J Prod Econ 141(1):112–126

Chapter 10
Energy-Efficient Process Parameters Optimizing Decision Method

10.1 Introduction

In Chap. 3, the introduction of the discount factor avoids the data saturation of the additional loading loss factor during identification and separates the cutting power to obtain the energy efficiency of the stand-alone equipment of the discrete manufacturing system. By obtaining machine efficiency and defining the energy level of the workshop, it is beneficial for the management to optimize the workshop and production reasonably. However, there is a problem that needs to be solved urgently, which is low machine efficiency. It is not enough to study the amount of energy efficiency. It is also necessary to explore methods and techniques for energy conservation. Known by the first chapter, according to the discussion in the first chapter, energy saving (lightweight design, optimization of processing parameters, etc.) can be carried out technically, and energy saving can be performed from the control (reasonable scheduling, reduction of standby time, etc.). Studies have shown that [1], in the discrete manufacturing industry, saving energy based on the optimization of cutting parameters, tools and tool trajectory has reached 6–40%.

Many experts and scholars at home and abroad have studied the problem of energy-saving optimization in discrete machine manufacturing systems.

Rajemi et al. [1]. studied the optimization method of optimal turning conditions under the lowest energy consumption target. Carmita et al. [2] applied the response surface method to optimize the cutting parameters to achieve the lowest energy consumption and the best processing quality, and then, the research results show that the feed rate and depth of cut are the processing parameters that affect the most energy consumption, and the feed rate is the most factor which is leading to surface roughness. Yang et al. [3] used differential evolution algorithm and non-dominated sorting genetic algorithm to optimize the turning parameters, choose tool wear and metal removal rate as optimization targets and use a non-dominated sorting preference for the obtained solution set. Li et al. [4] studied the turning optimization method based on genetic algorithm to study the minimum time and cost of the manufacturing system

Y. Wang et al., *Quantitative Analysis and Optimal Control of Energy Efficiency in Discrete Manufacturing System*, https://doi.org/10.1007/978-981-15-4462-0_10

during the turning process. Li et al. [5] solved milling parameter optimization problems by introducing maximum productivity as the optimization targets, and finally, they proposed a cell particle swarm optimization algorithm to optimize the model. Therefore, it is particularly important to study energy-saving optimization methods for discrete machine tool manufacturing systems to reduce energy consumption and improve energy efficiency.

Under the constraints of the actual machining environment, this chapter establishes a multi-objective optimization model for processing parameters of discrete manufacturing systems with energy efficiency as the optimization goal by establishing optimization variables under cutting speed, feed rate and depth of cut. After that, a random walk multi-objective particle swarm algorithm (RWA-MOPSO) was proposed to solve process parameter optimization problems. The algorithm uses particle swarm optimization as a global search method and introduces the random walk method into the particle swarm algorithm as a local search method to improve the local search ability of particles and improve the convergence speed. The non-dominated sorting and crowding distance strategies are used to obtain Pareto optimality. Finally, a method which is called analysis hierarchy process (AHP) is used to select the optimal processing parameters from the Pareto frontier solution, and the energy saving is achieved by optimizing the processing parameters. In the actual processing environment, the effectiveness and superiority of the proposed algorithm in solving the energy-saving optimization problem of discrete manufacturing system processing parameters are verified.

10.2 An Optimization Decision Model Facing Energy Efficiency Process Parameters

In discrete machine manufacturing system, the process parameters of individual stand-alone production nodes are same, such as such as cutting speed, feed rate and depth of cut et al. And the difference is the production equipment and the upper and lower boundaries of the constraint variables. Therefore, this book takes the process of single-machine equipment as the research object in discrete manufacturing system to conduct energy-saving optimization modelling.

10.2.1 Processing Energy Function

As mentioned earlier, when discrete machine manufacturing system is processing, generally, there are four processes of starting, standby, no-load and processing, then energy consumption model is

$$E = E_{st} + E_{s-s} + E_{ie} + E_c \tag{10.1}$$

where E is the total energy consumption, E_{st} is the machine start energy consumption, E_{s-s} is the machine standby energy consumption, E_{ie} is the machine no-load energy consumption, and E_c is the machining cutting energy consumption.

The starting energy consumption of the machine tool is generally changeless and is determined by the performance of the machine itself. The standby and no-load energy consumption are also constant after the machine is started and is obtained by the method of obtaining the energy-related parameters in the third chapter. The most energy-consuming and main energy consumption is cutting energy.

Cutting energy consumption represents the energy consumed to cut off the workpiece material:

$$E_c = \int_0^{t_c} P_c \mathrm{dt} \tag{10.2}$$

where $P_c(t)$ indicates the machine tool's cutting power, t_c indicates process time. During the turning process [6],

$$P_c(t) = \frac{1}{6 \times 10^4} C_{FC} a_{sp}^{x_{FC}} f^{y_{FC}} v_C^{n_{FC}} K_{FC} v_C \tag{10.3}$$

where v_c indicates cutting speed, f indicates the amount of feed, a_{sp} indicates cutting depth, C_{FC}, x_{FC}, y_{FC}, n_{FC}, K_{FC} indicate coefficients related to part material and tool material, and the corresponding value can be found from the cutting manual.

Therefore, the energy consumption model of the process is

$$E = E_{st} + E_{s-s} + E_{ie} + E_c = E_{st} + E_{s-s} + E_{ie} + \int_0^{t_c} P_c(t)\mathrm{dt} \tag{10.4}$$

10.2.2 Processing Time Function

Introducing the machining time objective function as another optimization object, machine processing time generally includes cutting time, tool change time and process assist time [7, 8], so the time model can be expressed as follows:

$$T_p = t_c + t_{ct}\frac{t_c}{T} + t_{ot} \tag{10.5}$$

$$t_c = \frac{L_w \Delta}{n f a_{sp}} = \frac{\pi d_0 L_w \Delta}{1000 v_c f_{\text{afeed}} a_{sp}} \tag{10.6}$$

where t_c indicates the cutting time from the start of machining to the end of the part, t_{ot} indicates auxiliary time such as clamping, L_w indicates the length of the machined part, T indicates the life of the tool used for machining, t_{ct} indicates the time taken by the magazine to change the tool once, Δ indicates the margin left in the process, d_0 indicates the diameter of the machined part, n indicates the spindle speed of the machine, v_c indicates the cutting speed of the machine, a_{sp} indicates depth of cut, and f_{afeed} indicates the amount of feed.

Tool life T is determined by Taylor's generalized formula:

$$T = \frac{C_T}{v_c^{x_{\text{life}}} f_{\text{afeed}}^{y_{\text{life}}} a_{sp}^{z_{\text{life}}}} \tag{10.7}$$

where C_T indicates constant values related to machining conditions such as parts, tools and machine tools, x_{life}, y_{life}, z_{life} indicate the life factor of the tool.

Then, the processing time function is

$$T_P = \frac{\pi d_0 L_w \Delta}{1000 v_c f a_{sp}} + \frac{t_{ct} \pi d_0 L_w \Delta v_c^{x-1} f^{y-1} a_{sp}^{z-1} t_c}{1000 C_T} + t_{ot} \tag{10.8}$$

10.2.3 Constraints

In the actual machining process, subject to the constraints of the machine itself and the size of the blank or semi-finished product, the value of the cutting parameters must meet these corresponding constraints.

(1) **Cutting speed constraint**. The speed of the machine tool during machining needs to be between the maximum and minimum cutting speeds:

$$\frac{\pi d_0 n_{\min}}{1000} \le v_c \le \frac{\pi d_0 n_{\max}}{1000} \tag{10.9}$$

where $n_{\min}$, $n_{\max}$ indicate the minimum and maximum values of the spindle speed of the processing equipment.

(2) **Feed quantity constraint**. The feed rate f of the machine tool during machining needs to be between the maximum and minimum feed rates.

$$f_{\min} \le f \le f_{\max} \tag{10.10}$$

where $f_{\min}$, $f_{\max}$ indicate the minimum and maximum values of the feed allowed by the processing equipment.

(3) **Cutting force constraint**. The feed resistance of the machine tool during machining is less than or equal to the maximum cutting force that the feed mechanism can withstand.

$$C_F a_{sp}^x f^y v_c^n K_F \leq F\max \tag{10.11}$$

where $F\max$ indicates maximum cutting force, C_F, x, y, n, K_F indicate the coefficient associated with the machined workpiece and the cutting conditions.

(4) **Processing quality requirements**. The processing quality is represented here by the roughness of the surface of the machined part.

$$R_a = \frac{0.0312 f^2}{r_\varepsilon} \leq R\max \tag{10.12}$$

where r_ε indicates tool radius, $R\max$ indicates the maximum value required for the surface roughness of the part.

(5) **Power constraints**. The power of the machine tool should be less than or equal to the maximum cutting power indicated on the machine nameplate.

$$\frac{F_c v_c}{1000\eta} \leq P\max \tag{10.13}$$

where η indicates the total efficiency of the machine tool, $P\max$ indicates the maximum cutting power indicated on the machine nameplate, F_c indicates the cutting force when the machine is machined.

Based on the above analysis, the multi-objective optimization model of cutting parameters described in this book can be summarized as Eqs. (10.14) and (10.15):

$$\min : F\left(v_c, a_{sp}, f\right) = \left(\min E, \min T_p\right) \tag{10.14}$$

$$s.t \begin{cases} \frac{\pi d_0 n_{\min}}{1000} \leq v_c \leq \frac{\pi d_0 n_{\max}}{1000} \\ f_{\min} \leq f \leq f\max \\ C_F a_{sp}^x f^y v_c^n K_F \leq F\max \\ \frac{0.0312 f^2}{r_\varepsilon} \leq R\max \\ \frac{F_c v_c}{1000} \leq \eta_{\text{total}} P\max \\ a_{p\min} \leq a_p \leq a_p\max \end{cases} \tag{10.15}$$

10.3 RWA-MOPSO Optimization Solution Algorithm

The energy-efficient process parameter optimization decision problem is a discrete combinatorial optimization problem, and the particle swarm optimization algorithm is a continuous space optimization algorithm. It is not feasible to solve the scheduling problem directly by particle swarm optimization. Based on the basic particle swarm optimization algorithm, this book introduces the random walk method to improve the algorithm, and through the operation of particle coding, a RWA-MOPSO algorithm is proposed for the optimization of process parameters described by solving Eqs. (10.14) and (10.15) for decision discrete group optimization problem.

10.3.1 Basic Particle Swarm Optimization

Particle swarm optimization(PSO) [9, 10] is an emerging group intelligent optimization algorithm based on group evolution algorithm. Each particle in the algorithm represents a potential solution, and the velocity determined the direction and distance of the particle's movement. Supposed that $P_{population}$ is a population of n particles whose search space dimension is D. V_{id} indicates the speed of particle i in d-dimensional space, X_{id} indicates the position of particle i in the d-dimensional space,its individual extremum is $P_i(t)$, group extremum is P_g. At each iteration, the speed and position update formula of the particle swarm algorithm is as follows:

$$V_{id}(t+1) = \omega(t)V_{id}(t) + c_1r_1(P_i(t) - X_{id}(t)) + c_2r_2\big(P_g(t) - X_{id}(t)\big) \tag{10.16}$$

$$X_{id}(t+1) = X_{id}(t) + V_{id}(t+1) \tag{10.17}$$

where $\omega(t)$ is the inertia factor, set by the linear decreasing strategy, with an expression:

$$\omega(t) = \omega\text{max} - \frac{(\omega\text{max} - \omega\text{min})}{i_{\text{num}}} \times t \tag{10.18}$$

where ωmax and ωmin indicate the maximum and minimum values of the inertia factor, indicate particle dimensions, $1 \le d \le D$, i_{num} indicates the number of particles, $1 \le i_{\text{num}} \le n$, t and iter are the current number of iterations and the maximum number of iterations, c_1 and c_2 are non-negative constant, named acceleration factor, r_1 and r_2 are a random number distributed between [0, 1].

10.3.2 Improvement Measures

(1) **Parameter coding**

Let the number of particles in the population P be n, and the position vector of each particle is composed of three parameters of the processing parameters, that is, the dimension of the position vector of the individual is $D = 3$. This population can be represented by a matrix of $P \times D$:

$$P(n \times D) = \begin{bmatrix} K_{v_c}^1 & K_{a_{sp}}^1 & K_f^1 \\ K_{v_c}^2 & K_{a_{sp}}^2 & K_f^2 \\ & \cdots & \\ K_{v_c}^n & K_{a_{sp}}^n & K_f^n \end{bmatrix}$$

(2) **Random walk method**

Aiming at problem that the basic PSO converges slowly, a solution is proposed. After the PSO algorithm searches for the global optimal solution, the random walk method is used as the local search strategy to avoid the algorithm stopping the search due to the local optimum, which improves the convergence speed and accuracy of the PSO algorithm. The random walk method is an algorithm that uses random numbers to find the optimal solution, which increases the diversity of particles. The specific expression is

$$x_i = x_{i-1} + \lambda u_{i-1} \tag{10.19}$$

where x_i indicates the approximate minimum value obtained in $i - 1$ steps, λ is a constant whose size determines the scope of the search. u_i is a randomly generated unit vector. The specific process is as follows:

Step1: $P_{\text{population}}$ is the optimal solution found by the particle swarm optimization algorithm.$\lambda = 0.5$ is the step size, $\varepsilon = 0.05$ is the minimum step size, t is iteration number;
Step2: make $t = 1$;
Step3: If $t \leq 10$, perform the following steps:

(1) $x = P_{\text{population}}$, generate a set of random numbers $r_1, r_2, \cdots, r_n \in [0, 1]$, n is the dimension of the search space. Make $h = \left(r_1^2 + r_2^2 + \cdots + r_n^2\right)$, if $h > 1$, then regenerate the random number until $h \leq 1$;

(2) $u = \dfrac{1}{\sqrt{r_1^2 + r_2^2 + \cdots r_n^2}} \begin{Bmatrix} r_1 \\ r_2 \\ \vdots \\ r_n \end{Bmatrix}$;

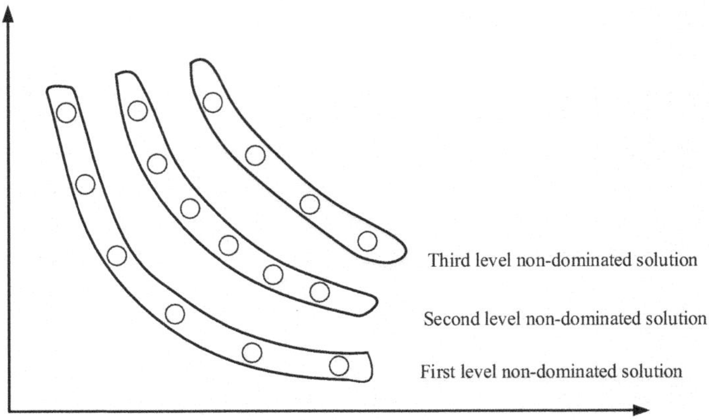

Fig. 10.1 Schematic representation of the non-dominated solution

(3) $x_1 = x + \lambda u$;

Step4: If x_1 is better than $P_{\text{optimization}}$, then $P_g = x_1, t = t + 1$; step Step3:
Step5: Make $\lambda = \lambda/2$ If $\lambda > \varepsilon$, step Step2, or output $P_{\text{optimization}}$.

(3) **Non-dominated sorting and crowding distance sorting**

Non-dominated sorting and crowding distance ordering selects the best solution of individuals from all particle individuals. The idea is as follows: First, identify all individuals in the population that are not dominated by the optimal solution and place them in the first rank and continue to rank the individuals in the remaining population according to the dominance relationship. Second, the above process is repeated until all the individuals in the solution set are classified. The hierarchical diagram of the non-dominated solution is shown in Fig. 10.1.

The crowded distance D_i is the sum of the particle i and the Euclidean distance of the particle in the same non-dominated solution level. The crowded distance of particle i can be calculated as follows:

$$D_i = \sum_{j=1}^{l} \sqrt{\sum_{h=1}^{p} (f_i^h - f_j^h)^2, i, j \in F(s), j \neq i} \tag{10.20}$$

where D_i indicates the crowded distance of the particles, $F(s)$ indicates a collection of all particles with a non-dominated solution level of s, f_j^h indicates the hth objective function representing the jth particle, a total of p objective functions, l indicates the number of particles in $F(s)$ set.

(4) **RAW-MOPSO algorithm flow**

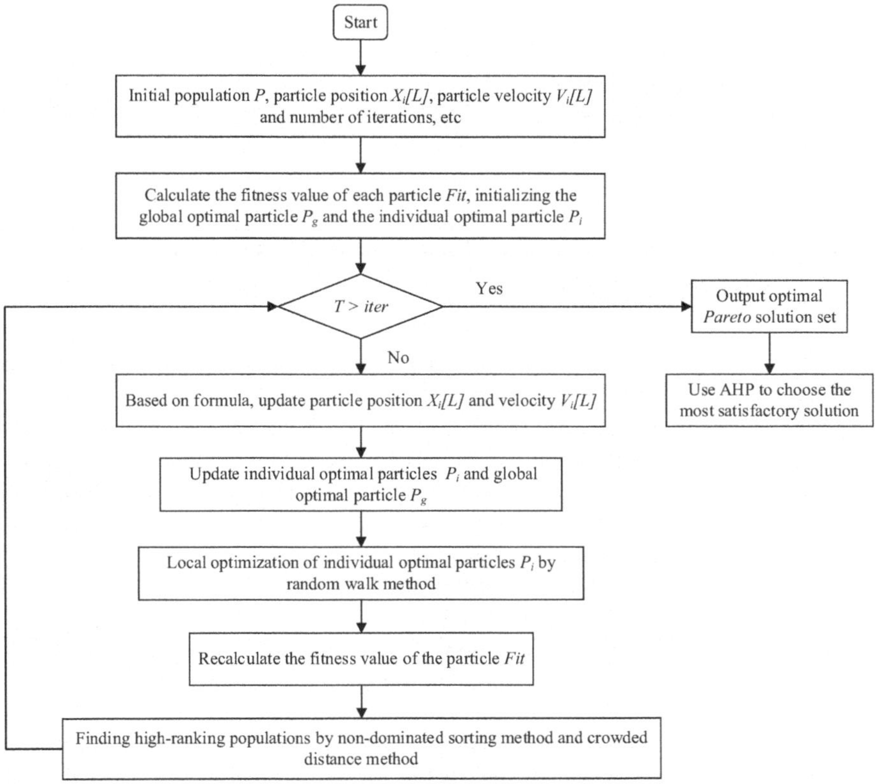

Fig. 10.2 RWA-MOPSO algorithm flow chart

According to the above analysis of the RWA-MOPSO hybrid algorithm, the specific steps of the improved algorithm to solve the processing parameter set are given as follows. The flow chart of the improved algorithm is shown in Fig. 10.2.

Step1: Initialize the population and parameters;
Step2: Calculating the fitness value of each particle, initializing the global optimal particle and the individual optimal particle;
Step3: Update the particles according to Formulas (10.16), (10.17);
Step4: Update individual optimal particles P_i and global optimal particle P_g;
Step5: Local optimization of individual optimal particles by random walk method;
Step7: Calculate the fitness value of the particles, compare them, and determine whether it is necessary to update the individual extremum P_i and the global extremum P_g of the population;
Step8: Finding high-ranking populations by non-dominated sorting method and crowded distance method;

Step9: Determining whether the current number of iterations is greater than a specified value is performing the next step 10, outputting the optimal solution set; or, executing the third step Step3;

Step10: The algorithm ends, and the optimal Pareto solution set is output.

10.3.3 Optimization Decision Based on AHP Analytic Hierarchy Process

After using the proposed algorithm to find a set of Pareto solutions for energy-saving optimization problems, it is necessary to objectively select a processing parameter from the set of solutions as the optimal solution. This book adopts the analytic hierarchy process. The basic idea of using the AHP decision-making method is to divide the complex solving problem into different levels. At the lowest level, the sub-objects are compared in pairs to obtain the weight coefficients of each sub-goal, and then calculate The weighting coefficient of each scheme to the total scheduling target and the scheme with the largest weighting coefficient are the final schemes. AHP decision-making effectively combines quantitative and qualitative analysis. The decision-making process requires less information, shorter time and simple implementation and is widely applied to discrete manufacturing production management.

For the decision of the final solution of the optimization problem, the AHP multi-level hierarchical structure is usually divided into three parts: the total target layer, the sub-target layer and the solution layer. The overall target layer is the overall goal of energy-saving optimization; the sub-target layer is the sub-goals to be considered for each selection. The processing time and energy consumption are considered in this book; the solution layer is a set of Pareto solution sets obtained by multi-objective algorithm. The specific hierarchical structure model is shown in Fig. 10.3.

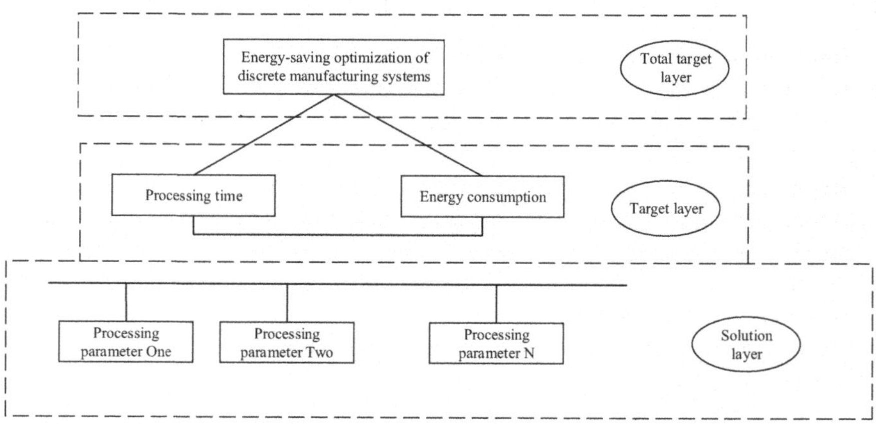

Fig. 10.3 AHP progressive structure for multi-objective decision-making

The importance of each goal in decision-making is in turn processing time and product energy consumption. Use the numbers 1–9 to indicate the degree of importance between the various objective functions, and obtain the judgment matrix:

$$A = \begin{bmatrix} & E_{\text{total}} & T_p \\ E_{\text{total}} & 1 & 1/5 \\ T_p & 5 & 1 \end{bmatrix} \tag{10.21}$$

The units (dimensions) of each indicator in the target layer are different. For example, the unit of processing time is seconds, and the unit of energy consumption is watt. There is no unified reference in the selection of optimal processing parameters, and it is impossible to carry out the relationship between the two comparisons. Therefore, before selecting the optimal processing parameter set, a one-step normalization process is performed to make each index one order of magnitude, which is convenient for comparison. The specific method is standardization of dispersion, linear transformation of the data of matrix A, then:

$$b_{ij} = \frac{a_j^{\max} - a_{ij}}{a_j^{\max} - a_j^{\min}} \tag{10.22}$$

The transformed data b_{ij} is summed according to the line:

$$\tilde{W}_j = \sum_{i=1}^{p_{\text{num}}} b_{ij}, q \geq j \geq 1 \tag{10.23}$$

Normalize the summed vector:

$$W_j = \frac{\tilde{W}_j}{\sum_{j=1}^{q} \tilde{W}_j} \tag{10.24}$$

where $a_j^{\max}$, $a_j^{\min}$ are the maximum and minimum values of the corresponding j columns of the judgment matrix A, p_{num} is the number of evaluation programs, q is the number of indicators, $i = 1, 2, \cdots p_{\text{num}}$, $j = 1, 2, \cdots, q$; a_{ij} is the value of the indicator in the Pareto program; the decision matrix of the scheme can be obtained from the above formula $B = (b_{ij})_{p \times q}$, then the decision matrix B is multiplied by the weight vector $W^{\mathrm{T}} = \left[W_1, W_2, \cdots, W_q\right]$ to calculate the optimal satisfaction matrix, that $D = (D_i) = BW^T = \sum_{j=1}^{q} b_{ij} W_j$, get evaluation plan satisfaction matrix.

Table 10.1 CNC lathe specifications

Name	Number
Minimum spindle speed (r/min)	80
Maximum spindle speed (r/min)	1400
Minimum feed (mm/r)	0.1
Maximum feed (mm/r)	3.5
Maximum cutting force (N)	900
Maximum effective cutting power (kW)	15

Table 10.2 Cutting force parameters

x	5	x_{FC}	1.0
y	1.75	y_{FC}	0.5
z	0.75	n_{FC}	-0.4
C_{FC}	2880	K_{FC}	1

10.4 Simulation Analysis

10.4.1 Test Conditions

Taking the single-machine equipment in a bearing grinding machine manufacturing system as an example, this book collects the data needed for the simulation experiment on a CNC lathe and verifies the effectiveness and superiority of the optimization algorithm. The specific parameters of the lathe are shown in Table 10.1.

In addition, the processed part is made of 45# steel rod. The quality standard of the machined parts is R_a cut less than or equal to 6.4 μm, cutting depth of the part is $a_{sp} = 1$ mm, the cutting fluid was used during the experiment. Find the cutting manual, the tool requirements are as follows: the material is cemented carbide material, the cutting edge angle is 5^0, leading angle is 45^0, front angle is 20^0, tool nose radius is $r_\varepsilon = 0.8$ mm, tool life is 64136 h, and the parameters related to the part material and machining tool are shown in Table 10.2.

10.4.2 Optimization Results and Analysis

Simulation experiment under MATLAB2010b, running on Windows7 operating system, single core Intel Core CPU, 2 GB memory. The simulation parameters are set as follows: $\omega = 1.3$, $c_1 = 1.6$, $c_2 = 1.6$, the maximum number of iterations is 200, initial population size $P = 60$.

The Pareto frontier solution obtained by the algorithm is shown in Fig. 10.3. In order to verify the effectiveness and superiority of the improved algorithm, it is compared with the multi-objective teaching–learning algorithm of the literature

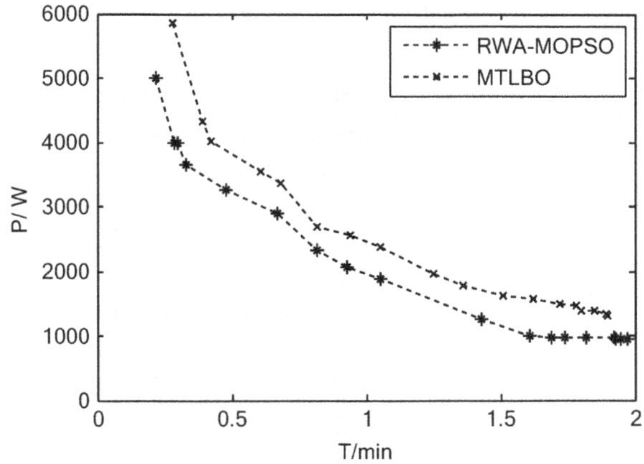

Fig. 10.4 Comparison of different optimization algorithms

[11]. Based on optimization, the Pareto frontier solutions of the two algorithms are obtained under the same processing conditions, as shown in Fig. 10.4.

Furthermore, the analytic hierarchy process AHP is used to objectively and practically select the optimal processing parameters from the obtained Pareto frontier solution. The calculated weight coefficient is $W^T = (0.167, 0.833)$, then $D = (D_i)_p = BW^T = \sum_{j=1}^{q} b_{ij} W_j$, then get evaluation plan satisfaction matrix $(D_i)_p$. Make $D_L = \max D = D_9$, that is, the processing parameters of the ninth group are the most reasonable. The corresponding processing parameters are $n = 800\ r/\text{min}$, $f = 0.20$ mm/r, $a_{sp} = 0.5$ mm, the corresponding energy consumption at this time is 1864 W, and processing time is 58 s.

The algorithm was randomly tested 20 times, and the average value and corresponding processing parameters were compared with the literature [70], as shown in Table 10.3. The empirical method refers to the processing parameters selected by the operator according to his own years of processing experience and reference manuals.

It can be seen from Table 10.3 that compared with the other two methods, the energy consumption and energy saving of the algorithm are shown in Table 10.4.

Table 10.3 Comparison of the algorithm proposed in this book with other algorithms

Algorithm	MTLBO	RWA-MOPSO	Empirical method
Average	65 s/1964 w	58 s/1853 w	75 s/2200 w
n	900 r/min	800 r/min	1200 r/min
f	0.19 mm/r	0.25 mm/r	0.16 mm/r
a_{sp}	0.5 mm	0.5 mm	0.5 mm

Table 10.4 Energy-saving percentage of the algorithm compared to other algorithms

Algorithm	MTLBO	Empirical method
Energy-saving ratio	15.8%	34.9%

10.5 Conclusion

Under the premise of the determined processing route, the process parameters of the equipment processing have a great influence on the energy consumption of the equipment. In order to reasonably select the process parameters of the discrete machine tool manufacturing system and reduce the energy consumption, this chapter aims to optimize the energy efficiency of the process parameters in the discrete machine tool manufacturing system and establish the energy-efficient process parameter optimization decision model of the discrete manufacturing system to energy efficiency and processing. Time is the goal, the improved multi-objective particle swarm optimization algorithm is used to optimize, and the optimal cutting parameter combination is obtained. The optimal set of processing parameters is selected by AHP analytic optimization decision. The simulation results show that the proposed algorithm is more advantageous in solving the parameter optimization problem.

References

1. Rajemi MF, Mativenga PT, Aramcharoen A (2010) Sustainable machining: Selection of optimum turning conditions based on minimum energy considera-tions. J Clean Prod 18:1059–1065
2. Carmita CN (2015) Optimization of cutting parameters using Response Surface Method for minimizing energy consumption and maximizing cutting quality in turning of AISI 6061 T6 aluminum. J Clean Prod 91:109–117
3. Yang SH, Natarajan U (2010) Multi-objective optimization of cutting parameters in turning process using differential evolution and non-dominated sorting genetic algorithm-II approaches. Int J Adv Manuf Technol 49(5–8):773–784
4. Li JG, YaoY X, Liu CQ (2006) Cutting parameters optimization in turning based on genetic algorithm. Comput Integr Manuf Syst 12(10):1651–1656
5. Li XP, Zhang CY, Gao L et al (2014) NC cutting parameter optimization based on cellular particle swarm optimization algorithm. Comput Eng Appl 50(2):252–257
6. Sheng M (2007) Research on cutting parameters optimization on NC turning. Harbin, Harbin Institute of Technology
7. Jiang ZG, Zhou F, Zhang H et al (2015) Optimization of machining parameters considering minimum cutting fluid consumption. J Clean Prod 108:183–191
8. Li AP, Gu ZY, Zhu J et al (2015) Optimization of cutting parameters for multi-pass hole machining based on low carbon manufacturing. Comput Integr Manuf Syst 21(6):1516–1521
9. Pinkey CC, Millie P, Kusum D (2015) Parameter optimization of multi-pass turning using chaotic PSO. Int J Mach Learn Cybernet 6(2):319–337
10. Mao ZH, Wang Y, Ji ZC (2015) Memetic non-dominated sorting particle swarm optimization algorithm for solving the multi-objective flexible job shop scheduling problem. Comput Syst Appl 24(10):155–161
11. Zhou Z, Zhang C, Xie Y et al (2015) Cutting parameters optimization for processing energy and effificiency in CNC lathe. Comput Integr Manuf Syst 21(09):2410–2418

Chapter 11
Design and Application of Energy Efficiency Optimization Control Software System

11.1 Introduction

The previous chapters have carried out detailed research on the energy efficiency modelling and optimization methods of discrete manufacturing systems. Based on ISOK industrial operation platform, this chapter will integrate energy efficiency real-time monitoring, quantitative analysis, process parameter optimization decision, static and dynamic optimization scheduling algorithm, research and development of energy efficiency optimization control software system and application verification on multi-variety flexible bearing processing automation production line. Firstly, this chapter introduces the development environment and overall architecture of the energy efficiency optimization control software system. Then, the functional modules of the system are introduced. Finally, the actual implementation results of the system are given.

11.2 System Development Environment

1. System development platform and tools

When it comes to development platforms, .net is Microsoft's latest platform, with C#.net, ASP.net and ADO.net as the main development tools. The programming environment of the system is MATLAB 2011b platform, whose programming language is M language, and the access to the database is realized by Java Database Connectivity (JDBC).

2. Software environment

   (a) Database server: win12 server;
   (b) Application server: win12 server;
   (c) Client terminal: winAll;
   (d) Database: oracle11g;

Y. Wang et al., *Quantitative Analysis and Optimal Control of Energy Efficiency in Discrete Manufacturing System*, https://doi.org/10.1007/978-981-15-4462-0_11

Table 11.1 Sever configuration indicator

Web server	Lenovo ThinkSever Server/RD450/DVD[E5-2620v3*2/Memory 64G/Hard Disk/600G*3]
Oracle server	Lenovo ThinkSever Server/RD450/DVD[E5-2620v3*2/Memory 64G/Hard Disk/2T*3]

(e) Application system requirements: The client requires IE6.0 or above, and MATLAB needs 2010 or above.

3. Operating environment

B/S; C/S + WebService; MATLAB 2011b.

4. Hardware environment

The system needs some conditions, as follows:

(a) The system needs one web server and one database server. The server configuration is shown in Table 11.1.
(b) Client: Based on 10 M/100 M Ethernet, 1500 M main frequency, 512 M memory or more.

11.3 Overall Framework for System Development

The system frame of the energy efficiency optimization of discrete manufacturing systems is shown in Fig. 11.1. The application system consists of four layers: data acquisition layer, database management layer, software development layer and scheduling application layer. Among them, the data acquisition layer is responsible for using the hardware (power sensor, smart metre, RFID), manual input collect application system basic information data at the bottom of the workshop. The database management layer is responsible for the information data frame design of the system software, including all the underlying data structures of the application system, the application data storage and the association between the upper interface data. This layer is the basis for energy efficiency optimization application software development. The software development layer is developed by C# programming language, PL/SQL Developer compiler, MATLAB, WEB front end and other related knowledge tools. The software development layer and the database layer exchange basic information data in two directions, and use the scheduling algorithm to realize scheduling according to the current basic data of the workshop scheduling. The shop scheduling application layer belongs to the interaction layer. The interaction layer responses for displaying the scheduling scheme generated by the scheduling algorithm in the scheduling report mode, which is used to guide the workshop scheduling for human–computer interaction, and can analyse and optimize the optimization result according to the basic data energy consumption and manufacturing data statistics.

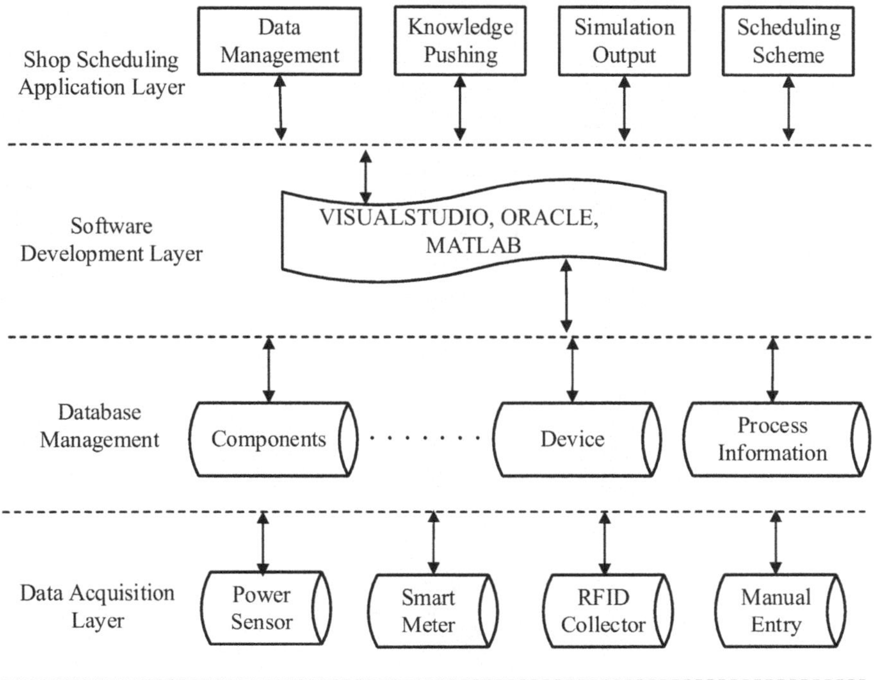

Fig. 11.1 System overall development framework

The energy efficiency optimization scheduling system is based on the energy efficiency optimization scheduling model and algorithm of the previous chapters, combined with the characteristics of the application workshop energy efficiency optimization problem, adaptively solves the scheduling model, and uses the C# programming language to write the scheduling programme. The data collected by the workshop is configured on the basis of ISOK industrial operation platform. At the same time, the interface between the scheduler and the ORACLE database is designed to facilitate the real-time writing of the scheduling results into the database to realize the optimal scheduling of the workshop work. According to the scheduling results, the manufacturing workshop processes and produces. When the rescheduling time is triggered, the scheduler determines the type of rescheduling through real-time production and energy consumption data and runs the rescheduling programme under the relevant type according to different rescheduling types, finally generate optimal scheduling solution feedbacking to production line execution and update database and knowledge base. The basic architecture of the optimized scheduling system is shown in Fig. 11.2.

1. The scheduling algorithm module is compiled by MATLAB programme, and the scheduling data in the Oracle database is read by OJDBC. The scheduling programme is completed to generate scheduling data and then transmitted to the Oracle database through OJDBC. Finally, the interactive platform displays the

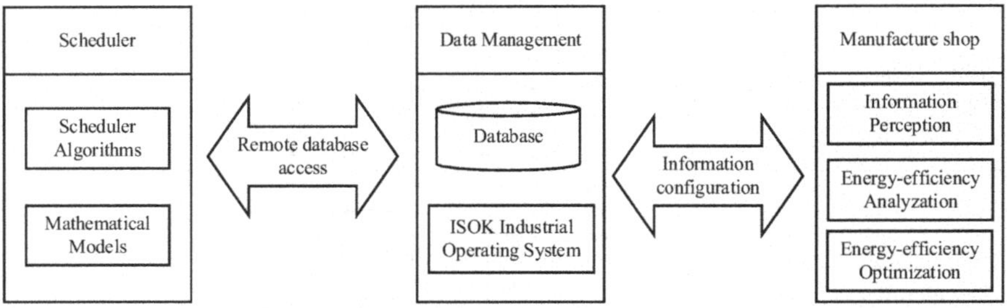

Fig. 11.2 Basic architecture of the optimal scheduling system

Fig. 11.3 Scheduling system function module

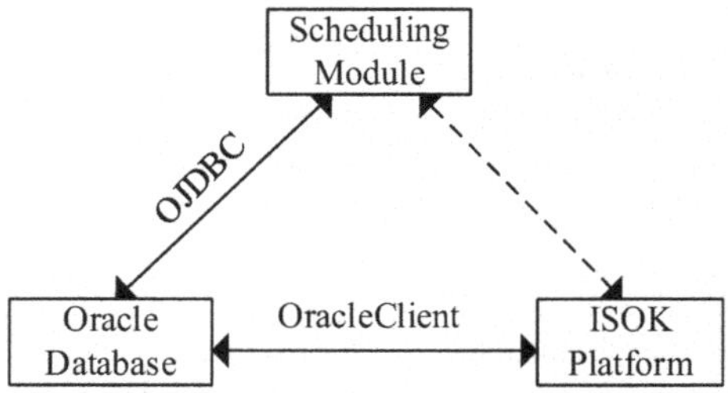

scheduling data on the scheduling platform through the OracleClient interface class.

2. According to the mathematical model in the previous chapters of this book, when converting actual scheduling problems into discrete models in a discrete manufacturing plant (Fig. 11.3), it is necessary to establish a mapping relational database in the Oracle database (scheduling information base table and scheduling result output table). The scheduling information base table is a part processing detail table, and the basic information in the table includes information such as part order number, quantity, routing and energy consumption of a single operation. The dispatch result output table includes the details of the workpiece scheduling arrangement generated by MATLAB calculation. The basic information in the table includes information such as the expected start time of the part, the estimated end time, the expected machine tool and the estimated energy consumption.
3. System function driving mechanism and algorithm in Table 11.2.

11.4 Application Object Description

The application line is the processing process of four batches of deep groove ball bearings, double row deep groove bearings, angular contact ball bearings and one-way stop bearings of a rolling bearing Co. Ltd. and the original energy efficiency level before the optimization of the workshop and its application. The energy efficiency levels of the optimized scheduling method are compared to verify the effectiveness

Table 11.2 System function drive mechanism and algorithm

Technical goal	Implementation/drive
Real-time dynamic monitoring of energy efficiency	Sensor sensing/non-intrusive sensing
Statistical analysis of energy efficiency	Three-layer energy efficiency quantitative evaluation index system and evaluation method
Real-time monitoring of production information	RFID/data collection intelligent terminal collaborative sensing
Equipment layer energy efficiency optimization	1. Process parameter optimization 2. Dynamic optimization scheduling triggered by bottleneck point
Task layer energy efficiency optimization	1. Static multi-objective optimization scheduling of knowledge fusion 2. Periodic scheduling 3. Emergency-driven dynamic optimization scheduling (fault, insert)
Production line comprehensive energy efficiency optimization	Energy saving based on equipment layer and task layer optimization integration

of the proposed optimization algorithm and energy efficiency optimization control software system in the application of discrete manufacturing systems. The energy efficiency data of 5 devices on a production line that performs processing tasks during the test were measured. Based on the same processing task, the energy efficiency statistics of the 5 devices of the production line were compared before the

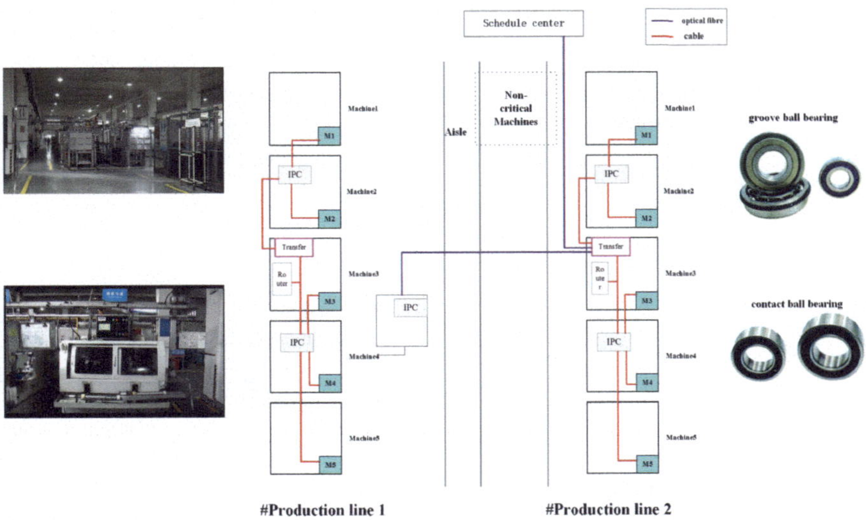

Fig. 11.4 Machine shop production line machine distribution and product map

Table 11.3 Test processing task status table

Product code	Product name	Employee ID	Process name	Processing content
AH5001	Deep groove ball bearing	1	Grinding channel	Inner groove trimming 12 times
		2	Grinding hole	Inner circle trimming 6 times
		3	Superfine channel	Internal super left consumption of oil stone 1000, right consumption oil stone 2000
		4	External communication mill	Outer groove trimming 5 times
		5	Grinding outer channel	External consumption of oil stone 300
AH5002	Double row deep groove bearing	1	Grinding channel	Inner groove trimming 12 times
		2	Grinding hole	Inner circle trimmed 5 times
		3	Superfine channel	Internal consumption of oil stone 1000
		4	External communication mill	Outer groove trimming 4 times
		5	Grinding outer channel	External consumption of oil stone 390
AH5003	Angular contact ball bearings	1	Grinding channel	Inner groove trimming 16 times
		2	Grinding hole	Inner circle trimming 3 times
		3	Superfine channel	Internal consumption of oil stone 550
		4	External communication mill	Outer groove trimming 5 times
		5	Grinding outer channel	External consumption of oil stone 250
AH5004	One-way stop bearing	1	Grinding channel	Inner groove trimming 15 times

(continued)

Table 11.3 (continued)

Product code	Product name	Employee ID	Process name	Processing content
		2	Grinding hole	Inner circle trimmed 5 times
		3	Superfine channel	Inside Super Left consumes 800 stone, right consumes 800 stone
		4	External communication mill	Outer groove trimming 6 times
		5	Grinding outer channel	External consumption of oil stone 390

implementation of the technology to verify the energy efficiency improvement level. The product line and machine distribution of the processing workshop are shown in Fig. 11.4, and the processing tasks are shown in Table 11.3.

11.5 System Main Function Module

The energy efficiency optimization control system realizes the collection and monitoring of energy consumption data, the basic information management of energy consumption, the energy consumption classification and statistics of key energy-consuming machine tools and the energy efficiency evaluation, etc., to achieve dynamic monitoring, statistics, management and energy consumption of the entire machine tool manufacturing workshop. The energy efficiency optimization control system is divided into six modules: energy basic information monitoring, real-time monitoring of production progress, energy efficiency index calculation, energy efficiency quantitative analysis, energy electronic accounting and parts process energy consumption specification and scheduling optimization.

1. **Energy Consumption Basic Information Monitoring**

In order to obtain real-time energy consumption information of machine tools, we installed smart metres and industrial computers. The smart metre is installed in the electric cabinet of the processing machine tool to monitor the power information and energy consumption of the processing machine. The main data parameters are phase voltage, phase current, line voltage, line current, combined apparent power, power consumption and other information (see Fig. 11.5). The smart metre obtains the machine current information through the transformer and connects with the three-phase voltage to obtain real-time three-phase voltage information. The data of the smart metre is transmitted to the data collector through the RS485 signal line. One end of the data collector is connected with the RS485 to USB module, and the RS485 to USB driver is installed. After successful communication, set the metre and configuration software to read ABC three-phase voltage, ABC three-phase current,

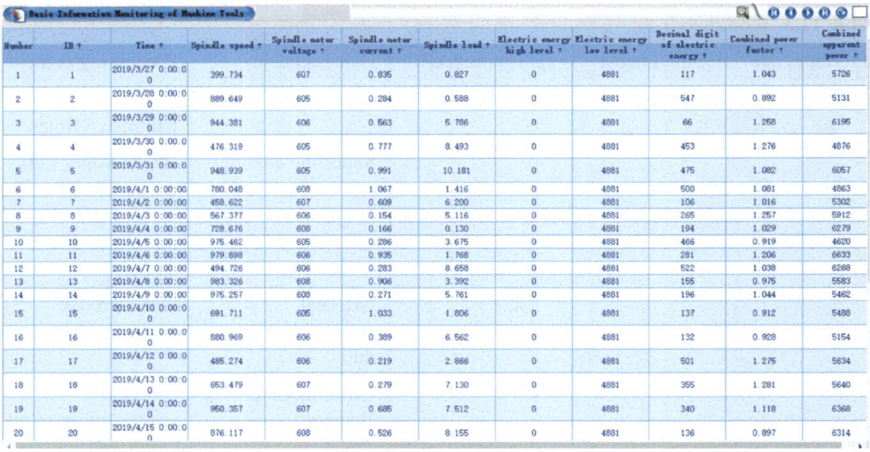

Number	ID	Time	Spindle speed	Spindle motor voltage	Spindle motor current	Spindle load	Electric energy high level	Electric energy low level	Decimal digit of electric energy	Combined power factor	Combined apparent power
1	1	2019/3/27 0:00:00	399.734	607	0.835	0.827	0	4881	117	1.043	5726
2	2	2019/3/28 0:00:00	889.649	605	0.284	0.588	0	4881	547	0.892	5131
3	3	2019/3/29 0:00:00	944.381	606	0.563	5.786	0	4881	66	1.258	6195
4	4	2019/3/30 0:00:00	476.319	605	0.777	8.493	0	4881	453	1.276	4876
5	5	2019/3/31 0:00:00	948.939	605	0.991	10.181	0	4881	475	1.082	6057
6	6	2019/4/1 0:00:00	780.048	608	1.067	1.416	0	4881	500	1.081	4863
7	7	2019/4/2 0:00:00	458.622	607	0.609	6.200	0	4881	106	1.016	5302
8	8	2019/4/3 0:00:00	567.377	606	0.154	5.116	0	4881	265	1.257	5912
9	9	2019/4/4 0:00:00	728.676	608	0.166	0.130	0	4881	194	1.029	6279
10	10	2019/4/5 0:00:00	975.462	605	0.286	3.675	0	4881	466	0.919	4620
11	11	2019/4/6 0:00:00	979.898	606	0.935	1.768	0	4881	281	1.206	6633
12	12	2019/4/7 0:00:00	494.726	606	0.283	8.658	0	4881	522	1.038	6268
13	13	2019/4/8 0:00:00	983.326	608	0.906	3.392	0	4881	155	0.975	5583
14	14	2019/4/9 0:00:00	975.257	608	0.271	5.761	0	4881	196	1.044	5462
15	15	2019/4/10 0:00:00	691.711	605	1.033	1.806	0	4881	137	0.912	5488
16	16	2019/4/11 0:00:00	880.969	606	0.389	6.562	0	4881	132	0.928	5154
17	17	2019/4/12 0:00:00	485.274	606	0.219	2.866	0	4881	501	1.275	5634
18	18	2019/4/13 0:00:00	653.479	607	0.279	7.130	0	4881	355	1.281	5640
19	19	2019/4/14 0:00:00	950.357	607	0.685	7.512	0	4881	340	1.118	6368
20	20	2019/4/15 0:00:00	876.117	608	0.526	8.155	0	4881	136	0.897	6314

Fig. 11.5 Power consumption information of each machine tool

combined apparent power, power consumption, etc., at a certain frequency, and store the read data into SQL Server, then pass the data mining software to the server database.

2. **Real-Time Monitoring of Production Progress**

In the interface of Fig. 11.6, green indicates that the processing task has been completed, and the workshop staff can arrange the work schedule reasonably, and ensure the processing task is completed regularly by checking the processing task completion situation.

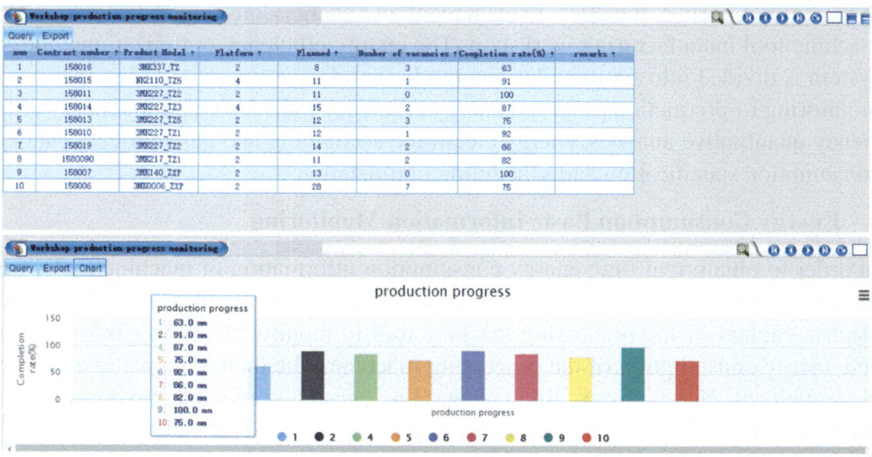

num	Contract number	Product Model	Platform	Planned	Number of vacancies	Completion rate(%)	remarks
1	158016	3MK337_TZ	2	8	3	63	
2	158015	MK2110_TZ5	4	11	1	91	
3	158011	3MK227_TZ2	2	11	0	100	
4	158014	3MK227_TZ3	4	15	2	87	
5	158013	3MK227_TZ5	2	12	3	75	
6	158010	3MK227_TZ1	2	12	1	92	
7	158019	3MK227_TZ2	2	14	2	86	
8	1580090	3MK217_TZ1	2	11	2	82	
9	158007	3MK140_ZXP	2	13	0	100	
10	158006	3MB0006_ZXP	2	28	7	75	

Fig. 11.6 Real-time monitoring of production progress

3. **Calculation of Energy Efficiency Indicators**

Calculate the energy efficiency by analysing the basic data, adding an algorithm and calculating the daily energy consumption and daily total energy consumption. The cutting energy consumption and other energy consumption in each row of data are represented in the graphical interface (Fig. 11.7) and expressed in the form of a bar graph, which can directly observe the efficiency of the machine tool.

4. **Energy Efficiency Quantitative Analysis**

Calculate the energy efficiency by analysing the basic data, adding an algorithm and calculating the daily energy consumption and daily total energy consumption. By cutting the energy consumption of each row of data (Fig. 11.8a) and other energy consumption in the graphical interface (Fig. 11.8b), you can directly observe the efficiency of the machine.

5. **Energy Electronic Ledger and Parts Process Energy Consumption Specification**

See Fig. 11.9.

6. **Optimized Scheduling**

The scheduling module starts four types of mechanism drivers: (1) generating an initial scheduling scheme, and when the new task is reached, starting to invoke a static optimization scheduling algorithm and a process parameter optimization algorithm; (2) periodic scheduling, scheduling time is scheduled by the dispatcher or system, and rescheduling is performed periodically; (3) machine fault scheduling, when the equipment fails, it is necessary to register the faulty machine, and read the workpiece being processed by the faulty machine and reschedule the production. (4) Emergency order scheduling, when an order requires early delivery or emergency

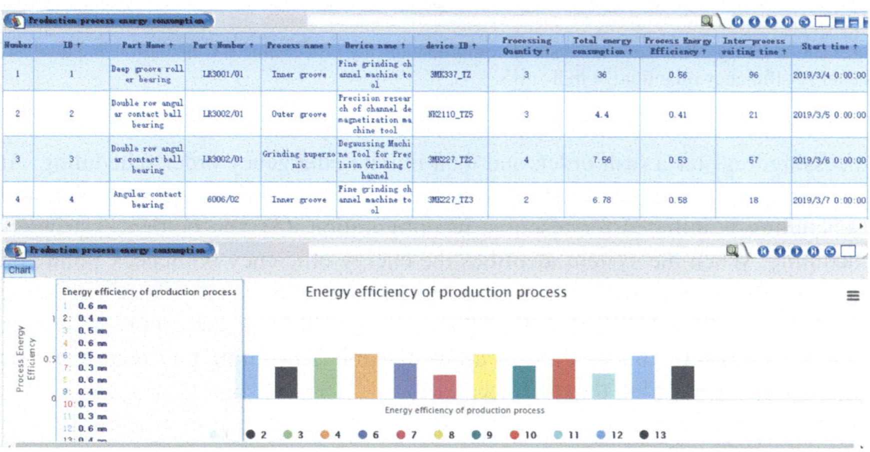

Number	ID	Part Name	Part Number	Process name	Device name	device ID	Processing Quantity	Total energy consumption	Process Energy Efficiency	Inter-process waiting time	Start time
1	1	Deep groove roller bearing	LR3001/01	Inner groove	Fine grinding channel machine tool	3MK337_TZ	3	36	0.56	96	2019/3/4 0:00:00
2	2	Double row angular contact ball bearing	LR3002/01	Outer groove	Precision research of channel demagnetization machine tool	NK2110_TZ5	3	4.4	0.41	21	2019/3/5 0:00:00
3	3	Double row angular contact ball bearing	LR3002/01	Grinding supersonic	Degaussing Machine Tool for Precision Grinding Channel	3MK227_TZ2	4	7.56	0.53	57	2019/3/6 0:00:00
4	4	Angular contact bearing	6006/02	Inner groove	Fine grinding channel machine tool	3MK227_TZ3	2	6.78	0.58	18	2019/3/7 0:00:00

Fig. 11.7 Calculation of energy efficiency indicators

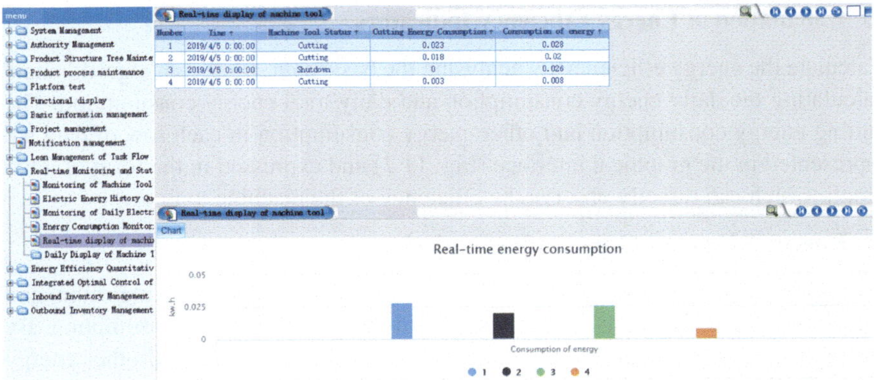

(a) Energy efficiency analysis of equipment during processing

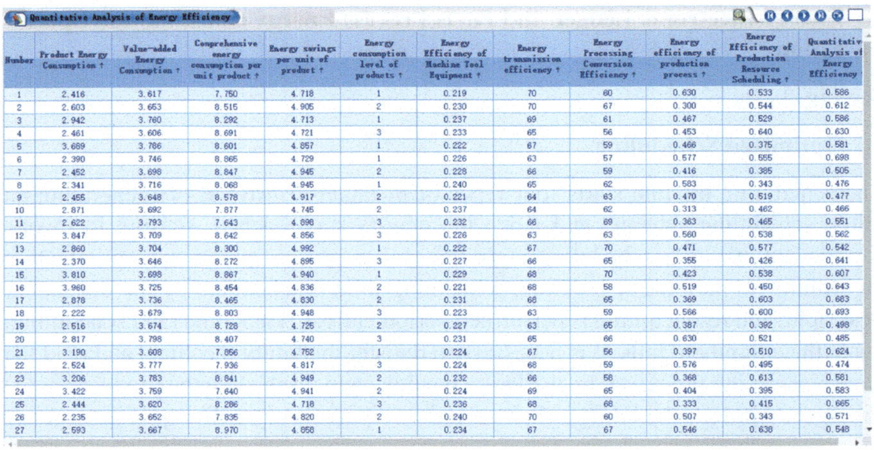

Number	Product Energy Consumption	Value-added Energy Consumption	Comprehensive energy consumption per unit product	Energy savings per unit of product	Energy consumption level of products	Energy Efficiency of Machine Tool Equipment	Energy transmission efficiency	Energy Processing Conversion Efficiency	Energy efficiency of production process	Energy Efficiency of Production Resource Scheduling	Quantitativ Analysis of Energy Efficiency
1	2.416	3.617	7.750	4.718	1	0.219	70	60	0.630	0.533	0.586
2	2.603	3.653	8.515	4.905	2	0.230	70	67	0.300	0.544	0.612
3	2.942	3.760	8.292	4.713	1	0.237	69	61	0.467	0.529	0.586
4	2.461	3.606	8.691	4.721	3	0.233	65	56	0.453	0.640	0.630
5	3.689	3.786	8.601	4.857	1	0.222	67	59	0.466	0.375	0.581
6	2.390	3.746	8.865	4.729	1	0.226	63	57	0.577	0.555	0.698
7	2.452	3.698	8.847	4.945	2	0.228	66	59	0.416	0.385	0.505
8	2.341	3.716	8.068	4.945	1	0.240	65	62	0.583	0.343	0.476
9	2.455	3.648	8.578	4.917	2	0.221	64	63	0.470	0.519	0.477
10	2.871	3.692	7.877	4.745	2	0.237	64	62	0.313	0.462	0.466
11	2.622	3.793	7.643	4.898	3	0.232	66	69	0.363	0.465	0.551
12	3.847	3.709	8.642	4.856	3	0.226	63	63	0.560	0.538	0.562
13	2.860	3.704	8.300	4.992	1	0.222	67	70	0.471	0.577	0.542
14	2.370	3.646	8.272	4.895	3	0.227	66	65	0.355	0.426	0.641
15	3.810	3.698	8.867	4.940	1	0.229	68	70	0.423	0.538	0.607
16	3.960	3.725	8.454	4.836	2	0.221	68	58	0.519	0.450	0.643
17	2.878	3.736	8.465	4.830	2	0.231	68	65	0.369	0.603	0.683
18	2.222	3.679	8.803	4.948	3	0.223	63	59	0.566	0.600	0.693
19	2.516	3.674	8.728	4.725	2	0.227	63	65	0.387	0.392	0.498
20	2.817	3.798	8.407	4.740	3	0.231	65	66	0.630	0.521	0.485
21	3.190	3.608	7.856	4.752	1	0.224	67	56	0.397	0.510	0.624
22	2.524	3.777	7.936	4.817	3	0.224	68	59	0.576	0.495	0.474
23	3.206	3.783	8.841	4.949	2	0.232	66	58	0.368	0.613	0.581
24	3.422	3.759	7.640	4.941	2	0.224	69	65	0.404	0.395	0.583
25	2.444	3.620	8.286	4.718	3	0.236	68	68	0.333	0.415	0.665
26	2.235	3.652	7.835	4.820	2	0.240	70	60	0.507	0.343	0.571
27	2.593	3.667	8.970	4.858	1	0.234	67	67	0.546	0.638	0.548

(b) Calculation and analysis of energy efficiency quantitative indicators

Fig. 11.8 **a** Energy efficiency analysis of equipment during processing. **b** Calculation and analysis of energy efficiency quantitative indicators

processing, register a rush order, and then initiate emergency order scheduling. At this time, priority is given to processing the contents of the emergency order, and rescheduling is initiated for resource reconfiguration. (5) The bottleneck triggers scheduling. When the system identifies the energy efficiency bottleneck point and location of the production line through quantitative analysis, it drives the rescheduling mechanism to optimize the resource allocation and eliminate the energy waste link as much as possible. In Fig. 11.10, various dynamic scheduling information can be dynamically displayed in the interface.

Monthly Energy Consumption Query of Machine Tools

Number	ID ↑	Machine number ↑	Date ↑	Month ↑	Monthly power consumption ↑	Monthly average daily consumption ↑
1	1	1003	2019/3/1 0:00:00	3	791.874	31.924
2	2	1001	2019/3/2 0:00:00	3	949.659	22.059
3	3	1001	2019/3/3 0:00:00	3	332.273	34.607
4	4	1001	2019/3/4 0:00:00	3	673.566	26.687
5	5	1001	2019/3/5 0:00:00	3	442.575	22.815
6	6	1001	2019/3/6 0:00:00	3	834.666	30.724
7	7	1001	2019/3/7 0:00:00	3	770.472	32.887
8	8	1001	2019/3/8 0:00:00	3	173.083	36.623
9	9	1001	2019/3/9 0:00:00	3	862.947	3.018
10	10	1001	2019/3/10 0:00:00	3	989.943	32.425
11	11	1001	2019/3/11 0:00:00	3	517.823	23.827
12	12	1001	2019/3/12 0:00:00	3	885.091	36.658
13	13	1002	2019/3/13 0:00:00	3	590.910	20.941
14	14	1002	2019/3/14 0:00:00	3	160.669	19.304
15	15	1002	2019/3/15 0:00:00	3	205.464	30.246
16	16	1002	2019/3/16 0:00:00	3	411.106	10.747
17	17	1002	2019/3/17 0:00:00	3	750.465	19.935
18	18	1002	2019/3/18 0:00:00	3	826.805	33.629
19	19	1002	2019/3/19 0:00:00	3	791.433	22.539
20	20	1002	2019/3/20 0:00:00	3	323.295	31.736
21	21	1002	2019/3/21 0:00:00	3	537.326	28.114

(a) Energy electronic ledger

Specification for Energy Consumption of Parts

Query Add Export

NUM	Part Name ↑	Process name ↑	Energy consumption limits ↑	average value ↑	Error range ↑
1	Deep groove ball bearing	Inner groove	185.69–201.17(W)	193.385	(−4.7%) − (+5.2%)
2	Deep groove ball bearing	Inner groove	60.619–80.87(W)	70.89	(−14%) − (+10%)
3	Deep groove ball bearing	Inner groove	153.169–185.67(W)	169.64	(−10%) − (+9.6%)
4	Deep groove ball bearing	Inner groove	256.54–289.56(W)	273.05	(−6%) − (+6%)
5	Angular contact ball bearing	Inner groove	123.57–130.46(W)	127.22	(−12%) − (+12%)
6	Angular contact ball bearing	Inner groove	110.54–132.214(W)	125.98	(−5%) − (+5.04%)
7	Angular contact ball bearing	Inner groove	170.23–185.45(W)	178.09	(−7%) − (+7.9%)
8	Angular contact ball bearing	Inner groove	245.24–267.98(W)	254.67	(−11%) − (+11.9%)
9	Angular contact ball bearing	Inner groove	167.67–175.59(W)	171.43	(−5%) − (+5%)
10	Deep groove ball bearing	Inner groove	130.69–170.17 (W)	150.328	(−10.7%) − (+12%)

(b) Part process energy consumption specification

Fig. 11.9 Calculation and analysis of energy efficiency quantitative indicators

11.6 Comparison of Process Parameters and Machine Energy Efficiency Before and After Optimization

11.6.1 Original Process Plan and Scheduling Plan

The process parameters of the four batch processing tasks of the original production line processing deep groove ball bearings, double row deep groove bearings, angular contact ball bearings and one-way stop bearings are shown in Table 11.4. Before the implementation of the technology, the manual scheduling and sequential scheduling schemes are mainly adopted.

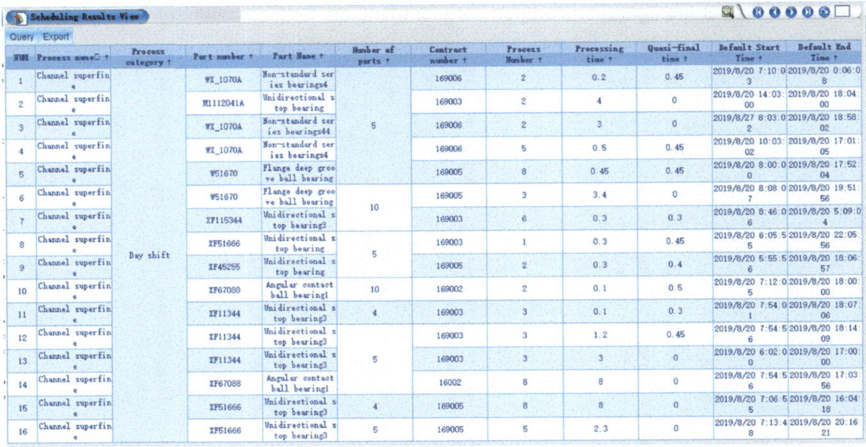

NUM	Process name	Process category	Part number	Part Name	Number of parts	Contract number	Process Number	Processing time	Quasi-final time	Default Start Time	Default End Time
1	Channel superfine	Day shift	WX_1070A	Non-standard series bearings4	5	169006	2	0.2	0.45	2019/8/20 7:10:03	2019/8/20 0:06:08
2	Channel superfine		M1112041A	Unidirectional stop bearing		169003	2	4	0	2019/8/20 14:03:00	2019/8/20 18:04:00
3	Channel superfine		WX_1070A	Non-standard series bearings44		169006	2	3	0	2019/8/27 8:03:02	2019/8/20 18:58:02
4	Channel superfine		WX_1070A	Non-standard series bearings4		169006	5	0.5	0.45	2019/8/20 10:03:02	2019/8/20 17:01:05
5	Channel superfine		W51670	Flange deep groove ball bearing		169005	8	0.45	0.45	2019/8/20 8:00:00	2019/8/20 17:52:04
6	Channel superfine		W51670	Flange deep groove ball bearing	10	169005	3	3.4	0	2019/8/20 8:08:07	2019/8/20 19:51:56
7	Channel superfine		XF115344	Unidirectional stop bearing3		169003	6	0.3	0.3	2019/8/20 8:46:06	2019/8/20 5:09:04
8	Channel superfine		XF51666	Unidirectional stop bearing	5	169003	1	0.3	0.45	2019/8/20 6:05:55	2019/8/20 22:05:56
9	Channel superfine		XF45255	Unidirectional stop bearing		169005	2	0.3	0.4	2019/8/20 5:55:56	2019/8/20 18:06:57
10	Channel superfine		XF67088	Angular contact ball bearing1	10	169002	2	0.1	0.5	2019/8/20 7:12:05	2019/8/20 18:00:00
11	Channel superfine		XF11344	Unidirectional stop bearing3	4	169003	3	0.1	0.3	2019/8/20 7:54:01	2019/8/20 18:07:06
12	Channel superfine		XF11344	Unidirectional stop bearing3	5	169003	3	1.2	0.45	2019/8/20 7:54:56	2019/8/20 18:14:09
13	Channel superfine		XF11344	Unidirectional stop bearing3		169003	3	3	0	2019/8/20 6:02:00	2019/8/20 17:00:00
14	Channel superfine		XF67088	Angular contact ball bearing1		16002	8	8	0	2019/8/20 7:54:56	2019/8/20 17:03:56
15	Channel superfine		XF51666	Unidirectional stop bearing3	4	169005	8	8	0	2019/8/20 7:06:55	2019/8/20 16:04:18
16	Channel superfine		XF51666	Unidirectional stop bearing3	5	169005	5	2.3	0	2019/8/20 7:13:48	2019/8/20 20:16:21

Fig. 11.10 Example of equipment scheduling plan

1. The Original Machine Energy Efficiency Level

The energy consumption optimization control system is randomly selected to test the monthly energy consumption data of 5 sets of equipment before the application, and the energy consumption of the first four types of processing tasks is optimized. Figure 11.11 and the energy efficiency level of the key machines before optimization are shown in Table 11.5. As shown in Table 11.6 and as can be seen from the chart, the energy efficiency of the machine energy efficiency and processing tasks is less than 75%.

11.6.2 Process Parameters and Energy Efficiency Levels After Energy Efficiency Optimization

1. Process Plan and Scheduling Plan

After using the static optimization scheduling, dynamic optimization scheduling and process parameter optimization algorithm proposed in the previous chapter, the process parameters of the four batch processing tasks of deep groove ball bearings, double row deep groove bearings, angular contact ball bearings and one-way stop bearings can be found in Table 11.7.

2. Equipment Layer Energy Efficiency Level

After using the equipment energy efficiency optimization driving mechanism shown in Table 11.2, deep groove ball bearings, double row deep groove bearings, angular contact ball bearings, one-way stop bearings, four batch processing tasks, five

Table 11.4 Details of the process before the implementation of the technology

Product code	Product name	Process name	Employee ID	Process parameters (before optimization)				
				Fast	Fine grinding amount	Compensation amount	Feed rate	Rewind
AH5001	Deep groove ball bearing	Grinding channel	1	123	123	123	123	123
		Grinding hole	2	222	120	222	222	120
		Superfine channel	3	238	238	220	135	118
		External communication mill	4	183	183	183	183	183
		Grinding outer channel	5	140	140	115	165	140
AH5002	Double row deep groove bearing	Grinding channel	1	110	110	120	100	120
		Grinding hole	2	222	213	222	213	213
		Superfine channel	3	138	140	132	138	130
		External communication mill	4	120	120	120	130	130
		Grinding outer channel	5	135	123	135	120	118
AH5003	Angular contact ball bearings	Grinding channel	1	240	50	25	125	70
		Grinding hole	2	265	67	235	210	114
		Superfine channel	3	140	66	214	260	174
		External communication mill	4	184	85	165	240	134
		Grinding outer channel	5	304	105	158	210	136
AH5004	One-way stop bearing	Grinding channel	1	305	241	185	230	85
		Grinding hole	2	77	165	232	147	188
		Superfine channel	3	255	314	257	314	58
		External communication mill	4	59	65	149	254	110
		Grinding outer channel	5	340	260	56	146	184

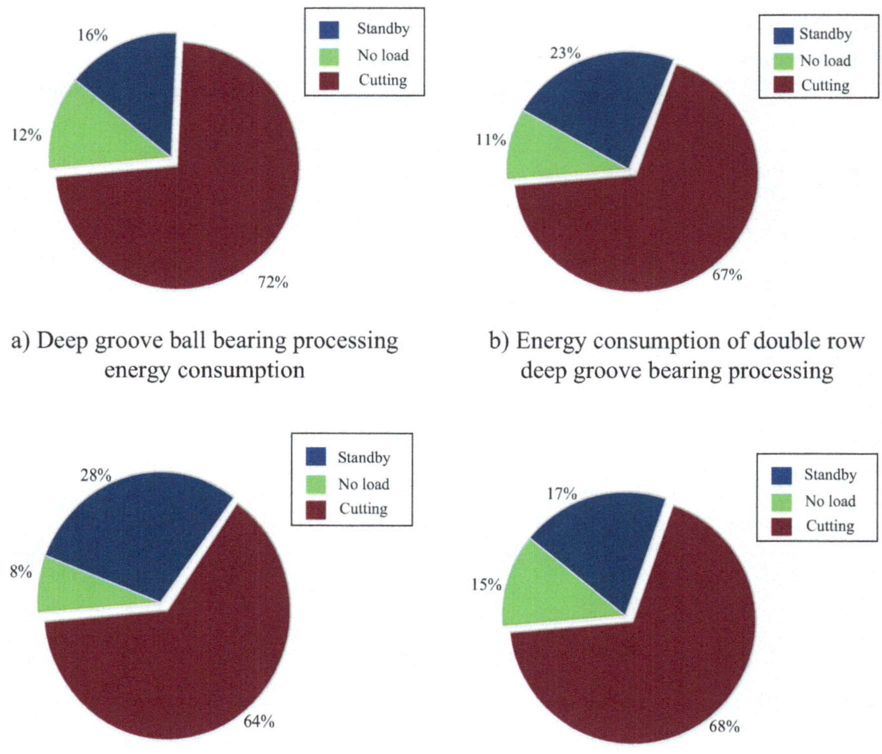

a) Deep groove ball bearing processing energy consumption

b) Energy consumption of double row deep groove bearing processing

c) Contact ball bearing processing energy consumption

d) One-way stop bearing processing energy composition

Fig. 11.11 Optimizing the energy consumption of the first four types of processing tasks

Table 11.5 Energy efficiency level of key machines before optimization (unit of energy consumption: KW.H)

Serial number	Machine number	Processing energy consumption	Total energy consumption	Efficiency (%)
1	1	164.76	222.65	**73.99**
2	2	126.75	230.45	**55**
3	3	182.07	300.12	**60.67**
4	4	169.57	256.32	**66.15**
5	5	116.58	180.35	**64.64**

machine tool processing energy consumption, 5 sets of equipment, the energy consumption of no-load and standby is greatly reduced. The energy efficiency level of the machine tool is shown in Table 11.8. The average energy efficiency level is improved by more than 10%, and the optimization effect is obvious.

Table 11.6 Energy efficiency level of original processing tasks (time unit: hour)

	Serial number			
	AH5001	AH5002	AH5003	AH5004
Processing time	9.5	11	6	11.5
Waiting time	3	3	2	3
Standby energy consumption	1.25	2.5	2.8	2.5
Processing energy	5.52	7.4	6.3	10.2
Total energy consumption	7.69	11.1	9.9	14.9
Efficiency (%)	**71.78**	**66.67**	**63.63**	**68.46**

Table 11.7 Process details after technical implementation (unit: mm)

Product code	Product name	Process name	Employee ID	Process parameters (before optimization)				
				Fast	Fine grinding amount	Compensation amount	Feed rate	Rewind
AH5001	Deep groove ball bearing	Grinding channel	1	123	123	123	123	123
		Grinding hole	2	222	120	222	222	120
		Superfine channel	3	238	238	220	135	118
		External communication mill	4	183	183	183	183	183
		Grinding outer channel	5	140	140	115	165	140
AH5002	Double row deep groove bearing	Grinding channel	1	110	110	120	100	120
		Grinding hole	2	222	213	222	213	213
		Superfine channel	3	138	140	132	138	130
		External communication mill	4	120	120	120	130	130
		Grinding outer channel	5	135	123	135	120	118
AH5003	Angular contact ball bearings	Grinding channel	1	240	50	25	125	70
		Grinding hole	2	265	67	235	210	114
		Superfine channel	3	140	66	214	260	174
		External communication mill	4	184	85	165	240	134
		Grinding outer channel	5	304	105	158	210	136

(continued)

Table 11.7 (continued)

Product code	Product name	Process name	Employee ID	Process parameters (before optimization)				
				Fast	Fine grinding amount	Compensation amount	Feed rate	Rewind
AH5004	One-way stop bearing	Grinding channel	1	305	241	185	230	85
		Grinding hole	2	77	165	232	147	188
		Superfine channel	3	255	314	257	314	58
		External communication mill	4	59	65	149	254	110
		Grinding outer channel	5	340	260	56	146	184

Table 11.8 Equipment layer energy efficiency level after technology implementation (energy unit: KW.H)

Serial number	Machine number	Processing energy consumption	Total energy consumption	Efficiency (%)
1	1	189.14	230.66	81.99
2	2	198.30	266.12	74.51
3	3	196.59	271.69	72.35
4	4	163.52	213.22	76.69
5	5	145.69	188.63	77.24

3. Task-Level Energy Efficiency Level After Technology Implementation

After using the process parameter optimization and scheduling optimization shown in Table 11.2, the energy efficiency levels corresponding to the four batch processing tasks of deep groove ball bearings, double row deep groove bearings, angular contact ball bearings and one-way stop bearings are shown in Table 11.9. The energy consumption corresponding to the task is shown in Fig. 11.12.

Table 11.9 Energy efficiency level of processing tasks after technology implementation (energy unit: KW.H)

Serial number	AH5001	AH5002	AH5003	AH5004
Processing time	9.5	11	6	11.5
Waiting time	2	1.5	1	2
Standby energy consumption	0.55	0.71	0.84	0.95
Processing energy	5.5	7.2	6.1	9.42
Total energy consumption	6.76	8.8	8.4	11.9
Efficiency (%)	**81.36**	**81.81**	**72.62**	**81.52**

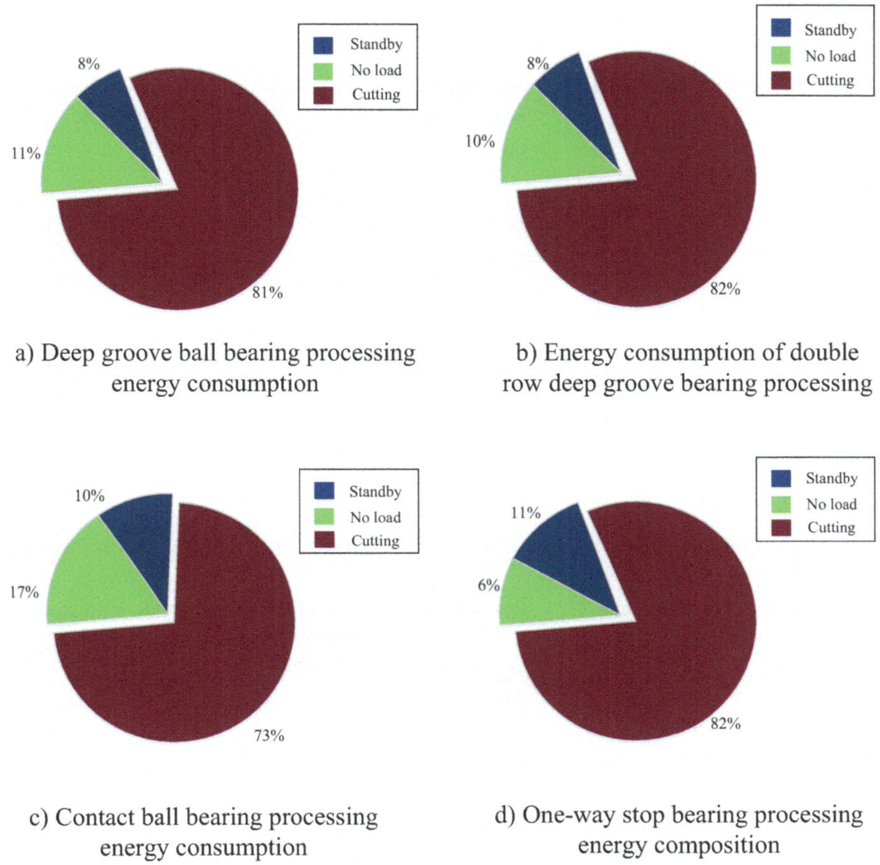

a) Deep groove ball bearing processing energy consumption

b) Energy consumption of double row deep groove bearing processing

c) Contact ball bearing processing energy consumption

d) One-way stop bearing processing energy composition

Fig. 11.12 Energy consumption of four types of processing tasks after technology implementation

11.6.3 Comparison of Energy Efficiency Levels Before and After Optimization

1. Energy efficiency comparison of equipment layers (Table 11.10).
2. Energy efficiency comparison of task layer (Table 11.11).

11.7 Conclusion

Based on the theory of multi-objective flexible job shop scheduling problem in the previous chapter, the mould production workshop of a plastic product factory was taken as the application object. According to the actual situation of the production

Table 11.10 Comparison of energy efficiency levels of equipment layers before and after technology implementation

Serial number	Machine number	Before optimized machine energy efficiency (%)	After optimized machine energy efficiency (%)
1	5	73.99	81.99
2	9	55	74.51
3	10	60.67	72.35
4	11	66.15	76.69
5	13	64.64	77.24
Average energy efficiency of equipment		**64.09**	**76.56**
Energy efficiency improvement after technology implementation		**12.47%**	

The bold represents the average enery and energy efficiency improvement

Table 11.11 Comparison of energy efficiency levels of task layer before and after technology implementation

Serial number	Product name	Product code	Before optimized machine energy efficiency (%)	After optimized machine energy efficiency (%)
1	Deep groove ball bearing	AH5001	71.78	81.36
2	Double row deep groove bearing	AH5002	66.67	81.81
3	Angular contact ball bearings	AH5003	63.63	72.62
4	One-way stop bearing	AH5004	68.46	81.52
Average energy efficiency of equipment (%)			**67.64**	**79.33**
Energy efficiency improvement after technology implementation			**11.69%**	

The bold represents the average enery and energy efficiency improvement

Note The average energy efficiency of the equipment before and after the implementation of the technology is the average of the energy efficiency of the five processing machines

The average energy efficiency of the tasks before and after the implementation of the technology is the average of the energy efficiency of the four types of processing tasks

workshop, the research and analysis, combined with the theoretical research results, developed the mould production. Multi-mode scheduling optimization scheduling system. This chapter describes the development environment, architecture, main functional modules and actual operation results of the system.

The manufacturer's authorised representative in the EU is Springer Nature Customer Service Centre GmbH, Europaplatz 3, 69115 Heidelberg, Germany. If you have any concerns regarding our products, please contact ProductSafety@springernature.com

Printed and bound by CPI Group (UK) Ltd, Croydon, CR0 4YY
30/09/2026
02225557-0005